The Handbook of Safety Engineering

Principles and Applications

Frank R. Spellman and Nancy E. Whiting

GOVERNMENT INSTITUTES
An imprint of

THE SCARECROW PRESS, INC.
Lanham • Toronto • Plymouth, UK
2009

 **Government Institutes**

Published by Government Institutes
An imprint of The Scarecrow Press, Inc.
A wholly owned subsidiary of The Rowman & Littlefield Publishing Group, Inc.
4501 Forbes Boulevard, Suite 200, Lanham, Maryland 20706
http://www.govinstpress.com

Estover Road, Plymouth PL6 7PY, United Kingdom

British Library Cataloguing in Publication Information Available

Library of Congress Cataloging-in-Publication Data

Spellman, Frank R.
  The handbook of safety engineering principles and applications / Frank R. Spellman and Nancy E. Whiting.
    p. cm.
  Includes bibliographical references and index.
  ISBN 978-1-60590-671-3 (cloth) — ISBN 978-1-60590-672-0 (electronic)
  1. Industrial safety—Handbooks, manuals, etc. 2. System safety—Handbooks, manuals, etc. I. Whiting, Nancy E. II. Title.
  T55.S6442 2010
  620.8′6—dc22   2009025336

∞™ The paper used in this publication meets the minimum requirements of American National Standard for Information Sciences—Permanence of Paper for Printed Library Materials, ANSI/NISO Z39.48-1992. Printed in the United States of America

For Jo Ann Garnett-Chapman (aka Wonder Woman)

Contents

PART III: SAFETY ENGINEERING CONCEPTS

Preface

We are loath to let others do unto us what we happily do to ourselves.

—Chauncey Starr

Safety is all about fear.

—Frank R. Spellman

Safety professionals know that the best solution to preventing accidents in the workplace boils down to engineering out the hazards. If there isn't any hazard or exposure, there can't be any accident. If you accept the premise that the ultimate method for protecting workers on the job requires the removal or engineering out of hazards in the workplace, this text is for you.

Handbook of Safety Engineering: Principles and Applications evolved out of the original first and second editions of our text, *Safety Engineering: Principles and Practices*, which served as a standard safety text for the past decade. These texts proved to be excellent guides and sources of information for students and practitioners. However, we found through practice, use, and constructive criticism that this resource could better serve students of safety and practitioners in the field (especially those studying for professional certification examinations) if we shifted the focus, placing more emphasis on engineering aspects and less on regulatory and administrative requirements. This is exactly what we have done with this new and completely reworked text.

One thing is certain: To attain the lofty but attainable (in the practice of safety, we should never admit that the impossible is possible) goal of engineering out hazards, safety practitioners must be thoroughly trained in the principles and applications of safety engineering. They must also have basic skill sets best described as "all encompassing." In short, the safety practitioner, commonly called the "safety engineer" in the real working world, must be a generalist; that is, he or she must be well trained in the sciences, cyber operations, math operations, mechanics, fire science (water hydraulics, etc.), electrical safety, and the technical and administrative aspects of the safety profession. In light of this fact, this text provides basic engineering principles that are accessible to the new (rookie) safety professional as well as the seasoned professional.

The text has been written for all three levels of on-the-job (real-world) competence and experience: fundamental, intermediate, and advanced levels. As before, this new text is designed as a standard college-level textbook, as a desk reference, and most importantly, as a hands-on field guide for safety professionals and practitioners. It is a practical handbook for general readers, students, professors, industrial hygienists, senior-level undergraduate and graduate students in safety and industrial engineering, science and engineering professionals, safety researchers, engineering designers, human-factor specialists, and all other safety practitioners.

How do we know what the safety professional requires? We know based on practical experience gained during more than forty years of on-the-job safety practice. Again, ongoing confrontation with safety problems has shown that the safety practitioner needs exposure to and/or training on basic and advanced mathematics. Moreover, basic mechanics is an important area that is fundamental to those studying for the kinds of mathematical/mechanical problems confronted by test takers on the professional licensure examinations [Certified Safety Professional (CSP)/Certified Industrial Hygienist (CIH) certification exams]. We also included extensive coverage of fire science, dealing with water hydraulics and math. Additionally, because of the major impact the tragic events of 9/11 have had on the safety profession (i.e., safety and security in the workplace are now synonymous), we have included an extensive section on security devices and apparatus that safety engineers not only need to be familiar with but also may need to procure and install to upgrade security in his/her workplace. Another important section of this text deals with the fundamentals of electricity (Ohm's Law through AC Theory) and electrical safety.

The text begins with an extensive introduction to the practice of safety engineering and what it entails; the rest deals with basic and advanced math/engineering operations and applications. Note that we have included fundamental sections—prior to our coverage of more advanced concepts—of the basics of mathematics, scientific principles, and engineering. This is necessary because in the real world of safety practice many safety practitioners are assigned the lofty and prestigious job title "Safety Engineer," but many of these folks are not degreed engineers or trained safety practitioners. Instead, they are often in-house personnel assigned to safety management as a collateral duty or (sometimes) as full-time positions. Unfortunately, many of these in-house personnel face this demanding and difficult job never having been trained in any aspect of safety and safety engineering. This book is designed to aid such personnel. We're not going to make engineering simplistic; we're going to make it simple. Degreed safety engineers will consider the basic introductory sections of this book as review material; the advanced/applied sections are designed for their perusal.

PURPOSE OF TEXT

One of the critically important lessons we've learned is that, just as providing your car with regular maintenance service can prevent large, nasty, unexpected repair bills, and just as careful engineering and design work of cars works to protect you in an accident, through safety engineering, we can work to eliminate or reduce on-the-job hazards by preventive attention and imagination directed toward the safety of workers in the workplace. Prevention through engineering and management is what this text is all about.

AUDIENCE

While the primary intent of this handbook is for use as a handheld field manual and a textbook in safety engineering, practitioners and students are far from the only ones who will find this book useful. The list of who can use this text includes:

- Students in safety engineering and systems safety
- Employer safety directors/managers
- Corporate managers, supervisors, forepersons
- Inspectors exposed to on-the-job hazards
- Architects
- Consulting engineers
- Building managers
- Contractors
- Owner contract administrators
- Equipment and prefab manufacturers
- Plaintiff lawyers
- Defense lawyers
- OSHA officials
- Standards writers and construction specifiers
- Justice and court officials
- Quality management personnel tasked with ensuring OSHA compliance
- Insurance company loss control engineers
- Union officials and workers
- Security directors/personnel

SCOPE OF TEXT

In the past (and even to a degree in the present), very little discussion or emphasis was developed on industrial safety—and even less emphasis on developing a consciousness for safety as an integral part of engineering practice. However, this trend has been changing—as it should. One of the most significant changes has taken place in colleges and universities. Safety and safety-related courses and curriculums have been added to pertinent fields of study; in such programs, safety engineering is usually approached as a science with well-defined goals and objectives, not as an exercise in lip service and sloganeering. With the new demands on industry and critical infrastructure for homeland security, the demand for safety professionals with security backgrounds has also increased.

Still, a problem exists. Many of these relatively recent courses and curriculums are focused on the purely theoretical and scientific aspects of safety and related topics. Many of the safety new-hires we meet are recent graduates from these programs, and a few minutes of conversation tells us all we need to know about their level of training and understanding. Hiring a highly educated safety graduate, one well-grounded in logic and logic systems such as Boolean algebra, systems analysis, and design for safety is not unusual—however, this same student may lack what is really required: a fundamental grounding in the concepts of real-world safety practice. In other words, a gap exists between what our undergraduate and graduate students are typically taught and in what is utterly necessary in the work world—the real world. *Handbook of Safety Engineering: Principles and Applications* is designed to fill this gap.

Clearly written in everyday English, with an understandable, accessible, and direct style, this text provides easy access to a wealth of practical and substantial information. It emphasizes developing a consciousness for safety as an integral part of engineering practice in the work world and addresses industry responsibility for homeland security needs. Deliberate sabotage, technological failures, and preventable disasters wreak havoc on workers and the public, whose lives and health depend on engineering expertise for protection—for proactive prevention versus reactionary measures. The principles in this book (if conscientiously applied) can prevent the devastating effects of improper or unsafe practices in the creation and delivery of work outputs and work activities.

Throughout this text, the authors stress the need for professionalism, scientific analysis of risks and safety measures, concern for human and environmental needs, and use of real-world and effective safety engineering.

NOTE TO READER

The argument is often made in the modern safety profession that the title *safety engineer* is a misnomer, implying that the person filling the position is a degreed engineer with formal education or specialized training in the safety field. We disagree with this point of view because typically the safety engineer (whomever this might be) is responsible not only for managing the overall program but also for preventing mechanical injuries; trip, slip, and fall injuries; impact and acceleration injuries; thermal injuries; fire-related accidents; electrical accidents, and so on. The bottom line: The title *safety engineer* is appropriate for anyone responsible for preventing and mitigating such incidents/occurrences.

A second requirement calls for educational emphasis upon quantitative and logical problem solving. The nontechnician whose mathematical ability breaks down at simple algebra is not likely to acquire the necessary quantitative expertise without great effort. Along with mathematics, the safety engineer must have a good foundation in mechanics and structures. An education that does not include a foundation in the study of forces that act on buildings, machines, processes, and humans leaves the safety practitioner in the same position as a thoracic surgeon with incomplete knowledge of gross anatomy—he or she simply has to feel their way to the target, groping in the dark, leaving a lot to be desired (especially for the patient). *Handbook of Safety Engineering: Principles and Applications* is a light leading the way through the darkness, focusing its beam on the safety, security, and health of workers.

Frank R. Spellman, CSP (Retired)
Norfolk, VA

Nancy E. Whiting
Columbia, PA

I

INTRODUCTION

1

Introduction

Safety often is viewed largely as a simple matter of applying specific routines. In many cases the routines are repeated regularly despite obvious signs of their inadequacies. Greatly needed is an understanding that the sources of harm, which the safety specialty should be able to control, have basic origins although their consequences will differ in character and severity. This view furnishes the realization that hazards are not simply the agents most closely identified with injuries. Merely regulating them is not the sure way to limit their effect.

—Grimaldi and Simonds (1989, 9)

In the practice of safety, if the responsible person in charge of safety does not have the total, absolute support of the senior executive in charge, from fresh start to exhausted and frustrated finish, the safety practitioner is doomed to fail in his/her assigned duties.

—Spellman and Whiting (1999)

SAFETY ENGINEERING?

When we ask new acquaintances standard questions (where they live, where they are from), sooner or later in the conversation, we'll ask the question, "What do you do for a living?" If the answer is "I'm an engineer," we might be impressed, but our question isn't completely answered. Engineer? What type of engineer? Many different engineering disciplines or specialties are possible, from locomotive operators to "flight engineers" to custodial personnel ("sanitation engineers").

So what kind of engineer is our engineer? Unless he or she answers with a specific engineering specialty, the natural tendency will be for the questioner to ask, "What kind of engineer are you?" or "What type of engineering do you do?"

The respondent may offer an answer we might expect—"I am a mechanical engineer" or maybe a civil, aeronautical, chemical, electrical, industrial, or environmental engineer. Or the respondent might answer, "I am a safety engineer." Unless you are familiar with safety engineering, you might be taken aback by this statement. Safety engineer? What is that?

What a safety engineer is and does is what this text is all about. This text works to complete (and to thoroughly explain) the opening statement by Grimaldi and Simonds. For now, we will simply continue their statement by adding our own words:

To regulate their effect (hazards) it is necessary to employ a means of removing or controlling the causes responsible for the presence of injurious agents. This is the essence of the practice of safety engineering.

SAFETY ENGINEERING: A CLOSER LOOK

Simply put, the safety engineer is a Jack or Jill of many engineering, scientific, and other professions. The safety engineer is a multidisciplinary professional with a broad array of knowledge in preventing accidents. Does this mean that the safety engineer is an expert in all engineering, scientific, and the other disciplines? Good question. The short answer is no. The long answer depends on your definition of *expert*. If you define *expert* as does *Webster* (expert: a person with a high degree of skill in or knowledge of a certain subject), we can answer yes, because safety engineers should possess a high degree of skill in or knowledge of safety engineering. But this really doesn't answer our question, does it? To better answer the original question, "Is the safety engineer an expert in all engineering, scientific, and other disciplines," two words in this question need to be changed to render a more accurate answer. The words *expert* and *all* should be changed—*expert* status denotes that the safety engineer knows everything about safety—something no one could do. To accomplish this amazing feat, the safety engineer would have to be an expert in law, engineering, technical equipment, manufacturing processes, behavioral sciences, security, management, health sciences, risk, finance, and insurance—as they relate to safety issues—to name just a few fields. While true that the safety engineer is typically an expert in a particular area of safety, it is also true that he or she can't be expected to be an expert in everything. This point is amplified even further when we remember the following famous quote: "An expert is someone who has found all the mistakes in a very narrow field" (Niels Bohr as quoted by Edward Teller). At this moment, we are unfamiliar with anyone who has found all the mistakes in the practice of safety engineering.

Let's take a look at what the safety engineer really is.

The safety engineer is knowledgeable in many aspects of engineering, scientific, and other related disciplines—a Jack or Jill of many engineering, scientific, and other disciplines. While a safety engineer isn't an expert in all these diverse fields, he or she should be expert in how safety issues affect these many fields and in how to eliminate, reduce, or control hazards for these fields.

The safety engineer is devoted to the application of engineering and scientific principles and methods to the elimination and control of hazards. He or she must know a lot about many different engineering and scientific fields. The safety engineer specializes in the recognition and control of hazards through use of knowledge and skill related to engineering, scientific, and other disciplines.

In short, the safety engineer must be a "generalist" with a firm grasp on how the many fields with safety concerns connect with his or her area of responsibility and influence.

WHY A GENERALIST?

Again, based on our experience, we feel that the safety engineer must be a generalist—the importance of achieving an interdisciplinary education mixed with experience can't be too highly stressed. In addition to a "generalized" education honed smooth by years of on-the-job experience, the safety engineer must know what role other organizational professionals, practitioners, and employees play in the organization and be able to properly identify hazards, develop effective solutions to safety problems, and achieve safe operations and systems.

As a generalist, the safety engineer functions to coordinate and facilitate the actions of other knowledgeable personnel in applying safety engineering principles to particular problems.

EDUCATION AND THE SAFETY ENGINEER

What should a training package for a course of study in safety engineering include? Although professional education (college education) provides the minimum standard for professional credibility, a degree certainly does not indicate that someone with such a degree is smarter—or more effective professionally—than someone who lacks a degree. What it does mean is that the degree holder has accomplished a series of standardized tasks by which he or she may be measured. What a college education really does is provide the graduate with the tools he or she needs to do research—"to find answers." When you get right down to it, this is a worthy goal of any good advanced training course—to teach the students how to problem-solve and to provide them with the tools that allow them to find pertinent answers independently.

So, why is a college education necessary? Can't an individual learn everything they need to know about safety engineering from on-the-job experience?

We can all learn from on-the-job experience—absolutely nothing can replace real experience. But the question was: "Can't someone learn *everything* they need to know about safety engineering from on-the-job experience?" A seasoned safety professional with many years of experience in the safety profession would probably answer yes to this question. But what that safety professional is really saying is that he or she learned all he or she "needed to know" about safety engineering via on-the-job experience—on their way up through their profession—and often as the material they were learning was being discovered. Thirty or forty years ago, this statement had much validity. Today, however, in most cases, the demands of the safety engineering profession are vastly different than they were thirty years ago—thus training requirements are also much different.

Why?

The requirements for a more demanding, classroom-based, formal education with an accompanied professional safety internship combined with rigorous on-the-job training and years of experience are necessary simply because times have changed. With this change, the complexity of problems related to industrial safety has increased exponentially.

What kinds of changes? Complexity of what problems?

Anyone who pays attention to news programs on television or radio or has scanned newspapers or the Internet in recent years should have no problem recognizing the changes we are referring to.

Let's take a look at a few of these changes and the corresponding industrial safety and health concerns they have generated.

Major Accidents and Disasters

Major advances (Herculean advances in some respects) in technology have been made in recent decades. These include advances in nuclear power, electronics, chemical processing, transportation, management information systems, medicine, space exploration, and manufacturing processes, to name just a few of hundreds of growth industries and processes. Modern society's attempts to ensure "the good life" for us all have, at the same time, generated many perils that are also responsible for many of the woes that beset us. Why? With technological advances come technological problems, many of which have direct or indirect impact, not only on the health and safety of employees but also on the public and the environment.

When a process or operation has been correctly engineered (properly designed and constructed), it will show clear and extensive evidence that the design included attention to safety. Such a process or operation, under the watchful and experienced eye of a competent safety engineer, can continue to operate in a way that reduces the chance of occupational injuries, illnesses, public exposure, and damage to the environment.

A safety engineer, one well versed in the basics of civil, industrial, mechanical, electrical, chemical, and environmental engineering, takes information pertinent to safety engineering from each of these disciplines. Exactly what will such exposure afford the safety engineer, our environment, and us? Let's take a look at some examples.

- Civil engineering exposure—Safety engineers who have had exposure to this field understand the need for safe and sanitary handling, storage, treatment, and disposal of wastes. They have knowledge of the controls needed for air and water pollution. They understand the need for structural integrity of bridges, buildings, and other constructed facilities. They understand the planning required to build highways that are safe to use.
- Industrial engineering exposure—Industrial engineers try to fit tasks to people, rather than people to tasks. By doing this, they make work methods and environments physically more comfortable and safer. Safety engineers need to understand the concepts involved with human-factor engineering and ergonomics.
- Mechanical engineering exposure—The student of safety engineering soon discovers that the mechanical engineering field really got the ball rolling toward incorporating safety requirements for machines such as boilers, pumps, air compressors, and elevators, and many other types of mechanized equipment and facilities. The safety engineer must understand the operation and limitations of mechanized machines and ancillary processes.
- Electrical engineering exposure—Electrical engineers are concerned with the design of electrical safety devices such as interlocks and ground fault interrupters and other items. Electrocution and fire caused by faulty electrical circuitry in the workplace are more common than you might think. The organizational safety engineer must be cognizant of these potential hazards and know what needs to be done to prevent and correct such deficiencies.

- Chemical engineering exposure—Chemical engineers apply system safety techniques to process design through the design of less hazardous processes, using less hazardous chemicals, chemicals that produce less waste, or that produce waste that can be easily reclaimed.
- Environmental engineering exposure—Like safety engineering, environmental engineering is a broad-based, interrelated discipline that incorporates the use of environmental science, engineering principles, and societal values. The safety engineer must have a good background in environmental science and environmental engineering practices, primarily because protecting the environment from pollution is often one of the safety engineer's chief duties.

Preventing major accidents and disasters is a primary responsibility of the safety engineer. Major conflagrations, explosions, catastrophic failures of equipment such as boilers or airplanes, and prevention of chemical releases and spills are just a few of the important responsibilities included in safety engineering.

Major accidents and disasters are terrible, obviously, to those directly affected. Serious injury, illnesses, or death resulting from such incidents take their toll; they have impact in ways we cannot imagine until we actually experience them ourselves. The point is with proper, complete, careful planning—safety engineering—accidents and disasters can be prevented.

The results, consequences, impact, and horror involved when "things go wrong" (when safety engineering and safety management take a back seat to other "more important" concerns) can kill or maim all within reach of the site—and what is within reach of the site is dependent upon the nature of the disaster and the physical conditions of the site as well as weather—elements beyond our control.

The potential for risks involved with many chemical processing operations is extreme. Bhopal is not alone in the historical records detailing the horrors of major accidents and disasters (the Chernobyl and Three Mile Island Unit 2 nuclear power incidents are other famous examples). This text's intent is not to point the finger of blame at anyone for these incidents, but rather to point out that they do occur. More importantly for our concerns, these horrific events trumpet the need for safety engineering. What is hardest to bear in these tragic incidents (and in many others over the years) is that we recognize that they could have been prevented—and we fear that the same greed, or carelessness, or disregard for human life may someday affect us as well.

In one respect, the United States has been fortunate. Two of the world's most terrible industrial accidents—accidents that changed how nations handled the very concept of industrial safety—had U.S. counterparts. The difference? The accidents in the United States were on a far less serious scale. Only months after OSHA's post-Bhopal study on the U.S. chemical industry's chances of producing a similar incident (OSHA reported the possibility as very unlikely), a similar spill did occur, in Institute, West Virginia—one-hundred-plus people became ill, no deaths. The result? The chemical manufacturing industry swiftly cleaned up its act. The Three Mile Island event terrified the American public into a more-than-healthy respect for the potential harm a nuclear reactor represents. Even with no deaths or injuries from Three Mile Island, in the incident aftermath, no aspect of nuclear power is treated casually here. The United States again reassessed its regulatory standards after Chernobyl, using, as they should have, the hard-bought lessons in industrial safety Chernobyl had to teach.

Regulatory Influence

Over the years, several changes related to safety in the workplace in the United States have come about because of regulations enacted by the U.S. Congress (most of which, at their conception at least, were based on the British example). The impetus for congressional action in this area was prompted by increased pressure on legislators (by the public) to force businesses to adopt safety measures and to provide hazard-free workplaces. The strong influence of governmental authority in regulating the safety and health of workers in the workplace cannot be overlooked. One of the most important pieces of legislation that directly affected the push for a safer and healthier workplace was the advent of workers' compensation laws. The main intention of the proponents of workers' compensation legislation was to advance occupational safety programs, which in turn sparked the need for highly educated, informed, and trained safety engineers to design, implement, and manage them.

Legal Ramifications

When laws are enacted, soon the legal profession is called in. New regulation means eventual testing of the law through the court system—for safety regulation, this is meant to provide legal services for workers' complaints against employers for hazarding their safety and health, and other related on-the-job concerns. This, of course, is what has happened, and it had a profound effect in the workplace. In the old days the employer hired and expected his employee to work long, hard hours without getting injured. If the employee was injured, the employer simply showed him the door and that was that. Workers now have a vehicle for pursuing legal action against dangerous conditions in the workplace and the employer. Times have changed.

Thus the goal of most employers quickly became to prevent such legal actions—because they are costly in both money and time. The primary way in which employers avoid expensive litigation is by ensuring compliance with applicable laws and regulations. They most often accomplish this by assigning the responsibility for ensuring workers' safety and health on the job to a designated safety official; often this designated safety official is a fully qualified safety engineer. Thus, the safety engineer must be fully educated on and cognizant of new and pending safety regulations.

Recommended Training Program for Safety Engineers

For several years there was (and to a lesser degree still is) considerable confusion and disagreement about exactly what type of preliminary training, if any, a typical safety engineer should have. Much of the confusion and disagreement stems from certain practices fostered by various industries throughout the industrial complex even to this day. Let's take a look at a few of these practices as they evolve, based on organizational attitudes.

Often a company considers safety within the organization as being the full responsibility of the employee. Management in these companies feels (obviously, they aren't thinking) that expensive safety programs, safety equipment, and safety professionals to administer these programs are not necessary. "Safety is expensive—safety doesn't contribute to my bottom line—it takes money away from it." Have you heard this statement or a version of it before? It expresses a view of safety that's more superficial than a paper cut. Unfortunately, this attitude is much more common than you might think.

Even today in this age of regulation and enforcement (namely by the Occupational Safety and Health Administration—OSHA) the scenario described above is still practiced in various industries—at least until such an organization is cited by OSHA or pays the expense brought about by litigation filed by workers and unions against the organization for violation of safe work practices.

When an organization is "hit in the head" by such actions, various (often predictable) reactions are to be expected. Typical reactions include:

- When cited by OSHA, an organization may decide to pay any associated fines and then go back to business as usual. There actually exists a mind-set out there in the real work world (held by certain company owners and/or managers) that it is easier and cheaper to pay any fines levied by OSHA or other state and local authoritative bodies than to fully comply with their "costly" regulatory requirements. This attitude and practice is unfortunate, shortsighted, and illegal—but until recently, when Congress stiffened the penalties and dramatically increased the fines levied, many industries simply ignored OSHA and its requirements that provide a safe and healthy workplace for workers. Unfortunately, many still do.
- Another typical reaction employed by company owners and managers of companies cited by OSHA is to comply by assigning a worker from within the organization as "the safety person" for the organization. In some cases, this practice has actually worked. For example, if the in-house employee (who probably has little or no safety experience) is a highly motivated, active, and dedicated worker, he or she can successfully complete assigned safety duties in a manner that both provides safety and health protection to the workers and may bring the company into compliance with the regulators.

However, let's take a look at what happens, more often than not, when a company chooses a general worker for the safety officer position.

In industry, often the opinion is held that with the appointment of a "safety person," the company's responsibility for controlling hazards becomes the safety person's sole responsibility. Upper management and managers, all the way down the chain of command, then "brush off" (delegate) their responsibility for safety onto the safety person. Not only is this notion an example of convoluted thinking but also it is decidedly wrong—the "ultimate" responsibility for the safety of workers cannot be delegated. Experience clearly shows that this practice ultimately destroys any real chance for safety achievement. Why? Let's take a closer look.

If a company follows the procedure described above and simply selects some worker to become the organization's "safety person," the success of the designated safety person in accomplishing his or her safety duties is directly related to the amount of upper management support he or she receives. If the safety person is simply pulled from the ranks of a workforce, assigned the safety person's duties, given no formal training in safety and zero support from upper management, unless that person is exceptionally dedicated to safety goals and has the brains, guts, and backbone to push for enforcement, he or she is doomed for failure—and thus so is the organization's safety program.

Another problem with pulling "someone" from the existing workforce to fill in as the company's safety person position is made readily apparent in the selection process itself. Who do you suppose is the most likely candidate to be chosen for the position? Let's take a look at three prospective candidates. Will it be Harry from the assembly line, who can outwork any three employees on any given day? Will it be JoAnn from quality control with an eagle eye for defects who saves the company big bucks? Or Frank, who is slower than slow, uncooperative, and basically just occupies the rarefied air in which he seems to hover?

You don't need to be a rocket scientist to determine that the likely candidate is often going to be Frank. The managerial mind-set might be: "Waste Harry or JoAnn on this? Safety's too expensive to begin with without taking away my best workers. Frank's not working well where he is—he needs something different to keep him busy—to keep him out of the way. The safety job (that no one else really wants) is just perfect for Frank." Thus Frank fills the company safety person position, and you can judge for yourself the potential for Frank's ultimate success. Popular culture has given us a sterling example of such a safety officer—one instantly recognizable if you watch *The Simpsons* on Fox.

Like Homer Simpson as safety officer of a nuclear power plant, Frank as the company safety person seems ridiculous and almost comical, doesn't he?—until you think about the potential for risk he represents. Incompetence in the safety officer's position is ultimately not very funny, is it? But what would have happened if the company had decided to "promote" Harry or JoAnn to the safety person position?

If either Harry or JoAnn (for the sake of simplicity, let's stick with Harry) had been chosen as the company's safety person, one thing is certain: the choice was a good one in respect to attitude and work ethic. But the bigger question is: What exactly is expected of Harry? Or as Harry would certainly put it: What am I supposed to do? Where do I start?

Where do I start? Does this sound familiar? This is a natural question for the new "safety person" to ask. This question is complicated, of course, by other considerations—other questions. For example, Harry is going to want to know if he is assigned the safety person position as full time or as a collateral duty. If assigned the safety person's duties as a separate full-time job, then Harry probably has the initiative and drive needed to perform the safety person's duties in a credible manner, though the task will require Harry to learn as he goes (difficult, but not impossible—it's been done before).

Harry might find out, however, that he is expected to continue his full-time duties on the assembly line (employees who can do the work of three people are difficult to replace) and that the safety work is a collateral duty. Again, this is a typical practice in many industries, especially in companies with a workforce of less than twenty people or so, for example.

If Harry is assigned the duties of company safety person, he is going to find out (if he is determined and conscientious—which he is) what several thousand other folks thrown into such positions in the past have already found out. That is, finding a more challenging or more mind-boggling collateral duty assignment than that of "the safety person job" approaches impossible. This statement may seem strange to those managers who view safety as a duty that only requires "someone" (anyone) to keep track of accident statistics, to conduct safety meetings, and perhaps place an occasional safety notice or poster on the company bulletin boards. The fact is, in this age of highly technical safety standards, government regulations, and insurance requirements, the safety person not only has much more to do than place posters on bulletin boards, he or she has more responsibility—much more. Our friend Harry will soon find out that safety is a full-time job. Another safety person scenario in common practice today (typical for companies with twenty-five or more employees) is to hire a professional full-time safety engineer as the company's safety person. This full-time position may be titled Safety Director, Safety Manager, Safety Administrator, Safety Officer, Safety Official, Safety Coordinator, Safety Specialist, or some other variation. The title is not important; the job classification requirements or duties are.

Typically, a safety professional is expected to organize, stimulate, and guide the company's safety program as well as keep up on safety subjects and regulations and function as a resource person for all those involved in the work. Normally, in this capacity, the safety professional, as a staff person, holds very few—or no—administrative powers over the operating components of the company. Is this a good practice? It depends. When you factor in the responsibility of company line officials who are primarily responsible for the operating components of the company, then yes, it is a good practice (in most cases).

Are you confused? If so, consider this. If the company safety person is designated as "the person" solely (the keyword is *solely*) responsible for the safety and health of workers involved in the operating components of the company, then instead of line management, which through chain of command holds the managerial power to control unsafe work practices and conditions, management often will relegate (or abdicate) this responsibility solely to the safety person. Obviously, this is not only an unhealthy situation—it could have titanic consequences (remember Bhopal, Chernobyl, or Three Mile Island? How could we forget?).

Just what responsibility—and enforcement power—does or should the safety professional have? What can (and does) a safety officer do in those situations when he or she observes, for example, an imminently dangerous work situation or practice in progress? Should the safety professional just stand there and let the line manager in charge continue a dangerous and life-threatening operation? No, the safety professional must take action.

What action? The potency of the safety professional lies in the capacity to use well-marshaled facts to persuade the line managers to act in safety's behalf. In other words, the safety professional who observes any unsafe act or unsafe condition should immediately point this out to the manager in charge.

What if the manager in charge is not available? What is the safety professional supposed to do then? Good question—one that has been asked for decades (the answers received have not always been what you might expect). A company safety professional should be given enough authority to stop work when required—when danger to the safety and health of workers, the public, or the environment is imminent.

Caution is advised, however. While the company safety professional should have the authority to stop work because of unsafe conditions or unsafe work practices, this authority must not be abused. Interfering with line operations and stopping work as a show of power over nickel-and-dime trouble or for dreamed-up causes is not recommended. Simply, the safety engineer should never forget that his or her duties require him or her to be an advocate for safety and not a Gestapo agent.

Getting back to the education requirements for today's safety professionals, common practice today (mainly because of OSHA, insurance company requirements, and fear of litigation) means larger companies hire a qualified safety professional—they don't promote from within (that is, unless a qualified safety professional is available in-house).

This brings us back to explaining exactly what qualifications the safety engineer is typically expected to have today. Along with experience, the professional safety engineer is expected to be a college graduate with a degree in safety and health management, industrial hygiene, environmental health, engineering, business administration (with safety course completion), or some other affiliated field.

The National Safety Council publishes every other year a list of colleges and universities that offer safety training (the list grows in length each year). In addition, the Board of Certified Safety Professionals [an organization that sets the standards for registering safety engineers as Certified Safety Professionals (CSP)] has suggested a standard baccalaureate curriculum for the safety professional. Note that the Board of Certified Safety Professionals does not infer that their suggested curriculum is the "only" academic program that can successfully prepare individuals who plan to enter the safety profession; instead, the content included in their proposed curriculum is minimal for meeting the educational requirements for becoming a CSP.

Let's take a look at some of the minimum college requirements.

- Calculus I and II
- Statistics
- Biology I and II
- Chemistry I and II
- Technical Writing
- Composition
- Physics I and II
- Graphics

- Economics
- Psychology
- Public Speaking
- Applied Mechanics
- Materials and Processes
- Several Engineering electives
- Safety and Health I and II
- Industrial Hygiene
- Toxicology
- Computer Science
- Management
- Safety Engineering
- Fire Prevention and Protection
- Global Positioning System
- Control of Physical Hazards
- Several other safety electives

Certified Safety Professional (CSP)

Does completing the minimum educational requirements automatically qualify a candidate for certification as a safety professional? Not exactly, but the candidate is headed in the right direction.

Why is it important for the safety professional to become a Certified Safety Professional (CSP)? According to the Board of Certified Safety Professionals, "certification provides a credential showing that a safety professional has achieved a professional level of qualification as judged by peers" (BCSP 2006). Based on personal experience (of the male coauthor; retired CSP), BCSP (2006) is correct when it states the "certification provides credibility with employers, clients, peers, and the public."

To qualify for and obtain certification as a certified safety professional (CSP), certain requirements must be completed. These requirements include:

- Graduation from an accredited college or university with a bachelor's degree in engineering, or some other field such as occupational safety and health.
- Completion of at least four years of professional safety experience. Graduate degrees from accredited institutions can be applied toward professional safety experience.
- Achievement of a passing score on both the core examination (safety fundamentals) and the comprehensive practice exam.
- To maintain certification, a person holding the CSP credential must pay an annual renewal fee and also meet Continuance of Certification requirements. This program, which operates in five-year cycles, helps ensure that those with the CSP credential keep current in the safety profession through reexamination, continuing education, professional writing, or other approved means (BCSP 2006).

If you are familiar with the requirements for registration and licensing for a Professional Engineer (PE), you will notice that the qualification requirements for CSP designation are very similar, especially regarding the two-tiered testing process. In achieving PE status, an engineer must first successfully complete a fundamentals examination, which leads to designation as an EIT (engineer-in-training). For CSP certification, candidates must also successfully complete a fundamentals (or core) examination, which leads to designation as an ASP (associate safety professional). This is not a recognized certification as such, but the designation does indicate a candidate's progression (as does the engineering EIT designation) toward certification.

The Safety Fundamentals and Comprehensive Practice Examinations are administered semiannually by the Board of Certified Safety Professionals (BCSP), Savoy, Illinois. The examinations are closed book and consist of approximately two hundred multiple-choice questions.

The subject areas covered in the safety fundamentals examination are weighted as follows:

The topics for the major areas given by the BCSP for the Safety Fundamentals and Comprehensive Practice Examination include:

Table 1.1 BCSP Topics for Safety Fundamentals and Comprehensive Practice Exams

Domain 1 Safety, Health, and Environmental Management	37%
Domain 2 Safety, Health, and Environmental Engineering	25%
Domain 3 Safety, Health, and Environmental Information Management and Communications	33%
Domain 4 Professional Conduct and Ethics	5%

1. Basic and Applied Sciences
- Mathematics: algebra, trigonometry, calculus, statistics, and symbolic logic
- Physics: mechanics, heat, light, sound, electricity, magnetism, and radiation
- Chemistry: atomic structure, bonding, states of matter, chemical energetics and equilibrium, and chemical kinetics, addressing both inorganic and organic chemistry
- Biological sciences: heredity, diversity, reproduction, development, structure and function of cells, organisms, and populations, with emphasis on human biology
- Behavioral sciences: individuals' differences, attitudes, learning, perception, and group behavior
- Ergonomics: the capabilities and limitations of humans and machines, simulation for design and training, principles of symbolic and pictorial displays, static and dynamic forces on the human frame, response to environmental stress, and vigilance and fatigue
- Engineering and technology: applied mechanics, properties of materials, electrical circuits and machines, principles of engineering design, and computer science
- Epidemiology: the application of epidemiological techniques to the study and prevention of injuries and occupational illnesses

2. Program Management and Evaluation
- Organizations, planning, and communication: general principles of management applicable to the organization of safety function, safety program planning, and communication
- Legal and regulatory considerations: federal, state, and local safety and health legislation, workers' compensation laws, regulatory agencies, contractual relationships, and public/product liability
- Program evaluation: the identification, selection, and application of qualitative and quantitative measures to evaluate and document the need for specific control programs
- Disaster and contingency planning: the identification of types of emergencies, plan-of-action considerations, training, and equipment needs
- Professional conduct and methods: principles that define the rights and responsibilities of safety professionals in their relationships with each other and with other parties

3. Fire Prevention and Protection
- Fire prevention: inspection procedures, hot-work permits, flammable liquids handling and storage considerations, and other factors intended to control sources of ignition and fuel
- Structural design standards: emergency egress design, fire resistive construction, and fire and smoke containment
- Detection and control systems and procedures: principles and design of fire detection and warning systems, fixed and portable extinguishing systems, fire and emergency drills, and fire brigade organization and training

4. Facilities and Equipment Design
- Facilities and equipment design: methods, standards, and principles of design for the control of injury-causing conditions associated with facilities and equipment in general
- Mechanical hazards: identification and control of injury-causing conditions associated with facilities and equipment in general

- Pressures: identification and control of exposures associated with compressed gases, pressure vessels, and high-pressure systems
- Electrical hazards: the nature of electrical shock, grounding, insulation, ground fault circuit interrupters (GFCIs), and inspections
- Transportation: identification and control of hazards associated with transportation of personnel and materials
- Materials handling: identification and control of injury-causing conditions associated with both manual and mechanical materials handling, hoisting apparatus, conveyors, and powered industrial trucks
- Illumination: the assessment of the quality and quantity of illumination with respect to accident prevention

5. **Environmental Aspects**
   - Toxic materials: the identification and assessment of hazards associated with chemicals and other materials producing or capable of producing a harmful effect on biological mechanisms
   - Environmental hazards: the identification, assessment, and control of potentially harmful effluents
   - Noise: the principles and methods of measurement, assessment, and control of noise exposures
   - Radiation: the principles and methods of measurement, assessment, and control of radiation exposures, both ionizing and nonionizing
   - Thermal hazards: the identification and assessment of exposures associated with temperature extremes
   - Control methods: respiratory protection, protective clothing, ventilation, shielding, and process controls

6. **System Safety and Product Safety**
   - Product/System design considerations: product or system design considerations associated with the prevention of product liability losses.
   - Product liability administrative aspects: administrative and legal considerations associated with the prevention of product liability losses.
   - Reliability, quality control, and systems safety: general concepts of product and system safety as they relate to reliability and quality control

Upon successful completion of the safety fundamentals examination, the ASP designation is awarded. Upon successful completion of the Comprehensive Practice Examination, the CSP designation is awarded.

Did You Know?

Although the three Es of safety have been around for some time, recently safety professionals in industry have begun to refer to the four Es of safety. The fourth E ranges from the terms *encouragement* to *emergency response*, depending on the emphasis applied by the industry involved.

THE THREE Es PARADIGM

Throughout this chapter, we have stressed the importance of employing a qualified safety person to be responsible for the controlling, coordinating, and managing of the company's safety program. The keyword we have stressed throughout is *qualified*. By now you should have a good feel for why "qualified" is important.

Creating a safe work environment is the primary goal of safety officers—and that safe work environment can be created, controlled, coordinated, and managed by applying the three Es of safety: engineering, education, and enforcement.

We have pointed to the importance of engineering in safety—to the design of control systems to enhance occupational safety and health. Again, the requirements of OSHA, increased employer liability, and heightened worker awareness of the implications of unsafe and unhealthy work environments all contribute to this need. Experience has shown that the vast majority of engineers concerned with design and operation have not been adequately trained to recognize or control existing or potential safety and health hazards. Worse, the past tendency for safety considerations has been to call in the "engineers" only after equipment or process design and installation have been completed. This has been and is, obviously, a serious shortcoming. In this text we not only stress the need to "engineer out a hazard" but also to do so in the preliminary process—in the design phase. Safety engineers should be given an early design of a system (any system) to analyze it to find what faults can occur, and then propose safety requirements in design specifications up-front and changes to existing systems to make the system safer.

In actual practice, however, it is too often the case that rather than actually influencing the design, safety engineers are assigned to prove that an exiting, completed design is safe. If a safety engineer then discovers significant safety problems late in the design process, correcting them can be very expensive. As a case in point, consider, for example, the design and construction of a chemical storage building. The design and construction personnel did not run the design plans past the resident safety engineer. After construction was completed and the internal chemical storage tanks were filled with listed hazardous chemicals, the resident safety engineer inspected the new installation. Almost immediately the safety engineer recognized that there was no installed emergency eyewash/deluge shower installed within or immediately outside the building as per OSHA requirements. Later, when the company attempted to install the eyewash/deluge showers, they discovered that there was no convenient potable water line anywhere close to the new structure. In the end, more than $30,000 was spent to bring the chemical storage structure into compliance with applicable regulations. Obviously, this type of error caused the company to waste a large sum of money. Later, it was found that the installation of the eyewash/deluge shower and the associated piping and hookup to an approved, potable water line would have cost less than $4,000.

We stress the proactive mode (developing preventive actions before accidents occur), not the reactive mode (deriving preventive actions from accidents), because fixing a problem before it becomes a problem just makes good common sense and can result in the saving of precious lives.

Engineering includes virtually every preventive action necessary to correct or eliminate a physical hazard (including such simple steps as maintaining good order and discipline within the workplace and maintaining housekeeping measures such as cleaning up spills, removing obstructions, etc.). Major preventive actions (i.e., "engineering out" a hazard) cover a broad range of possible steps and require the application of physical sciences to control hazards in the workplace, including designing out hazards, modifying processes, and incorporating fail-safe devices. In other instances, "engineering" may be devoted largely to more routine actions—substituting less hazardous materials, reducing the inventory of hazardous materials, and, as a last resort, prescribing protective equipment (Grimaldi and Simonds 1989).

Grimaldi and Simonds also point out that the scope and order of actions applied as engineering steps for the control of hazards usually include:

Step 1. Evaluate the process or operation and identify its harmful agents (if any).
Step 2. Eliminate the harmful agents by redesign, or substitute a less harmful material, arrangement, and so on.
Step 3. Shield or guard the hazard.
Step 4. Isolate the hazards.
Step 5. Use ventilation, wet processing, and other techniques to dilute the harmful effect.
Step 6. When steps 2 through 5 do not furnish the level of control needed, provide personal protective equipment (PPE).

Education (employee training), the second of the three Es of safety, is one of the most important preventive actions the safety engineer incorporates into the company safety program. The reason for this is not difficult to understand and appreciate. Simply put, employees can't be expected to perform their assigned functions unless they have been properly trained to perform them. Without proper training, the potential for injury or death is magnified many times. As an extreme example, if a manager tells an employee to operate a dangerous machine (say, a bulldozer or front-end loader) without ensuring that the employee is qualified (trained) on how to operate the machine, this is inviting trouble—disaster, in fact.

Safety engineers are routinely called upon to conduct safety training at all levels of organizational strata, but safety training usually begins with the supervisor. This makes sense because the supervisor is primarily responsible for ensuring that his or her workers are qualified (trained) to do their jobs.

Note that safety engineers can enhance the company's overall safety profile by employing the expertise of supervisors in the establishment and maintenance of the company's safety program. Experience has shown that no matter how hard the safety engineer works to improve the company's safety program and no matter how conscientious he or she is, if the supervisors don't buy into the program, it is doomed to failure.

Exactly what training should the safety engineer provide to supervisors? Generally, the safety engineer needs to train supervisors to recognize hazards and explain the appropriate actions to be taken. More specifically, supervisor safety training should focus on four tenets:

- Develop safe working conditions and safe working practices.
- Conduct employee safety training that stresses the real need for safety. This is best accomplished by personalizing the training (i.e., relate the training to workers' jobs and their functions within the company, show how it can and will benefit them).
- In personalizing safety, what we are really talking about is employee participation. We have stated that it is important for the safety engineer to have the supervisors "buy in" on the company's safety program—and this is true—but it should be stressed that worker safety is what safety engineering is all about. Make the workers part of the company's safety program. In our experience, safety engineers can learn a lot by just listening to workers—they have good ideas. Look at it this way—the workers face the on-the-job hazards you are trying to eliminate or reduce, each and every day—their value to the safety engineer can't be overstressed.
- Implement safety rules. While the safety engineer usually has a good "feel" for what an organization needs in the way of safety rules, safety rules often need to be tailored to the type of work performed. For example, if the company's primary work deals with material handling operations, the focus of safety rules will more than likely be on rules dealing with safe material handling operations.

In regard to training workers, it is important to point out that whether you provide the training yourself or hire experts in the field, all the training in the world is worthless unless it is properly documented. Having sat through several civil lawsuit cases in various courts of law where a company or a company's officials were being sued for any number of inadequate safety provisions, the one glaring deficiency that I noticed in all of these cases was either the lack of training or the lack of documentation to prove that the training was conducted. In the eyes of OSHA and the law, training without documentation is training that was never accomplished.

Enforcement, the last of the three Es of safety, is more than just achieving compliance with company rules and procedures. This is especially the case today when company (and thus employee) practices must be in compliance with federal, state, and local laws and regulations and consensus standards.

When a company puts together a set of safety rules, safe work practices, and procedures—and fails to enforce them—the rules, practices, and procedures are not worth the paper they are written on. Enforcement normally implies punishment; however, the safety engineer must not be regarded as the wielder of such. The safety engineer makes a serious mistake whenever he or she becomes the company "enforcer" (often viewed by workers and supervisors as the company's Gestapo agent). Instead, the primary lesson the company safety engineer must learn to be successful is to be an advocate for safe work practices in his or her company, not just a regulator of safe work practices.

A significant factor included in the enforcement element (one often overlooked) is the use of "praise" when a good safety effort emerges. The safety engineer who does not recognize employees who complete their tasks in a correct, safe manner loses a golden opportunity to use the positive to reinforce what is commonly viewed as the negative: enforcement. When an employee is singled out by the company safety engineer and supervisors for outstanding performance of some function conducted safely—in full view and earshot of his coworkers—the effectiveness of the company's safety program is often elevated to a higher level, one that is otherwise often difficult to attain.

THE SAFETY ENGINEER'S DUTIES

The company safety engineer, as the organizer, stimulator, and guide of the safety program, performs a number of significant tasks. Probably the easiest way to illustrate what is typically required of a safety engineer (and clearly reinforce our belief that a safety engineer must be a Jack or Jill of all trades) is to take a look at an example job description for such a position.

Company Safety Engineer—Examples of Duties

- Conducts independent work activities based on broad directives.
- Manages safety division and safety/environmental health activities for company departments.
- Promotes awareness among company employees of the benefits of safe operation.
- Plans, coordinates, and supervises safety activities, evaluates company safety rules, policies, safe work practices, and training procedures and recommends changes as appropriate.
- Coordinates implementation of safety rules, policies, safe work practices, and procedures.
- Provides liaison and coordination between departmental safety officers/committees concerning safety activities; reviews and inspects operations and related maintenance procedures and recommends actions to insure safety of company employees.
- Investigates safety accidents and recommends policies/procedures and/or other actions to correct problems or prevent future accidents.
- Plans, coordinates, supervises, and reviews recording and reporting of safety data and information.
- Assists with the procurement of contractual services or materials and supplies for safety activities.
- Supervises safety training activities for all company departments.
- Manages OSHA, SARA Title III, CERCLA, and other regulatory requirements.
- Directs development of trial or pilot safety programs.
- Reviews and critiques drawings, contract specifications, and OSHA 200/300 logs of perspective outside contractors, ensuring compliance with 29 CFR 1910.119 process safety management requirements, including preconstruction safety briefs for contractor supervisory personnel.
- Continuously improves knowledge of general and specific information pertaining to safety equipment, regulatory requirements, programs, and procedures.
- Collects, analyzes, and arranges information in such a manner as to fulfill management's decision-making needs.
- Assists various levels of management by originating and developing policy recommendations, operating plans, programs, measures, and controls that will permit the effective accident-free use of human and material resources.
- Manages the company medical surveillance program, including CDL physical exam requirements; pulmonary function testing; audiometric testing; and environmental sampling/testing programs.
- Manages training program for HAZWOPER compliance.
- Publishes annual safety training/medical testing/safety inspection schedule.
- Investigates and mitigates employee reports of unsafe conditions.

The S in Safety Engineer

September 11, 2001: Four commercial airplanes are hijacked on the East Coast. American Airlines Flight 11 out of Boston crashes into the North Tower of the World Trade Center with ninety-two people on board. United Airlines Flight 172, also out of Boston, crashes into the South Tower of the World Trade Center with sixty-five passengers on board. American Airlines Flight 77 out of Washington Dulles crashes into the Pentagon with sixty-four on board. Flight 93 out of Newark, New Jersey, crashes into a rural, western Pennsylvania field with forty-four people on board. An estimated additional 2,666 people on the ground (many of them fire and rescue workers) die in the World Trade Center and another 125 are dead in the Pentagon. Over 3,000 people die in one day as a result of terrorist attacks (U.S. Department of State 2002).

Are safety engineers responsible for preventing terrorism in the workplace? While no one will hold safety engineers responsible for preventing passenger planes from being deliberately crashed into buildings (that kind of responsibility takes place on a higher level—actually, some would say at a lower level), the short answer is yes. The compound answer is absolutely yes.

Even though no written OSHA standard mandates that a written workplace terrorism program is required, it should be included in any comprehensive safety or loss control program. For example, the workplace security plan could be attached to the emergency response plan. The point is, because of 9/11, companies are expanding their programs to incorporate the possibility of terrorism and acts of subversion within the company. This makes sense when we consider that many company safety engineers are already responsible for addressing bomb threats and prevention of workplace violence. Adding security concerns to the wide range of responsibilities of safety engineers is the logical next step.

9/11 changed many elements of our society. One of the major changes has to do with the workplace itself—architecture, building plans, and company activities are now viewed in a different light. Now we must factor in a new "what if" scenario—what if our company comes under terrorist attack, from within or from outside the company?

Because of the need for increased security, the safety engineer's job has expanded exponentially. We discuss security concerns, prevention techniques, and detection devices in greater detail in chapter 14. For now we point out that the *s* in safety engineering takes on a new connotation—security. Based on our experience, we can say that 9/11 supersized the *s* in the practice of safety engineering. Earlier, we listed the duties of the company safety engineer. Here's an essential addition to the list:

- Planning, managing, and implementing a company security plan.

Reality Check

Does the preceding job description make the job of safety engineer seem like a task too great for one individual to handle? Don't let it scare you away. With the proper training, attention to duty and detail, good organizational skills, proper companywide delegation of safety responsibilities, and clearheaded thinking, the job of safety engineer, while always challenging, is achievable.

REVIEW QUESTIONS

1. In what disciplines do safety engineers need expertise?
2. In what ways are safety engineers "generalists"?
3. What is the safety engineer's role in industry?
4. Why are both college degree study and hands-on experience important for safety engineering professionals today?
5. Why is "prevention" more important than "reaction" in safety engineering?
6. Aspects of civil, industrial, mechanical, electrical, chemical, and environmental engineering are critical to the safety engineering profession. Why, and what do these areas provide for the safety engineer?
7. What do safety engineers learn from studying the Bhopal case?
8. What can safety engineers learn from studying 9/11?
9. When new safety regulations are enacted, what is the general industry response?
10. How have safety and health regulations affected the average workplace?
11. Who is (or should be) responsible for safety in industrial facilities?
12. What are the pros and cons of promoting from within versus hiring a safety professional?
13. What problems are associated with assigning someone safety officer functions as collateral duty?
14. What is a safety engineer's duty if he sees workers involved in unsafe work practices?
15. Discuss the CSP certification process.
16. List, define, and discuss the three Es of safety. Why and how is each element important?
17. What are the six parts of the engineering steps for control of hazards?
18. What four tenets should supervisor safety training focus on?
19. What role does praise play in safety training?
20. On our list of possible safety engineering duties, we have twenty entries. Take five, and define and discuss what the duties entail.
21. Discuss the most recent addition to the list of duties for the safety engineer. Why is this so important?

REFERENCES AND RECOMMENDED READING

Allison, W. W. 1991. Other voices: Are we doing enough? *Professional Safety* (February): 31–32.

Banks, W. W., and F. Cerven. 1985. Predictor displays: The application of human engineering in process control systems. *Hazard Prevention* (January/February): 26–32.

BCSP. 2006. *Comprehensive practice self-assessment examination.* 4th ed. (revised). Savoy, IL: Board of Certified Safety Professionals.

Benner, L., Jr. 1990. Safety training Achilles' heel. *Hazard Prevention* 1: 6112.

CoVan, J. 1995. *Safety engineering.* New York: John Wiley and Sons, Inc.

Fawcett, H. K., and W. S. Wood, eds. 1982. *Safety and accident prevention in chemical operations.* New York: John Wiley and Sons, Inc.

Friend, M. A., and J. P. Kohn. 2003. *Fundamentals of occupational safety and health.* 3rd ed. Rockville, MD: ABS Consulting/Government Institutes.

Grimaldi, J. V., and R. H. Simonds. 1989. *Safety management.* 5th ed. Homewood, IL: Irwin.

McCormick, E. J., ed. 1970. *Human factors engineering.* 3rd ed. New York: McGraw-Hill.

Spellman, F. R. 1996. *Safe work practices for wastewater treatment plants.* Lancaster, PA: Technomic Publishing Company.

Spellman, F. R. 1998. *Surviving an OSHA audit.* Lancaster, PA: Technomic Publishing Company.

Spellman, F. R., and N. E. Whiting. 1999. *Confined space entry.* Lancaster, PA: Technomic Publishing Company.

U.S. Department of State. 2003. *September 11, 2001: Basic facts.* Washington, DC: U.S. Department of State.

❷

Safety: A Historical Perspective

Even though life has improved and been extended, citizens of the United States pay a high price for their high-technology life-style. Each year there are over 100,000 accidental deaths and nearly 11 million disabling injuries. The cost of all accidents in the United States is about $100 billion annually, excluding some indirect costs and the value resulting from pain and suffering. Accidents are the leading cause of death for people aged 1 through 44. For those aged 45 or older the accidental death rate increases rapidly with age. Only heart disease, cancer and stroke exceed it. For the total population, the two leading causes of accidental death are motor vehicles and falls. Although the accidental death rate in the United States has declined from about 85–90 per 100,000 population in 1910 to below 50 today, the actual number of accidental deaths has increased over the same period.

—Brauer (1994, 4)

If a builder builds a house for someone, and does not construct it properly, and the house which he built falls in and kills its owner, then that builder shall be put to death.

—Code of Hammurabi

INTRODUCTION

Since the dawn of human existence when we chased down the woolly mammoth and the mastodon to eat or ran in fear from Smilodon (saber-toothed cat), cave bear, and other hungry creatures, we have always had the desire to be safe and secure—reckless, yes, but safe and secure, always.

In regard to the shift from hunting and gathering to working for a living, it has long been known that working conditions, occupations, some diseases, and shortened life expectancy are connected. Lead poisoning, for example, was recognized as early as the fourth century BC. In the first century AD, Pliny the Elder, in his *Historia Naturalis*, pointed out that dust from mercury ore grinding and fumes from lead are hazardous. In more recent times, exposure to a myriad of on-the-job hazards has increased in many occupations. The connection between job type, disease, and life expectancy is made clear by Cohen (1984), who pointed out that the difference in average life expectancy in U.S. industries varies dramatically based on occupation. For example, workers employed in clothing manufacturing and communication typically outlive (up to five years longer on average) firefighters, police officers, truckers, and coal miners.

Few would dispute that technological advancements have created many changes in society from the fourth century to the present. Disputing the fact that technological advances have improved society would be equally difficult. In the workplace, the introduction of robotics in manufacturing, for example, has increased productivity and quality of products produced. For most of us, automation and industrial safety improvements have made our lives easier. Medical advances made possible through microminiaturization sometimes seem miraculous, allowing doctors to save the lives of people who, only a few years before, would have died.

These advances are not without cost, though. Technological advances have created many changes that have created new economic, social, political, environmental, and safety and health problems. The old saying goes: Every problem has a solution. Today we might add: Every solution creates new problems.

Probably the easiest way to understand the benefit and the "cost" of technological advancement (in terms of creating hazards) is to illustrate by example. For instance, the widespread use of electricity has occurred in the United States within the twentieth century. Few people today (especially those who remember what living without it was like) would argue against its merits. Electricity has been a godsend, touching every aspect of our lives, from modern refrigeration to thousands of products that we now take for granted—lighting, heating, cooling, machine operation, a source of power to operate most of our household appliances, to computers, BlackBerries and iPods. Few would want to do away with electricity, and in fact, when you're accustomed to life with electrical power, the learning curve for living without electricity is steep.

If you had to list all the benefits to humankind derived by the use of electricity, you'd have a long list—and you could never be sure you'd gotten everything. Explaining how our lives would be different without electricity would also be close to impossible because we simply have lived with it too long—we have gotten used to electricity and are dependent on it. We accept electricity as a fact of life, and take it for granted, ignoring the fact that millions of people in the world still live without it.

However, as a technological advance, electricity has a cost. While the introduction of electricity solved many problems for us and improved the quality of life in the United States, it also created certain problems.

Before the introduction of electricity, very few people (excepting those struck by lightning) died by accidental electrocution. Fires ignited by electricity from lightning storms were not uncommon, but not as common as fires from other ignition sources—now, electricity in the industrial work setting causes more fires than any other single source.

In the workplace, electrical hazards did not exist before the advent of electricity. Fires in the workplace caused by electrical current (as mentioned, a very common occurrence today) were not possible.

Electricity has been a boon to civilization as we know it today—no question about that. The point? While technological advancements can improve society, they can also be detrimental. We tend not to pay attention to the detrimental aspects when the benefits outweigh the costs.

As safety engineers, when we think about the monetary, physical, and environmental costs of technology, it seems rather ironic that the myriad of technological advancements that we have developed over the years were primarily developed via engineering, by engineers. We now call upon engineering and engineers to design and construct innovative technology that is not only safe to use but also safe to be around (e.g., nuclear power plants). In particular, we are concerned about the safety and health of workers in the workplace, since many of the technological advances made are designed for the workplace—designed to make work easier, faster, more productive, more economical, and hopefully, safer. As Grimaldi (1980) points out, "If two certainties in life are death and taxes, surely a third is daily exposure to hazards." True—but were workplaces any safer before?

PREINDUSTRIAL SAFETY

Most of us have seen movies or read books that describe the horrors of working in pre-Industrial-Revolution-era factories and sweatshops. Indeed, if it were not for movies and books detailing the type of workplaces our ancestors labored in before the Industrial Revolution, we would have little knowledge and/or understanding about the hazards they were exposed to regularly, at work or at home. For example, who knows all the hazards involved with transportation by horse? Most people who work with horses professionally don't rely on horses as their basic mode of transportation, with the exception of the Amish (who have the additional hazard of trying to travel by horse and buggy through modern traffic). Whether you get where you are going in a horse-drawn vehicle or by riding, horses present the possibility of constant risk. Horses can rear, buck, bite, kick, run away, roll, spook—and with a rider in the saddle or driver in a vehicle, any of these possibilities could, and did, lead to injury or death.

We take for granted a steady supply of hot water. We simply turn on the proper tap, and hot water flows out. Take yourself back before the time of the hot water heater. Water was heated over a fire, and someone had to carry the hot water to where it was needed. Obviously, scalding by hot water was a common hazard then. How many of us today have ever had to manufacture a bar of soap? Probably not many of us. Can you imagine the work that went into this process—working with hot, boiling lard and lye? Being severely burned while making soap was a hazard that was all too common before industrialization.

In short, hazards were everywhere in preindustrial society, too. We sometimes forget that life expectancy was shorter then, infant mortality was higher, and that because many hazards were associated with day-to-day living, death and disability were common—accepted as facts of everyday life. When we hear someone say that they wish they could go back in time and live in the "good old days" when everything was simpler, we wonder if these dreamers have any idea of what living in the "good old days" was really all about. In the United States, our lives are so protected from harm that it seems as a nation, we have lost our perspective on safety and personal responsibility. Consider this: a lawsuit over a spilled cup of McDonald's coffee made national headlines only a few years ago.

INDUSTRIAL SAFETY

During the 1800s and through the early stages of the Industrial Revolution, industrial safety was often ignored in the United States. The exception to this practice was in fire protection. Fire was a common problem (with devastating results) in factories of that day, and with a healthy push from the factory insurance companies, the National Board of Fire Underwriters was formed in 1866. Insurance companies understood the need for factory fire prevention inspections and employed their own teams of inspectors to ensure that factories were inspected periodically to improve protection—and reduce financial losses.

Massachusetts was the first state to recognize the need for pursuing safety and health guidelines for the workplace. Based on the British model, the Commonwealth passed factory acts covering the general provisions of the British laws. Included in these laws was the provision for the inspection of factories and public buildings. Also included was an important provision related to providing guards on dangerous machinery—specifically on shafting, gearing, belting, and drums. Interestingly, cleanliness and air quality was also addressed. Other provisions were provided for ensuring the safe operation of elevators and hoistways. All establishments of three or more stories in height were required to have fire escapes. Certain doors in manufacturing establishments (and in other places people congregate as well) were directed to open outward. Presently, regulatory provisions intended to protect workers and the public from safety and hazards had been enacted in every state (Grimaldi and Simonds 1989).

Actually, the fear of regulatory intervention by the government on business owners spurred on the early workplace safety and health movement. The last thing the business owners wanted was for the government to burden them with regulations—with controls.

Other groups worked to push safety and health issues for workers to the forefront. Along with the government, these included organized labor (e.g., in 1890, the United Mine Workers organized to demand shorter hours, increased pay, and improved safety conditions). Industry itself made a move toward safety and health in the workplace (they came to the realization that money could be saved if lives and limbs were protected). Voluntary safety organizations (such as the American Public Health Association) advanced the safety movement. Some of these reforms were successful, other were less so (see Case Study 2.1 following), but worker health and safety issues became important industrial considerations.

One of the most influential early books on industrial safety, *Industrial Accident Prevention*, was written by J. W. Heinrich and was published in 1931. Heinrich (the father of safety theory) maintained that accidents result from unsafe actions and unsafe conditions. Heinrich postulated that people cause far more accidents than unsafe conditions do and that accidents result from a sequence of events:

1. Social environment
2. Fault of a person
3. Unsafe condition or act
4. Accident
5. Injury

Though many modern safety professionals question some of Heinrich's ideas, his was the beginning effort to develop the conceptual, theoretical framework that was required in making society view safety as a true profession. Moreover, Heinrich's work is the basis for the theory of behavior-based safety, which holds that as many as 95 percent of all workplace accidents are caused by unsafe acts. Note, however, that the authors of this text do not hold with Heinrich or the behavior-based-safety-model view that workers and their actions are the main causal factors related to the occurrence of on-the-job injuries; more will be said on this controversial topic later in the text.

Case Study 2.1: A Stitch in Time

The working conditions of the seamstress in nineteenth-century England received a great deal of concern and attention in the media of the day, but actual reform came very slowly. Victorian women had little choice of career—nanny or governess positions, shop positions, servant occupations, certain kinds of factory work, and dressmaking were often the only legitimate jobs a woman might be qualified (or allowed) to do. Of these, dressmaking was frequently the only choice possible, because while educational effort was seldom spent on girls until mandatory educational requirements went into effect around the turn of the century, virtually every girl learned how to sew and do fancy needlework. Without a father, brother, or husband to protect or support her, a woman with limited education might be forced into a difficult choice—servant work, dressmaking, or prostitution.

Because these women all could sew, the potential labor force was huge. In London in 1851, there were between 15,000 and 20,000 seamstresses working. Since competition for jobs was steep and because people wanted to buy their goods as cheaply as possible, wages were often inadequate—too small to live on. Working conditions were generally dreadful (only the first-rate houses cared to treat their employees well—they hired top-quality workers and expected to keep them), and by the nature of the profession, many women gained little more than lifelong ill health from their labor.

Comfortable hand sewing requires only a few necessities: good lighting and ventilation; comfortable, straight furniture; and frequent chances for movement. But nineteenth-century custom houses frequently ignored these considerations. The custom houses fed, housed, and worked their apprenticed help as they pleased, in conditions that ranged from reasonable in first-class establishments to places with living quarters where the women slept five to a bed. At the height of the social season, when the houses were at their busiest, women would commonly work eighteen- to twenty-hour days for weeks on end to finish overbooked orders. To finish special projects, they might work around the clock and into the next day. With bad light, stuffy quarters, no chance for exercise, and little food, the health of these women suffered.

High fashion also created seamstress problems. When the sewing machine came into general use in the mid-1860s, a logical expectation might be that more could be accomplished faster. But as soon as the sewing machine came into use, it was used to create clothing with more fine detail—if not handwork, then decoration: pleats, ruffles, lace, braid, and trimmings—as a display of caste and wealth. Production time actually went up.

Their work caused many of the health problems prevalent within the group. Lack of fresh air (the gaslights used for illumination consumed as much of the available air as five adults—for each light), no exercise, the bent-over posture, as well as breathing lint and fiber particles caused many women to develop respiratory problems, sometimes compounded by organic poisoning from inhaled toxins. Tuberculosis was quite common, and internal distresses caused by bad posture, tight corseting, and bad food gradually dragged down the workers' health. Blindness, headaches, nervous exhaustion, anemia, faintness, and chronic fatigue were other side effects. In *The Ghost in the Looking Glass*, Christina Walkley quotes nineteenth-century medical experts, who agree that "no life style was better calculated to destroy health and induce early death" (32).

The upbringing that most Victorian women shared also created part of their problems. Brought up to depend on men, these women were terribly unsophisticated when dealing for money and for their rights. They had no organized lobby behind them. They really didn't know how to complain—or have anyone effective to complain to. Factories employed larger numbers of people and had a formal internal structure, with a recognizable chain of command. Custom houses were generally run by a single owner with only a few employees on site. Many seamstresses relied on piecework employment, and often the only way to find piecework would involve a middleman who could easily skim his portion of already meager wages. Since often this was a woman's only possible source of income and the job market was full, she had little recourse. She could work at the wages offered—or starve. Often a slave would have been physically better off, because a slave was valuable property. The seamstress was an expendable worker.

Over and over again, *Punch* brought the seamstress's plight to England's attention. In 1843, *Punch* published Thomas Hood's "The Song of the Shirt" (a poem that inspired much pity), but little toward reform could be accomplished. Between the private homes that housed the custom houses, the amount of untraceable piecework farmed out to individuals, and the fear of job loss for those who were working, concrete proof significant enough to pass legislation was difficult to obtain. Conditions did not become easier until long after the turn of the century, when World War I changed clothing styles radically enough to allow factory-made clothing to become the norm. By 1918, the Trade Boards Acts established hour and wage regulations. But even today, immigrant workers in the United States suffer many of the same sweatshop conditions as those Victorian seamstresses, with as little recourse. Sweatshop factories exist in most third-world countries, often using children to produce fashionable clothing inexpensively for companies in industrialized nations. The Nike brand and a popular clothing line associated with Kathie Lee Gifford have been tied to sweatshops and third-world children's labor, and stories about the plight of deaf Mexican immigrants in U.S. sweatshops have all made national news.

DISASTERS GIVE SAFETY A PUSH

When the public becomes concerned about their health and safety and the health and safety of their loved ones, something has to give. What usually ends up giving is the legislators who make the laws and regulations and other enforcement actions to ensure citizen safety. Nothing tweaks the public's attention or demands their concern more than a huge disaster, especially when such a disaster (whatever might it be) hits close to home or to loved ones. This is the reaction to the old Aesop fable: "He that hath been once beguiled by some other ought to keep him well from the same"—roughly meaning "once bitten, twice shy."

History is replete with chronicles of one major disaster after another. In the last four decades, we can find countless examples of multiple fatalities caused by some element or another of our industrialized society.

- In 1974 in São Paulo, Brazil, a fire in a twenty-five-story bank building killed 227 people.
- In 1976, 182 cases of a mysterious pneumonia occurred among members of the American Legion attending a convention in Philadelphia. After months of laboratory analysis, a newly discovered bacterial organism, Legionella pneumophila, was found responsible.
- In 1977, the explosions of two grain elevators (one in Louisiana and one in Texas within five days of each other) resulted in fifty-four deaths. In 1975, in India, a coal mine exploded and flooded, killing 431 miners.
- In 1982, in Washington, D.C., an Air Florida Boeing 727 crashed into the Potomac River Bridge during a snowstorm, killing seventy-eight people. In 1981, in Missouri, the Kansas City Hyatt Regency skywalk collapsed (the worse structural disaster in U.S. history), killing 114 people and injuring over 200 others.
- In 1986 (Chernobyl, Ukraine), a nuclear reactor exploded, and its overall death toll was estimated to be around 10,000 persons.
- On April 19, 1995, in Oklahoma City, Oklahoma, an explosion ripped through the Murrah Federal Building. It completely tore the roof off the north wall and left 168 dead, including nineteen children. The terrorists had purchased ammonium nitrate from a farm cooperative and nitromethane racing fuel at a racetrack, assembling the materials to construct the bomb. It was rigged in a rental truck and parked in front of the building. A denotation-delay device allowed the bomber to escape prior to the explosion. This was the most lethal terrorist attack to ever occur in the United States—until 9/11.

We could go on and on listing recent disasters and terrorist attacks and describe the gore and awful personal impact such events cause on the victims and their family members, but the point is clear. When disasters occur, we work hard to find out what happened and try to correct it. However, what we should really do is anticipate the accident and prevent it. Remember, prevention, not correction, is the safety engineer's mantra.

Let's get back to the evolution of safety in the workplace.

SAFETY EVOLVES VIA SAFETY-RELATED ACTS

One of the first safety-related acts was enacted in 1906—the Pure Food and Drug Act. Though this act primarily targeted the safety of goods, drugs, and cosmetics, it set the tone—opened the way, so to speak—for safety and health regulations that followed.

One of the generation of acts that followed was the Walsh-Healy Public Contract Act (1936). Although the purpose of this act was an attempt to make government contractors model employers, it was the first legislation that included occupational health standards. As originally conceived, the act required most government contractors to adopt an eight-hour day and a forty-hour week and to pay the prevailing minimum wage. One of its major focuses was on child labor. The act specifically eliminated use of child labor on government contracts by establishing minimum ages for males (sixteen) and for females (eighteen). The Walsh-Healy Act was amended in 1969, setting threshold limit values (TLVs) for many substances and possible worker exposures (including noise criteria), and limits for many toxic materials.

The Walsh-Healy Act left many safety and health stipulations untouched. But as time went on and pressure mounted, Congress passed one of the most far-reaching pieces of social legislation to date: The Fair Labor Standards Act of 1938. Though principally social legislation, it had a peripheral impact on workers' safety. While actually the act only applied to about 20 percent of the U.S. workforce, it did ban child labor, set the minimum wage, and set the maximum workweek at forty-four hours. If employers wanted workers to work beyond the forty-four-hour limit, they had to pay them time-and-a-half for overtime. Children under sixteen were banned from employment, and children under eighteen were prohibited from working in hazardous industries. Arguments over this legislation still continue in a minor way.

Case Study 2.2

Witnesses Cite Need for Legislation to Address Amish Youth Employment
WASHINGTON, D.C.—Witnesses before the House Workforce Protections Subcommittee today testified in favor of legislation (H.R. 1943) designed to allow Amish Americans to continue their traditional way of training their children in a craft or occupation while ensuring the safety of those who are employed in woodworking occupations. H.R. 1943, introduced by Rep. Joseph Pitts (R-PA), amends the 1938 Fair Labor Standards Act (FLSA) to allow Amish youth to continue working in businesses where machinery is used to process wood products under certain safety conditions.

—*News from the Committee on Education and the Workforce,* John Boehner, Chair, October 8, 2003

Amish children are finished with school at age fourteen. Federal workplace regulations do not allow those under the age of sixteen in some cases, eighteen in others, to work in certain industries, including sawmills and carpentry or woodworking shops. These are common businesses for the 150,000 Amish in the thirty-three states and 115 congressional districts with Amish populations. The Amish want to be able to teach their sons the family business in their time-honored, traditional way—learning by doing and apprenticeship.

Others disagree. They cite inexperience, smaller size, immaturity, and lack of training as extreme areas of risk for these potential workers. The Department of Labor is opposed to the proposed legislation and objects to a recent trend by Congress, which is exhibiting exceptions for younger workers instead of enforcing and standing behind legislation and regulations designed to protect worker health and safety.

Nicolas W. Clark, counsel for the international United Food and Commercial Workers Union, agrees. "Amish children have benefited from the same child labor protection afforded children of other faiths for over 60 years. Congress should not deny them those protections now simply because they are Amish." Clark says that the death rate for workers in sawmills or woodworking shops is five times the national average for all industries (Jordan 2003).

Before 1970, the primary responsibility for governmental action concerning the prevention of occupational injuries and disease in the United States rested with the states. After 1970, upon enactment of the Occupational Safety and Health Act (OSH Act), the federal government stepped in and required that workplaces be made safe and healthy. Without a doubt, the OSH Act has become the most pervasive safety law ever passed. It stipulated that no longer could workers be made to trade safety for employment. The law set the minimum standards for safety in all workplaces in the United States.

Another group of federal and state acts that greatly impacted industry, safety engineers (remember, safety engineers are most likely to be designated as responsible parties for ensuring compliance with these acts), and ultimately the safety and health of workers falls under various Environmental Protection Acts that were promulgated in the 1970s and 1980s: the Toxic Substances Control Act (TSCA) of 1976; the Clean Air and Clean Water Acts; Federal Insecticide, Fungicide, and Rodenticide Act (FIFRA); Resource Conservation and Recovery Act (RCRA); Comprehensive Environmental Response, Compensation, and Liability Act (CERCLA "Superfund"); Underground Storage Tank Regulation under RCRA; and Superfund Amendments and Reauthorization Act (SARA) Title III or "Community Right to Know." The real significance of these acts is that for the first time, violations of safety and environmental law could result in both civil and criminal prosecution.

Reality Check

Much progress has been made in the last fifty years to protect the safety, health, and well-being of workers. The changes have been significant. No doubt, laws, regulations, and the provisions for their enforcement have significantly eased the occurrence of on-the-job injuries and illnesses. However, the jury is still out on ruling whether the results have been satisfactory or not.

Much of the success of any company's safety program rests squarely on the shoulders of the safety engineer. Remember, many out there in the real world of work misunderstand the scope, import, and significance of the safety engineer's task. When we say that the safety engineer must be, and is, a Jack or Jill or all trades, we include in this succinct definition the qualities required of a peace negotiator as well. Rules and regulations are easy to enact. Convincing those responsible to comply is much more difficult.

REVIEW QUESTIONS

1. Some occupations and their working conditions are connected to two principal potential drawbacks—illness and decreased life expectancy. What are some of the occupations, and what are the related factors?
2. Explain "every solution creates new problems."
3. Explain "costs" in terms of technological advances and detrimental effects.
4. Discuss "hazards" through history from a modern safety-engineering point of view.
5. Discuss possible changes in safety management in light of the Oklahoma City and 9/11 attacks.
6. What was the initial safety focus of Industrial-Revolution-era safety regulation, and why?
7. What groups acted as agents of change to bring about safety regulation in industry?
8. Discuss Heinrich's accident sequence of events.
9. What social factors kept early garment workers (seamstresses) in unhealthy working conditions?
10. What popular novel of the early part of this century led to the Pure Food and Drug Act? Why was it so effective?
11. The Walsh-Healy Public Contract Act set important limits, as did the Fair Labor Standards Act. What limits did they set, and why were they so important?
12. Why is the issue of adolescent children in certain industrial environments again attracting regulatory attention?
13. The OSH Act took many aspects of industrial safety away from the states. Discuss how this affected and affects American workplaces.
14. How did the Environmental Protection Acts affect regulation?

REFERENCES AND RECOMMENDED READING

Ashford, N. A. 1976. *Crisis in the workplace.* Cambridge, MA: MIT Press.
Brauer, R. L. 1994. *Safety and health for engineers.* New York: Van Nostrand Reinhold.
Cohen, B. L. 1984. Risk and risk aversion in our society. *Hazard Prevention* (September/October).
Ferry, T., ed. 1985. *New directions in safety.* Park Ridge, IL: American Society of Safety Engineers.
Grimaldi, J. V. 1980. Hazards, harms, and hegemony. *Professional Safety* (October).
Grimaldi, J. V., and R. H. Simonds. 1989. *Safety management.* 5th ed. Homewood, IL: Irwin.
Heinrich, H. 1931. *Industrial accident prevention.* New York: McGraw-Hill.
Hoover, R. L., R. L. Hancock, K. L. Hylton, O. B. Dickerson, and G. E. Harris. 1989. *Health, safety, and environmental control.* New York: Van Nostrand Reinhold.

Jordan, L. J. 2003. Amish want federal labor laws for teenagers relaxed. *Associated Press* (October 20).

Sinclair, U. 1980. *The jungle.* New York: New American Library.

Walkley, C. 1981. *The ghost in the looking glass.* London: Peter Owen LTD.

3

Safety Terminology

Every branch of science, every profession, and every engineering process has its own language for communication. Safety engineering is no different. To work even at the edge of safety engineering, you must acquire a fundamental vocabulary of the components that make up the process of administering safety engineering. Administering safety engineering? Absolutely.

Remember the safety engineer is—and must be—a practitioner of safety—a specialized field.

INTRODUCTION

As Voltaire said: "If you wish to converse with me, define your terms." In this chapter, we define the terms or "tools" (concepts and ideas) used by safety engineers in applying their skills to make our technological world safer. We present these concepts early in the text rather than later (as is traditionally done in an end-of-book glossary), so you can become familiar with the terms early, before the text approaches the issues those terms describe. The practicing safety engineer or student of safety engineering should know these concepts—without them it is difficult (if not impossible) to practice safety engineering. Several other chapters contain vocabulary specific to those more specialized fields.

Safety engineering has extensive and unique terminology, most with well-defined meanings, but a few terms (especially safety, accident, injuries, and engineering [as used in the safety context]) often are not only poorly defined but also are defined from different and conflicting points of view. For our purpose, we present the definitions of key terms, highlighting and explaining those poorly defined terms (showing different views from different sides) where necessary.

We do not define every safety term—only those terms and concepts necessary to understand the technical jargon presented in this text. For those practicing safety engineers and students of safety engineering who want a complete, up-to-date, and accurate "dictionary" of terms used in the safety profession, we recommend the American Society of Safety Engineers (ASSE) text, *The Dictionary of Terms used in the Safety Profession*. This concise, informative, and valuable safety asset can be obtained from the ASSE at 1800 East Oakton Street, Des Plaines, Illinois 60018 (1-312-692-4121).

SAFETY TERMS

abatement period: The amount of time given an employer to correct a hazardous condition that has been cited.

　absorption: The taking up of one substance by another, such as a liquid by a solid or a gas by a liquid.

　accident: This term is often misunderstood and is often mistakenly used interchangeably with *injury*. The meanings of the two terms are different, of course. Let's look at the confusion caused by the different definitions supplied to the term *accident*. The dictionary defines an accident as "a happening or event that is not expected, foreseen or intended." Defined another way: "an accident is an event or condition occurring by chance or arising from an unknown or remote cause." The legal definition is "an unexpected happening causing loss or injury which is not due to any fault or misconduct on the part of the person injured, yet entitles some kind of legal relief."

Are you confused? Stand by. The following definition will help clear your head.

> With rare exception, an accident is defined, explicitly or implicitly, by the unexpected occurrence of physical or chemical change to an animate or inanimate structure. It is important to note that the term covers only damage of certain types. Thus, if a person is injured by inadvertently ingesting poison, an accident is said to have taken place; but if the same individual is injured by inadvertently ingesting poliovirus, the result is but rarely considered accidental. This illustrates a curious inconsistency in the approach to accidents as opposed to other sources of morbidity, one, which continues to delay progress in the field. In addition, although accidents are defined by the unexpected occurrence of damage, it is the unexpectedness, rather than the production and prevention of that damage per se, that has been emphasized by much of accident research. The approach is not justified by present knowledge and is in sharp contrast to the approach to the causation and prevention of other forms of damage, such as those produced by infectious organisms, where little, if any, attention is paid to the unexpectedness of the insults involved, and only their physical and biological nature is emphasized—with notable success.
>
> —Haddon et al. (1964, 28)

Now you should have a better feel for what an accident really is; however, another definition, perhaps one more applicable to our needs, is provided by safety experts—the authors of the ASSE *The Dictionary of Terms used in the Safety Profession*, 1998. Let's see how they define *accident*.

> An accident is an unplanned and sometimes injurious or damaging event which interrupts the normal progress of an activity and is invariably preceded by an unsafe act or unsafe condition thereof. An accident may be seen as resulting from a failure to identify a hazard or from some inadequacy in an existing system of hazard controls. Based on applications in casualty insurance, an event that is definite in point of time and place but unexpected as to either its occurrence or its results (1).

In this text we use the ASSE's definition of accident.

accident analysis: A comprehensive, detailed review of the data and information compiled from an accident investigation. An accident analysis should be used to determine causal factors only, and not to point the finger of blame at anyone. Once the causal factors have been determined, corrective measures should be prescribed to prevent recurrence.

accident prevention: The act of averting a circumstance that could cause loss or injury to a person.

accommodation: The ability of the eye to quickly and easily readjust to other focal points after viewing a video display terminal so as to be able to focus on other objects, particularly objects at a distance.

acoustics: In general, the experimental and theoretical science of sound and its transmission; in particular, that branch of the science that has to do with the phenomena of sound in a particular space such as a room or theater. Safety engineering is concerned with the technical control of sound and involves architecture and construction, studying control of vibration, soundproofing, and the elimination of noise in order to engineer out the noise hazard.

action level: Term used by OSHA and NIOSH (National Institute for Occupational Safety and Health—a federal agency that conducts research on safety and health concerns) and defined in the Code of Federal Regulations (CFR), Title 40, Protection of Environment. Under OSHA, "action level" is the level of toxicant that requires medical surveillance, usually 50 percent of the PEL (Personal Exposure Level). Note that OSHA also uses the action level in other ways besides setting the level of "toxicant." For example, in its hearing conservation standard, 29 CFR 1910.95, OSHA defines the action level as an eight-hour, time-weighted average (TWA) of 85 decibels measured on the A-scale, slow response, or equivalently, a dose of 50 percent. Under CFR 40 §763.121, action level means an airborne concentration of asbestos of 0.1 fiber per cubic centimeter (f/cc) of air calculated as an eight-hour, time-weighted average.

acute: Health effects that show up a short length of time after exposure. An acute exposure runs a comparatively short course, and its effects are easier to reverse than those of a chronic exposure.

acute toxicity: The discernible adverse effects induced in an organism with a short period of time (days) of exposure to an agent.

adsorption: The taking up of a gas or liquid at the surface of another substance, usually a solid (e.g., activated charcoal adsorbs gases).

aerosols: Liquid or solid particles so small they can remain suspended in air long enough to be transported over a distance.

air contamination: The result of introducing foreign substances into the air so as to make the air contaminated.

air pollution: Contamination of the atmosphere (indoor or outdoor) caused by the discharge (accidental or deliberate) of a wide range of toxic airborne substances.

air sampling: Safety engineers are interested in knowing what contaminants workers are exposed to and the contaminant concentrations. Determining the quantities and types of atmospheric contaminants is accomplished by measuring and evaluating a representative sample of air. The types of air contaminants that occur in the workplace depend upon the raw materials used and the processes used. Air contaminants can be divided into two broad groups, depending upon physical characteristics: 1. gases and vapors and 2. particulates.

allergens: Because of the presence of allergens on spores, all molds studied to date have the potential to cause allergic reaction in susceptible people. Allergic reactions are believed to be the most common exposure reaction to molds (Rose 1999).

ambient: Descriptive of any condition of the environment surrounding a given point. For example, ambient air means that portion of the atmosphere, external to buildings, to which the general public has access. Ambient sound is the sound generated by the environment.

asphyxiation: Suffocation from lack of oxygen. A substance (e.g., carbon monoxide) that combines with hemoglobin to reduce the blood's capacity to transport oxygen produces chemical asphyxiation. Simple asphyxia is the result of exposure to a substance (such as methane) that displaces oxygen.

atmosphere: In physics, a unit of pressure whereby 1 atmosphere (atm) equals 14.7 pounds per square inch (psi).

attenuation: The reduction of the intensity at a designated first location as compared with intensity at a second location, which is farther from the source (reducing the level of noise by increasing distance from the source is a good example).

audible range: The frequency range over which normal hearing occurs—approximately 20 Hz through 20,000 Hz. Above the range of 20,000 Hz, the term *ultrasonic* is used. Below 20 Hz, the term *subsonic* is used.

audiogram: A record of hearing loss or hearing level measured at several different frequencies—usually 500 to 6000 Hz. The audiogram may be presented graphically or numerically. Hearing level is shown as a function of frequency.

audiometric testing: Objective measuring of a person's hearing sensitivity. By recording the response to a measured signal, a person's level of hearing sensitivity can be expressed in decibels, as related to an audiometric zero or no-sound base.

authorized person: (see **competent person**) A person designated or assigned by an employer or supervisor to perform a specific type of duty or duties, to use specified equipment, and/or to be present in a given location at specified times (for example, an authorized or qualified person is used in confined space entry).

autoignition temperature: The lowest temperature at which a vapor-producing substance or a flammable gas will ignite even without the presence of a spark or flame.

baghouse: Term commonly used for the housing containing bag filters for recovery of fumes from arsenic, lead, sulfa, etc. It has many different trade meanings, however.

baseline data: Data collected prior to a project for later use in describing conditions before the project began. Also commonly used to describe the first audiogram given (within six months) to a worker after he or she has been exposed to the action level (85 dBA) to establish his or her baseline for comparison to subsequent audiograms for comparison.

behavior-based safety (BBS) management models: A management theory based on the work of B. F. Skinner, it explains behavior in terms of stimulus, response, and consequences. BBS refers to a wide range of programs that focus almost entirely on changing the behavior of workers to prevent occupational injuries and illnesses.

bel: A unit equal to 10 decibels (see **decibel**).

benchmarking: A process for rigorously measuring company performance vs. "best-in-class" companies and using analysis to meet and exceed the best in class.

biohazard: (biological hazard) Organisms or products of organisms that present a risk to humans.

biological aerosols: Naturally occurring, biologically generated, and active particles small enough to become suspended in air. These include mold spores, pollen, viruses, bacteria, insect parts, animal dander, etc.

boiler code: ANSI/ASME Pressure Vessel Code, whereby a set of standards prescribing requirements for the design, construction, testing, and installation of boilers and unfired pressure vessels.

Boyle's Law: The product of a given pressure and volume is constant with a constant temperature.

carcinogen: A cancer-producing agent.

carpal tunnel syndrome: An injury to the median nerve inside the wrist, frequently caused by ergonomically incorrect repetitive motion.

casual factor: (accident cause) A person, thing, or condition that contributes significantly to an accident or to a project outcome.

catalyst: A substance that alters the speed of, or makes possible, a chemical or biochemical reaction, but remains unchanged at the end of the reaction.

catastrophe: A loss of extraordinary large dimensions in terms of injury, death, damage, and destruction.

Charles's Law: The volume of a given mass of gas at constant pressure is directly proportional to its absolute temperature (temperature in Kelvin).

chemical change: Change that occurs when two or more substances (reactants) interact with each other, resulting in the production of different substances (products) with different chemical compositions. A simple example of chemical change is the burning of carbon in oxygen to produce carbon dioxide.

chemical hazards: Includes hazardous chemicals conveyed in various forms: mists, vapors, gases, dusts, and fumes.

chemical spill: An accidental dumping, leakage, or splashing of a harmful or potentially harmful substance.

chronic: Persistent, prolonged, repeated. Chronic exposure occurs when repeated exposure to or contact with a toxic substance occurs over a period of time, the effects of which become evident only after multiple exposures.

coefficient of friction: A numerical correlation of the resistance of one surface against another surface.

combustible gas indicator: An instrument that samples air and indicates whether an explosive mixture is present, and the percentage of the lower explosive limit (LEL) of the air-gas mixture that has been reached.

combustible liquid: Liquids having a flash point at or above 37.8°C (100°F).

combustion: Burning, defined in chemical terms as the rapid combination of a substance with oxygen, accompanied by the evolution of heat and usually light.

competent person: As defined by OSHA, one who is capable of recognizing and evaluating employee exposure to hazardous substances or to unsafe conditions and who is capable of specifying protective and precautionary measures to be taken to ensure the safety of employees as required by particular OSHA regulations under the conditions to which such regulations apply.

confined space: A vessel, compartment, or any area having limited access and (usually) no alternate escape route, having severely limited natural ventilation or an atmosphere containing less than 19.5 percent oxygen, and having the capability of accumulating a toxic, flammable, or explosive atmosphere or of being flooded (engulfing a victim).

containment: In fire terminology, restricting the spread of fire. For chemicals, restricting chemicals to an area that is diked or walled off to protect personnel and the environment.

contingency plan: (commonly called the **emergency plan**) Under CFR 40 § 260.10, a document that sets forth an organized, planned, and coordinated course of action to be followed in the event of an emergency that could threaten human health or the environment.

convection: The transfer of heat from one location to another by way of a moving medium, including air and water.

corrosive material: Any material that dissolves metals or other materials or that burns the skin.

cumulative injury: Any physical or psychological disability that results from the combined effects of related injuries or illnesses in the workplace.

cumulative trauma disorder: A disorder caused by the highly repetitive motion required of one or more parts of a worker's body, which in some cases, can result in moderate to total disability.

Dalton's Law of Partial Pressures: In a mixture of theoretically ideal gases, the pressure exerted by the mixture is the sum of the pressures exerted by each component gas of the mixture.

decibel (dB): A unit of measure used originally to compare sound intensities and subsequently electrical or electronic power outputs; now also used to compare voltages. In hearing conservation, a logarithmic unit used to express the magnitude of a change in level of sound intensity.

decontamination: The process of reducing or eliminating the presence of harmful substances such as infectious agents to reduce the likelihood of disease transmission from those substances.

density: A measure of the compactness of a substance; it is equal to its mass per unit volume and is measured in kg per cubic meter/LB per cubic foot (D = mass/Volume).

dermatitis: Inflammation or irritation of the skin from any cause. Industrial dermatitis is an occupational skin disease.

design load: The weight that can be safely supported by a floor, equipment, or structure, as defined by its design characteristics.

dike: An embankment or ridge of either natural or humanmade materials used to prevent the movement of liquids, sludges, solids, or other materials.

dilute: Adding material to a chemical by the user or manufacturer to reduce the concentration of an active ingredient in the mixture.

dose: An exposure level. Exposure is expressed as weight or volume of test substance per volume of air (mg/l), or as parts per million (ppm).

dosimeter: Measuring tool that provides a time-weighted average over a period of time such as one complete work shift.

dusts: Various types of solid particles produced when a given type of organic or inorganic material is scraped, sawed, ground, drilled, heated, crushed, or otherwise deformed.

electrical grounding: Precautionary measures designed into an electrical installation to eliminate dangerous voltages in and around the installation and to operate protective devices in case of current leakage from energized conductors to their enclosures.

emergency plan: See **contingency plan**.

emergency response: The response made by firefighters, police, health care personnel, and/or other emergency service workers upon notification of a fire, chemical spill, explosion, or other incident in which human life and/or property may be in jeopardy.

energized ("live"): The conductors of an electrical circuit. Having voltage applied to such conductors and to surfaces that a person might touch; having voltage between such surfaces and other surfaces that might complete a circuit and allow current to flow.

energy: The capacity for doing work. Potential energy (PE) is energy deriving from position; thus a stretched spring has elastic PE and an object raised to a height above the earth's surface, or the water in an elevated reservoir, has gravitational PE. A lump of coal and a tank of oil, together with oxygen needed for their combustion, have chemical energy. Other sorts of energy include electrical and nuclear energy, light, and sound. Moving bodies possess kinetic energy (KE). Energy can be converted from one form to another, but the total quantity stays the same (in accordance with the conservation of energy principle). For example, as an orange falls, it loses gravitational PE, but gains KE.

engineering: The application of scientific principles to the design and construction of structures, machines, apparatus, manufacturing processes, and power generation and utilization for the purpose of satisfying human needs. Safety engineering is concerned with control of environment and humankind's interface with it, especially safety interaction with machines, hazardous materials, and radiation.

engineering controls: Methods of controlling employee exposures by modifying the source or reducing the quantity of contaminants released into the workplace environment.

Epidemiological Theory: This theory holds that the models used for studying and determining epidemiological relationships can also be used to study causal relationships between environmental factors and accidents or diseases.

ergonomics: A multidisciplinary activity dealing with interactions between man and his total working environment, plus stresses related to such environmental elements as atmosphere, heat, light, and sound as well as all tools and equipment of the workplace.

etiology: The study or knowledge of the causes of disease.

exposure: Contact with a chemical, biological, or physical hazard.

exposure ceiling: The concentration level of a given substance that should not be exceeded at any point during an exposure period.

fall arresting system: A system consisting of a body harness, a lanyard, or lifeline, and an arresting mechanism with built-in shock absorber, designed for use by workers performing tasks in locations from which falls would be injurious or fatal or where other kinds of protection are not practical.

fire: A chemical reaction between oxygen and a combustible fuel.

flammable liquid: Any liquid having a flash point below 37.8°C (100°F).

flammable solid: A nonexplosive solid liable to cause fire through friction, absorption of moisture, spontaneous chemical change, or heat retained from a manufacturing process, or that can be ignited readily and when ignited, burns so vigorously and persistently as to create a serious hazard.

flash point: The lowest temperature at which a liquid gives off enough vapor to form ignitable moisture with air and produce a flame when a source of ignition is present. Two tests are used—open cup and closed cup.

foot-candle: A unit of illumination. The illumination at a point on a surface one foot from, and perpendicular to, a uniform point source of one candle.

fume: Airborne particulate matter formed by the evaporation of solid materials, e.g., metal fume emitted during welding. Usually less that 1 micron in diameter.

gas: A state of matter in which the material has very low density and viscosity, can expand and contract greatly in response to changes in temperature and pressure, easily diffuses into other gases, and readily and uniformly distributes itself throughout any container.

grounded system: A system of conductors in which at least one conductor or point is intentionally grounded, either solidly or through a current-limiting (current transformer) device.

ground-fault circuit interrupter (GFCI): A sensitive device intended for shock protection, which functions to de-energize an electrical circuit or portion thereof within a fraction of a second in case of leakage to ground of current sufficient to be dangerous to persons but less than that required to operate the overcurrent protective device of the circuit.

hazard: The potential for an activity, condition, circumstance, or changing conditions or circumstances to produce harmful effects. Also an unsafe condition.

hazard analysis: A systematic process for identifying hazards and recommending corrective action.

hazard assessment: A qualitative evaluation of potential hazards in the interrelationships between and among the elements of a system, upon the basis of which the occurrence probability of each identified hazard is rated.

Hazard Communication Standard (HazCom): An OSHA workplace standard found in 29 CFR 1910.1200 that requires all employers to become aware of the chemical hazards in their workplace and relay that information to their employees. In addition, a contractor conducting work at a client's site must provide chemical information to the client regarding the chemicals that are brought on to the work site.

Hazard and Operability (HAZOP) Analysis: A systematic method in which process hazards and potential operating problems are identified, using a series of guide words to investigate process deviations.

hazard identification: The pinpointing of material, system, process, and plant characteristics that can produce undesirable consequences through the occurrence of an accident.

hazard control: A means of reducing the risk from exposure to a hazard.

hazardous material: Any material possessing a relatively high potential for harmful effects upon persons.

hazardous substance: Any substance that has the potential for causing injury by reason of its being explosive, flammable, toxic, corrosive, oxidizing, irritating, or otherwise harmful to personnel.

hazardous waste: A solid, liquid, or gaseous waste that may cause or significantly contribute to serious illness or death or that poses a substantial threat to human health or the environment when the waste is improperly managed.

hearing conservation: The prevention of, or minimizing of, noise-induced deafness through the use of hearing protection devices, the control of noise through engineering controls, annual audiometric tests, and employee training.

heat cramps: A type of heat stress (a possible side effect of dehydration) that occurs as a result of salt and potassium depletion.

heat exhaustion: A condition usually caused by loss of body water from exposure to excess heat. Symptoms include headache, tiredness, nausea, and sometimes fainting.

heatstroke: A serious disorder resulting from exposure to excess heat. It results from sweat suppression and increased storage of body heat, characterized by high fever, collapse, and sometimes convulsions or coma.

Homeland Security: Federal cabinet-level department created to protect the United States of America and its citizens as a result of 9/11. The new Department of Homeland Security (DHS) has three primary missions: prevent terrorist attacks within the United States, reduce America's vulnerability to terrorism, and minimize the damage from potential attacks and natural disasters.

hot work: Work involving electric or gas welding, cutting, brazing, or similar flame or spark-producing operations.

human factor engineering: (used in the United States) or **ergonomics:** (used in Europe)—for practical purposes, the terms are synonymous and focus on human beings and their interaction with products, equipment, facilities, procedures, and environments used in work and everyday living. The emphasis is on human beings (as opposed to engineering, where the emphasis is more strictly on technical engineering considerations) and how the design of things influences people. Human factors, then, seek to change the things people use and the environments in which they use these things to better match the capabilities, limitations, and needs of people (Sanders and McCormick 1993).

ignition temperature: The temperature at which a given fuel bursts into flame.

illumination: The amount of light flux a surface receives per unit area. May be expressed in lumens per square foot or in foot-candles.

impulse noise: A noise characterized by rapid rise time, high peak value, and rapid decay.

incident: An undesired event that, under slightly different circumstances, could have resulted in personal harm or property damage; any undesired loss of resources.

indoor air quality (IAQ): The effect, good or bad, of the contents of the air inside a structure on its occupants. While usually temperature (too hot and cold), humidity (too dry or too damp), and air velocity (draftiness or motionless) are considered "comfort" rather than indoor air quality issues, IAQ refers to such problems as asbestosis, sick-building syndrome, biological aerosols, ventilation issues concerning dusts, fumes, and so forth.

industrial hygiene: The American Industrial Hygiene Association (AIHA) defines industrial hygiene as "that science and art devoted to the anticipation, recognition, evaluation, and control of those environmental factors or stresses—arising in the workplace—which may cause sickness, impaired health and well-being, or significant discomfort and inefficiency among workers or among citizens of the community."

ingestion: Entry of a foreign substance into the body through the mouth.

injury: A wound or other specific damage.

interlock: A device that interacts with another device or mechanism to govern succeeding operations. For example: an interlock on an elevator door prevents the car from moving unless the door is properly closed.

ionizing radiation: Radiation that becomes electrically charged (i.e., changed into ions).

irritant: A substance that produces an irritating effect when it contacts skin, eyes, nose, or respiratory system.

job hazard analysis: (also called **job safety analysis**) The breaking down into its component parts of any method or procedure to determine the hazards connected therewith and the requirements for performing it safely.

kinetic energy: The energy resulting from a moving object.

laboratory safety standard: A specific hazard communication program for laboratories, found in 29 CFR 1910.1450. These regulations are essentially a blend of hazard communication and emergency response for laboratories. The cornerstone of the Lab Safety Standard is the requirement for a written chemical hygiene plan.

lockout/tagout procedure: An OSHA procedure found in 29 CFR 1910.147. A tag or lock is used to "tag out" or "log out" a device, so that no one can inadvertently actuate the circuit, system, or equipment that is temporarily out of service.

Log and Summary of Occupational Injuries and Illnesses (OSHA-300 Log): A cumulative record that employers (generally of more than ten employees) are required to maintain, showing essential facts of all reportable occupational injuries and illnesses.

loss: The degradation of a system or component. Loss is best understood when related to dollars lost. Examples include death or injury to a worker, destruction or impairment of facilities or machines, destruction or spoiling of raw materials, and creation of delay. In the insurance business, loss connotes dollar loss, and we have seen underwriters who write it as LO$$ to make that point.

lower explosive limit (LEL): The minimum concentration of a flammable gas in air required for ignition in the presence of an ignition source. Listed as a percent by volume in air.

material safety data sheet (MSDS): Chemical information sheets provided by the chemical manufacturer that include information such as: chemical and physical characteristics; long- and short-term health hazards; spill-control procedures; personal protective equipment (PPE) to be used when handling the chemical; reactivity with other chemicals; incompatibility with other chemicals; and manufacturer's name, address, and phone number. Employee access to and understanding of MSDS are important parts of the HazCom program.

medical monitoring: The initial medical exam of a worker, followed by periodic exams. The purpose of medical monitoring is to assess workers' health, determine their fitness to wear personal protective equipment, and maintain records of their health.

metabolic heat: Produced within a body as a result of activity that burns energy.

mists: Minute liquid droplets suspended in air.

molds: The most typical forms of fungus found on earth, comprising approximately 25 percent of the earth's biomass (McNeel and Kreutzer 1996).

monitoring: Periodic or continuous surveillance or testing to determine the level of compliance with statutory requirements and/or pollutant levels in various media or in humans, animals, or other living things.

mycotoxins: Some molds are able to produce *mycotoxins,* natural organic compounds that are capable of initiating a toxic response in vertebrates (McNeel and Kreutzer 1996).

nonionizing radiation: That radiation on the electromagnet spectrum that has a frequency of 10^{15} or less and a wavelength in meters of 3 x 10^{-7}.

Occupational Safety and Health Act (OSH Act): A federal law passed in 1970 to assure, so far as possible, every working man and woman in the nation safe and healthful working conditions. To achieve this goal, the act authorizes several functions, such as encouraging safety and health programs in the workplace and encouraging labor–management cooperation in health and safety issues.

OSHA Form 300: Log and summary of occupational injuries and illnesses. Formerly OSHA Form 200.

oxidation: When a substance either gains oxygen or loses hydrogen or electrons in a chemical reaction. One of the chemical treatment methods.

oxidizer: Also known as an oxidizing agent, a substance that oxidizes another substance. Oxidizers are a category of hazardous materials that may assist in the production of fire by readily yielding oxygen.

oxygen deficient atmospheres: The legal definition of an atmosphere where the oxygen concentration is less than 19.5 percent by volume of air.

particulate matter: Substances (such as diesel soot and combustion products resulting from the burning of wood) released directly into the air; any minute, separate particle of liquid or solid material.

performance standards: A form of OSHA regulation standards that lists the ultimate goal of compliance but does not explain exactly how compliance is to be accomplished. Compliance is usually based on accomplishing the act or process in the safest manner possible, based on experience (past performance).

permissible exposure limit (PEL): The time-weighted average concentration of an airborne contaminant that a healthy worker may be exposed to eight hours per day or forty hours per week without suffering any adverse health effects. Established by legal means and enforceable by OSHA.

personal protective equipment (PPE): Any material or device worn to protect a worker from exposure to or contact with any harmful substance or force.

preliminary assessment: A quick analysis to determine how serious the situation is and to identify all potentially responsible parties. The preliminary assessment uses readily available information; for instance, forms, records, aerial photographs, and personnel interviews.

pressure: The force exerted against an opposing fluid or thrust, distributed over a surface.

radiant heat: The result of electromagnetic nonionizing energy that is transmitted through space without the movement of matter within that space.

radiation: Energetic nuclear particles, including alpha rays, beta rays, gamma rays, e-rays, neutrons, high-speed electrons, and high-speed protons.

reactive: A substance that reacts violently by catching on fire, exploding, or giving off fumes when exposed to water, air, or low heat.

reactivity hazard: The ability of a material to release energy when in contact with water. Also, the tendency of a material, when in its pure state or as a commercially produced product, to vigorously polymerize, decompose, condense, or otherwise self-react and undergo violent chemical change.

reportable quantity (RQ): The minimum amount of a hazardous material that, if spilled while in transport, must be reported immediately to the National Response Center. Minimum reportable quantities range from 1 pound to 5,000 pounds per twenty-four-hour day.

Resource Conservation and Recovery Act (RCRA): A federal law enacted in 1976 to deal with both municipal and hazardous waste problems and to encourage resource recovery and recycling.

risk: The combination of the expected frequency (event/year) and consequence (effects/event) of a single accident or a group of accidents; the result of a loss-probability occurrence and the acceptability of that loss.

risk assessment: A process that uses scientific principles to determine the level of risk that actually exists in a contaminated area.

risk characterization: The final step in the risk assessment process, it involves determining a numerical risk factor. This step ensures that exposed populations are not at significant risk.

risk management: The professional assessment of all loss potentials in an organization's structure and operations, leading to the establishment and administration of a comprehensive loss-control program.

safety: A general term denoting an acceptable level of risk of, relative freedom from, and low probability of harm.

safety factor: Based on experimental data, the amount added (e.g., thousand-fold) to ensure worker health and safety.

safety standard: A set of criteria specifically designed to define a safe product, practice, mechanism, arrangement, process, or environment, produced by a body representative of all concerned interests and based upon currently available scientific and empirical knowledge concerning the subject or scope of the standard.

secondary containment: A method using two containment systems so that if the first is breached, the second will contain all of the fluid in the first. For underground storage tanks, secondary containment consists of either a double-walled tank or a liner system.

security assessment: A security test intensified in scope and effort, the purpose of which is to obtain an advanced and very accurate idea of how well the organization has implemented security mechanisms, and to some degree, policy.

sensitizers: Chemicals that in very low doses trigger an allergic response.

short-term exposure limit (STEL): The time-weighted average concentration to which workers can be exposed continuously for a short period of time (typically fifteen minutes) without suffering irritation, chronic or irreversible tissue damage, or impairment for self-rescue.

silica: Crystalline silica (SiO_2) is a major component of the earth's crust and the cause of silicosis.

specific gravity: The ratio of the densities of a substance to water.

threshold limit value (TLV): The same concept as PEL, except that TLVs do not have the force of governmental regulations behind them but are based on recommended limits established and promoted by the American Conference of Governmental Industrial Hygienists.

time-weighted average (TWA): A mathematical average [(exposure in ppm x time in hours) 1/4 time in hours = time weighted average in ppm] of exposure concentration over a specific time.

total quality management (TQM): A way of managing a company that entails a total and willing commitment of all personnel at all levels to quality.

toxicity: The relative property of a chemical agent with reference to a harmful effect on some biologic mechanism and the condition under which this effect occurs. The quality of being poisonous.

toxicology: The study of poisons, which are substances that can cause harmful effects to living things.

unsafe condition: Any physical state that deviates from that which is acceptable, normal, or correct in terms of past production or potential future production of personal injury and/or damage to property; any physical state that results in a reduction in the degree of safety normally present.

upper explosive limit (UEL): The maximum concentration of a flammable gas in air required for ignition in the presence of an ignition source.

vulnerability assessment: A very regulated, controlled, cooperative, and documented evaluation of an organization's security posture from outside-in and inside-out, for the purpose of defining or greatly enhancing security policy.

workers' compensation: A system of insurance required by state law and financed by employers that provides payments to employees and their families for occupational illnesses, injuries, or fatalities incurred while at work and resulting in loss of wage income, usually regardless of the employer's or employee's negligence.

zero energy state: The state of equipment in which every power source that can produce movement of a part of the equipment, or the release of energy, has been rendered inactive.

Note: Although the preceding definitions are in no way "all inclusive," they do provide the student of safety with some fundamental concepts important for understanding the material to follow and should help when talking with others in the field and in solving safety problems.

REVIEW QUESTIONS

1. Define "accident" from a safety point of view.
2. Define "safety" from a safety point of view.
3. Define "injuries" from a safety point of view.
4. Define "engineering" from a safety point of view.

REFERENCES AND RECOMMENDED READING

ASSE. 1998. *Dictionary of terms used in the safety profession.* 3rd ed. Des Plaines, IL: American Society of Safety Engineers.

Bird, F. E., and G. L. Germain. 1966. *Damage control.* New York: American Management Association.

Boyce, A. 1997. *Introduction to environmental technology.* New York: Van Nostrand Reinhold.

CCPS. 2008. *Guidelines for hazard evaluation procedures.* 2nd ed. New York: American Institute of Chemical Engineers.

CFR 29 part 1910 1995. *Occupational safety and health standards for general industry.* Washington, DC: U.S. Department of Labor, OSHA.

CFR 40 1987. *Protection of environment.* Code of Federal Regulations Title 40.

Fletcher, J. A. 1972. *The industrial environment: Total loss control.* Ontario, Canada: National Profile Limited.

Haddon, W., Jr., E. A. Suchman, and D. Klein. 1964. *Accident research.* New York: Harper and Row.

McNeel, S., and R. Kreutzer. 1996. Fungi and indoor air quality. *Health and Environment Digest* 10:2 (May/June): 9–12.

Plog, B. A., ed. 2001. *Fundamentals of industrial hygiene.* 5th ed. Chicago, IL: National Safety Council.

Rose, C. F. 1999. *Antigens.* Cincinnati: American Conference of Governmental Industrial Hygienists. ACGIH Bioaerosols Assessment and Control, ch. 25, 25-1 to 25.11.

Sanders, M. S., and E. J. McCormick. 1993. *Human factors in engineering and design.* 7th ed. New York: McGraw-Hill.

II

MATHEMATICS

4

Conversion Factors and Units of Measurement

Basic mathematics and probability concepts are used to perform a variety of safety analyses.

INTRODUCTION

The units most commonly used by safety engineering professionals are based on the complicated English system of weights and measures. However, bench work is usually based on the metric system or the International System of Units (SI) due to the convenient relationship between milliliters (mL), cubic centimeters (cm^3), and grams (g).

The International System of Units (SI) is a modernized version of the metric system established by international agreement. The metric system of measurement was developed during the French Revolution and was first promoted in the United States in 1866. In 1902, proposed congressional legislation requiring the U.S. government to use the metric system exclusively was defeated by a single vote.

While we use both systems in this text, SI provides a logical and interconnected framework for all measurements in engineering, science, industry, and commerce. The metric system is much simpler to use than the existing English system since all its units of measurement are divisible by ten.

Before listing the various conversion factors commonly used in safety engineering, it is important to describe the prefixes commonly used in the SI system. These prefixes are based on the power 10. For example, a kilo means 1000 grams and a centimeter means one-hundredth of 1 meter. The twenty SI prefixes used to form decimal multiples and submultiples of SI units are given in table 4.1.

Note that the kilogram is the only SI unit with a prefix as part of its name and symbol. Because multiple prefixes may not be used, in the case of the kilogram the prefix names of table 4.1 are used with the unit name *gram* and the prefix symbols are used with the unit symbol *g*. With this exception, any SI prefix may be used with any SI unit, including the degree Celsius and its symbol °C.

Example 4.1

$$10^{-6} \text{ kg} = 1 \text{ mg (one milligram), but not } 10^{-6} \text{ kg} = 1\mu\text{kg (one microkilogram)}$$

Example 4.2
Consider the height of the Washington Monument (169 m or 555 ft). We may write h_w = 169,000 mm = 16,900 cm = 169 m = 0.169 km using the millimeter (SI prefix milli, symbol m), centimeter (SI prefix centi, symbol c), or kilometer (SI prefix kilo, symbol k).

CONVERSION FACTORS

Conversion factors are given below in alphabetical order (table 4.2) and in unit category listing order (table 4.3).

Table 4.1   SI Prefixes

Factor	Name	Symbol
10^{24}	yotta	Y
10^{21}	zetta	Z
10^{18}	exa	E
10^{15}	peta	P
10^{12}	tera	T
10^{9}	giga	G
10^{6}	mega	M
10^{3}	kilo	k
10^{2}	hecto	h
10^{1}	deka	da
10^{-1}	deci	d
10^{-2}	centi	c
10^{-3}	milli	m
10^{-6}	micro	μ
10^{-9}	nano	n
10^{-12}	pico	p
10^{-15}	femto	f
10^{-18}	atto	a
10^{-21}	zepto	z
10^{-24}	yocto	y

Example 4.3

Problem:

Find degrees in Celsius of water at 72°F.

Solution:

$$°C = (F - 32) \times 5/9 = (72 - 32) \times 5/9 = 22.2$$

Conversion Factors: Practical Examples

Sometimes we have to convert between different units. Suppose that a sixty-inch piece of pipe is attached to an existing six-foot piece of pipe. Joined together, how long are they?

Obviously, we cannot find the answer to this question by adding 60 to 6. Why? Because the two lengths are given in different units. Before we can add the two lengths, we must convert one of them to the units of the other. Then, when we have two lengths in the same units, we can add them.

In order to perform this conversion, we need a *conversion factor*. That is, in this case, we have to know how many inches make up a foot; that is, twelve inches is one foot. Knowing this, we can perform the calculation in two steps as follows:

1. 60 inches is really 60/12 = 5 feet
2. 5 feet + 6 feet = 11 feet

From the example above it can be seen that a conversion factor changes known quantities in one unit of measure to an equivalent quantity in another unit of measure.

In making the conversion from one unit to another, we must know two things:

1. the exact number that relates the two units
2. whether to multiply or divide by that number

When making conversions, confusion over whether to multiply or divide is common; on the other hand, the number that relates the two units is usually known and, thus, is not a problem. Understanding the proper methodology—the "mechanics"—to use for various operations requires practice and common sense.

Table 4.2   Alphabetical Listing of Conversion Factors

Factors	Metric (SI) or English Conversions
1 atm (atmosphere) =	1.013 bars 10.133 newtons/cm^2 (newtons/square centimeter) 33.90 ft of H_2O (feet of water) 101.325 kp (kilopascals) 1,013.25 mg (millibars) 13.70.70 psia (pounds/square inch—absolute) 760 torr 760 mm Hg (millimeters of mercury)
1 bar =	0.987 atm (atmospheres) 1 x 10^6 dynes/cm^2 (dynes/square centimeter) 33.45 ft of H_2O (feet of water) 1 x 10^5 pascals [nt/m^2] (newtons/square meter) 750.06 torr 750.06 mm Hg (millimeters of mercury)
1 Bq (becquerel) =	1 radioactive disintegration/second 2.7 x 10^{-11} Ci (curie) 2.7 x 10^{-8} mCi (millicurie)
1 BTU (British Thermal Unit) =	252 cal (calories) 1,055.06 j (joules) 10.41 liter-atmospheres 0.293 watt-hours
1 cal (calories) =	3.97 x 10^{-3} BTUs (British Thermal Units) 4.18 j (joules) 0.0413 liter-atmospheres 1.163 x 10^{-3} watt-hours
1 cm (centimeters) =	0.0328 ft (feet) 0.394 in (inches) 10,000 microns (micrometers) 100,000,000 Å =10^8 Å (Ångstroms)
1 cc (cubic centimeter) =	3.53 x 10^{-5} ft^3 (cubic feet) 0.061 in^3 (cubic inches) 2.64 x 10^{-4} gal (gallons) 52.18 ℓ (liters) 52.18 ml (milliliters)
1 ft^3 (cubic foot) =	28.317 cc (cubic centimeters) 1,728 in^3 (cubic inches) 0.0283 m^3 (cubic meters) 7.48 gal (gallons) 28.32 ℓ (liters) 29.92 qts (quarts)
1 in^3 =	16.39 cc (cubic centimeters) 16.39 ml (milliliters) 5.79 x 10^{-4} ft^3 (cubic feet) 1.64 x 10^{-5} m^3 (cubic meters) 4.33 x 10^{-3} gal (gallons) 0.0164 ℓ (liters) 0.55 fl oz (fluid ounces)

Table 4.2   Alphabetical Listing of Conversion Factors (continued)

Factors	Metric (SI) or English Conversions
1 m³ (cubic meter) =	1,000,000 cc = 10^6 cc (cubic centimeters)
	33.32 ft³ (cubic feet)
	61,023 in³ (cubic inches)
	264.17 gal (gallons)
	1,000 ℓ (liters)
1 yd³ (cubic yard) =	201.97 gal (gallons)
	764.55 ℓ (liters)
1 Ci (curie) =	3.7×10^{10} radioactive disintegrations/second
	3.7×10^{10} Bq (becquerel)
	1,000 mCi (millicurie)
1 day =	24 hrs (hours)
	1,440 min (minutes)
	86,400 sec (seconds)
	0.143 weeks
	2.738×10^{-3} yrs (years)
1°C (expressed as an interval) =	1.8°F = [9/5]°F (degrees Fahrenheit)
	1.8°R (degrees Rankine)
	1.0K (degrees Kelvin)
°C (degree Celsius) =	[(5/9)(°F − 32°)]
1°F (expressed as an interval) =	0.556°C = [5/9]°C (degrees Celsius)
	1.0°R (degrees Rankine)
	0.556K (degrees Kelvin)
°F (degree Fahrenheit) =	[(9/5)(°C) + 32°]
1 dyne =	1×10^{-5} nt (newton)
1 ev (electron volt) =	1.602×10^{-12} ergs
	1.602×10^{-19} j (joules)
1 erg =	1 dyne-centimeters
	1×10^{-7} j (joules)
	2.78×10^{-11} watt-hours
1 fps (feet/second) =	1.097 kmph (kilometers/hour)
	0.305 mps (meters/second)
	0.01136 mph (miles/hour)
1 ft (foot) =	30.48 cm (centimeters)
	12 in (inches)
	0.3048 m (meters)
	1.65×10^{-4} nt (nautical miles)
	1.89×10^{-4} mi (statute miles)
1 gal (gallon) =	3,785 cc (cubic centimeters)
	0.134 ft³ (cubic feet)
	231 in³ (cubic inches)
	3.785 ℓ (liters)
1 gm (gram) =	0.001 kg (kilogram)
	1,000 mg (milligrams)
	1,000,000 ng = 10^6 ng (nanograms)
	2.205×10^{-3} lbs (pounds)

Table 4.2 Alphabetical Listing of Conversion Factors (continued)

Factors	Metric (SI) or English Conversions
1 gm/cc (grams/cubic cent.) =	62.43 lbs/ft^3 (pounds/cubic foot) 0.0361 lbs/in^3 (pounds/cubic inch) 8.345 lbs/gal (pounds/gallon)
1 Gy (gray) =	1 j/kg (joules/kilogram) 100 rad 1 Sv (sievert) [unless modified through division by an appropriate factor, such as Q and/or N]
1 hp (horsepower) =	745.7 j/sec (joules/sec)
1 hr (hour) =	0.0417 days 60 min (minutes) 3,600 sec (seconds) 5.95 × 10^{-3} weeks 1.14 × 10^{-4} yrs (years)
1 in (inch) =	2.54 cm (centimeters) 1,000 mils
1 inch of water =	1.86 mm Hg (millimeters of mercury) 249.09 pascals 0.0361 psi (lbs/in^2)
1 j (joule) =	9.48 ×10^{-4} BTUs (British Thermal Units) 0.239 cal (calories) 10,000,000 ergs = 1 × 10^7 ergs 9.87 × 10^{-3} liter-atmospheres 1.0 nt-m (newton-meters)
1 kcal (kilocalories) =	3.97 BTUs (British Thermal Units) 1,000 cal (calories) 4,186.8 j (joules)
1 kg (kilogram) =	1,000 gms (grams) 2,205 lbs (pounds)
1 km (kilometer) =	3,280 ft (feet) 0.54 nt (nautical miles) 0.6214 mi (statute miles)
1 kw (kilowatt) =	56.87 BTU/min (British Thermal Units) 1.341 hp (horsepower) 1,000 j/sec (kilocalories)
1 kw-hr (kilowatt-hour) =	3,412.14 BTU (British Thermal Units) 3.6 × 10^6 j (joules) 859.8 kcal (kilocalories)
1 ℓ (liter) =	1,000 cc (cubic centimeters) 1 dm^3 (cubic decimeters) 0.0353 ft^3 (cubic feet) 61.02 in^3 (cubic inches) 0.264 gal (gallons) 1,000 ml (milliliters) 1.057 qts (quarts)

Table 4.2   Alphabetical Listing of Conversion Factors (continued)

Factors	Metric (SI) or English Conversions
1 m (meter) =	1×10^{10} Å (Ångstroms)
	100 cm (centimeters)
	3.28 ft (feet)
	39.37 in (inches)
	1×10^{-3} km (kilometers)
	1,000 mm (millimeters)
	1,000,000 µ = 1×10^{6} µ (micrometers)
	1×10^{9} nm (nanometers)
1 mps (meter/second) =	196.9 fpm (feet/minute)
	3.6 kmph (kilometers/hour)
	2.237 mph (miles/hour)
1 mph (mile/hour) =	88 fpm (feet/minute)
	1.61 kmph (kilometers/hour)
	0.447 mps (meters/second)
1 kt (nautical mile) =	6,076.1 ft (feet)
	1.852 km (kilometers)
	1.15 mi (statute miles)
	2,025.4 yds (yards)
1 mi (statute mile) =	5,280 ft (feet)
	1.609 km (kilometers)
	1,609.3 m (meters)
	0.869 nt (nautical miles)
	1,760 yds (yards)
1 miCi (millicurie) =	0.001 Ci (curie)
	3.7×10^{10} radioactive disintegrations/second
	3.7×10^{10} Bq (becquerel)
1 mm Hg (mm of mercury) =	1.316×10^{-3} atm (atmosphere)
	0.535 in H_2O (inches of water)
	1.33 mb (millibars)
	133.32 pascals
	1 torr
	0.0193 psia (pounds/square inch—absolute)
1 min (minute) =	6.94×10^{-4} days
	0.0167 hrs (hours)
	60 sec (seconds)
	9.92×10^{-5} weeks
	1.90×10^{-6} yrs (years)
1 nt (newton) =	1×10^{5} dynes
1 nt-m (newton-meter) =	1.00 j (joules)
	2.78×10^{-4} watt-hours
1 ppm (part/million-volume) =	1.00 ml/m³ (milliliters/cubic meter)
1 ppm [wt] (part/million-weight) =	1.00 mg/kg (milligrams/kilograms)
1 pascal =	9.87×10^{-6} atm (atmospheres)
	4.015×10^{-3} in H_2O (inches of water)
	1.01 mb (millibars)
	7.5×10^{-3} mm Hg (milliliters of mercury)

Table 4.2 Alphabetical Listing of Conversion Factors (continued)

Factors	Metric (SI) or English Conversions
1 lb (pound) =	453.59 gms (grams) 16 oz (ounces)
1 lb/ft^3 (pound/cubic foot) =	16.02 gms/l (grams/liter)
1 lb/ft^3 (pound/cubic inch) =	27.68 gms/cc (grams/cubic centimeter) 1,728 lbs/ft^3 (pounds/cubic feet)
1 psi (pound/square inch) =	0.068 atm (atmospheres) 27.67 in H$_2$O (inches or water) 68.85 mb (millibars) 51.71 mm Hg (millimeters of mercury) 6,894.76 pascals
1 qt (quart) =	946.4 cc (cubic centimeters) 57.75 in^3 (cubic inches) 0.946 ℓ (liters)
1 rad =	100 ergs/gm (ergs/gram) 0.01 Gy (gray) 1 rem [unless modified through division by an appropriate factor, such as Q and/or N]
1 rem =	1 rad [unless modified through division by an appropriate factor, such as Q and/or N]
1 Sv (sievert) =	1 Gy (gray) [unless modified through division by an appropriate factor, such as Q and/or N]
1 cm^2 (square centimeter) =	1.076 x 10^{-3} ft^2 (square feet) 0.155 in^2 (square inches) 1 x10^{-4} m^2 (square meters)
1 ft^2 (square foot) =	2.296 x 10^{-5} acres 9.296 cm^2 (square centimeters) 144 in^2 (square inches) 0.0929 m^2 (square meters)
1 m^2 (square meter) =	10.76 ft^2 (square feet) 1,550 in^2 (square inches)
1 mi^2 (square mile) =	640 acres 2.79 x 10^7 ft^2 (square feet) 2.59 x 10^6 m^2 (square meters)
1 torr =	1.33 mb (millibars)
1 watt =	3.41 BTl/hr (British Thermal Units/hour) 1.341 x 10^{-3} hp (horsepower) 18.18 j/sec (joules/second)
1 watt-hour =	3.412 BTUs (British Thermal Unit) 859.8 cal (calories) 3,600 j (joules) 35.53 liter-atmosphere

Table 4.2   Alphabetical Listing of Conversion Factors (continued)

Factors	Metric (SI) or English Conversions
1 week =	7 days 168 hrs (hours) 10,080 min (minutes) 6.048×10^5 sec (seconds) 0.0192 yrs (years)
1 yr (year) =	365.25 days 8,766 hrs (hours) 5.26×10^5 min (minutes) 3.16×10^7 sec (seconds) 52.18 weeks

Table 4.3   Conversion Factors by Unit Category

Units of Length

1 cm (centimeter) =	0.0328 ft (feet) 0.394 in (inches) 10,000 microns (micrometers) 100,000,000 Å = 10^8 Å (Ångstroms)
1 ft (foot) =	30.48 cm (centimeters) 12 in (inches) 0.3048 m (meters) 1.65×10^{-4} nt (nautical miles) 1.89×10^{-4} mi (statute miles)
1 in (inch) =	2.54 cm (centimeters) 1,000 mils
1 km (kilometer) =	3,280.8 ft (feet) 0.54 nt (nautical miles) 0.6214 mi (statute miles)
1 m (meter) =	1×10^{10} Å (Ångstroms) 100 cm (centimeters) 3.28 ft (feet) 39.37 in (inches) 1×10^{-3} km (kilometers) 1,000 mm (millimeters) 1,000,000 μ = 1×10^6 μ (micrometers) 1×10^9 nm (nanometers)
1 kt (nautical mile) =	6,076.1 ft (feet) 1.852 km (kilometers) 1.15 km (statute miles) 2.025.4 yds (yards)
1 mi (statute mile) =	5,280 ft (feet) 1.609 km (kilometers) 1.690.3 m (meters) 0.869 nt (nautical miles) 1,760 yds (yards)

Units of Area

1 cm² (square centimeter) =	1.076×10^{-3} ft² (square feet) 0.155 in² (square inches) 1×10^{-4} m² (square meters)

Table 4.3   Conversion Factors by Unit Category (continued)

Units of Length	

1 ft^2 (square foot) =	2.296 x 10^{-5} acres
	929.03 cm^2 (square centimeters)
	144 in^2 (square inches)
	0.0929 m^2 (square meters)
1 m^2 (square meter) =	10.76 ft^2 (square feet)
	1,550 in^2 (square inches)
1 mi^2 (square mile) =	640 acres
	2.79 x 10^7 ft^2 (square feet)
	2.59 x 10^6 m^2 (square meters)

Units of Volume

1 cc (cubic centimeter) =	3.53 x 10^{-5} ft^3 (cubic feet)
	0.061 in^3 (cubic inches)
	2.64 x 10^{-4} gal (gallons)
	0.001 ℓ (liters)
	1.00 ml (milliliters)
1 ft^3 (cubic foot) =	28,317 cc (cubic centimeters)
	1,728 in^3 (cubic inches)
	0.0283 m^3 (cubic meters)
	7.48 gal (gallons)
	28.32 ℓ (liters)
	29.92 qts (quarts)
1 in^3 (cubic inch) =	16.39 cc (cubic centimeters)
	16.39 ml (milliliters)
	5.79 x 10^{-4} ft^3 (cubic feet)
	1.64 x 10^{-5} m^3 (cubic meters)
	4.33 x 10^{-3} gal (gallons)
	0.0164 ℓ (liters)
	0.55 fl oz (fluid ounces)
1 m^3 (cubic meter) =	1,000,000 cc = 10^6 cc (cubic centimeters)
	35.31 ft^3 (cubic feet)
	61,023 in^3 (cubic inches)
	264.17 gal (gallons)
	1,000 ℓ (liters)
1 yd^3 (cubic yard) =	201.97 gal (gallons)
	764.55 ℓ (liters)
1 gal (gallon) =	3,785 cc (cubic centimeters)
	0.134 ft^3 (cubic feet)
	231 in^3 (cubic inches)
	3.785 ℓ (liters)
1 ℓ (liter) =	1,000 cc (cubic centimeters)
	1 dm^3 (cubic decimeters)
	0.0353 ft^3 (cubic feet)
	61.02 in^3 (cubic inches)
	0.264 gal (gallons)
	1,000 (milliliters)
	1.057 qts (quarts)
1 qt (quart) =	946.4 cc (cubic centimeters)
	57.75 in^3 (cubic inches)
	0.946 ℓ (liters)

Table 4.3   Conversion Factors by Unit Category (continued)

Units of Length

Units of Mass	
1 gm (gram) =	0.001 kg (kilograms)
	1,000 mg (milligrams)
	1,000,000 mg = 10^6 ng (nanograms)
	2.205 x 10^{-3} lbs (pounds)
1 kg (kilogram) =	1,000 gms (grams)
	2.205 lbs (pounds)
1 lb (pound) =	453.59 gms (grams)
	16 oz (ounces)

Units of Time	
1 day =	24 hrs (hours)
	1440 min (minutes)
	86,400 sec (seconds)
	0.143 weeks
	2.738 x 10^{-3} yrs (years)
1 hr (hour) =	0.0417 days
	60 min (minutes)
	3,600 sec (seconds)
	5.95 x 10^{-3} weeks
	1.14 x 10^{-4} yrs (years)
1 min (minutes) =	6.94 x 10^{-4} days
	0.0167 hrs (hours)
	60 sec (seconds)
	9.92 x 10^{-5} weeks
	1.90 x 10^{-6} yrs (years)
1 week =	7 days
	168 hrs (hours)
	10,080 min (minutes)
	6.048 x 10^5 sec (seconds)
	0.0192 yrs (years)
1 yr (year) =	365.25 days
	8,766 hrs (hours)
	5.26 x 10^5 min (minutes)
	3.16 x 10^7 sec (seconds)
	52.18 weeks

Units of the Measure of Temperature	
°C (degrees Celsius) =	[(5/9)(°F - 32°)]
1°C (expressed as an interval) =	1.8°F = [9/5]°F (degrees Fahrenheit)
	1.8°R (degrees Rankine)
	10°K (degrees Kelvin)
°F (degree Fahrenheit) =	[(9/5)(°C) + 32°]
1°F (expressed as an interval) =	0.556°C = [5/9]°C (degrees Celsius)
	1.0°R (degrees Rankine)
	0.556°K (degrees Kelvin)

Table 4.3   Conversion Factors by Unit Category (continued)

Units of Length

Units of Force

1 dyne =	1 x 10^{-5} nt (newtons)
1 nt (newton) =	1 x 10^5 dynes

Units of Work or Energy

1 BTU (British Thermal Unit) =
  252 cal (calories)
  1,055.06 j (joules)
  10.41 liter-atmospheres
  0.293 watt-hours

1 cal (calorie) =
  3.97 x 10^{-3} BTUs (British Thermal Units)
  4.18 j (joules)
  0.0413 liter-atmospheres
  1.163 x 10^{-3} watt-hours

1 ev (electron volt) =
  1.602 x 10^{-12} ergs
  1.602 x 10^{-19} j (joules)

0 erg =
  1 dyne-centimeter
  1 x 10^{-7} j (joules)
  2.78 x 10^{-11} watt-hours

1 j (joule) =
  9.48 x 10^{-4} BTUs (British Thermal Units)
  0.239 cal (calories)
  10,000,000 ergs = 1 x 10^7 ergs
  9.87 x 10^{-3} liter-atmospheres
  1.00 nt-m (newton-meters)

1 kcal (kilocalorie) =
  3.97 BTUs (British Thermal Units)
  1,000 cal (calories)
  4,186.8 j (joules)

1 kw-hr (kilowatt-hour) =
  3,412.14 BTU (British Thermal Units)
  3.6 x10^6 j (joules)
  859.8 kcal (kilocalories)

1 nt-m (newton-meter) =
  1.00 j (joule)
  2.78 x 10^{-4} watt-hours

1 watt-hour =
  3.412 BTUs (British Thermal Units)
  859.8 cal (calories)
  3,600 j (joules)
  35.53 liter-atmospheres

Units of Power

1 hp (horsepower) =
  745.7 j/sec (joules/sec)

1 kw (kilowatt) =
  56.87 BTU/min (British Thermal Units/minute)
  1.341 hp (horsepower)
  1,000 j/sec (joules/sec)

1 watt =
  3.41 BTU/hr (British Thermal Units/hour)
  1.341 x 10^{-3} hp (horsepower)
  1.00 j/sec (joule/sec)

Units of Pressure

1 atm (atmosphere) =
  1.013 bars
  10.133 newtons/cm² (newtons/square centimeters)
  33.90 ft. of H_2O (feet of water)

Table 4.3   Conversion Factors by Unit Category (continued)

Units of Length	

	101.325 kp (kilopascals)
	14.70 psia (pounds/square inch – absolute)
	760 torr
	760 mm Hg (millimeters of mercury)
1 bar =	0.987 atm (atmospheres)
	1 x 10⁶ dynes/cm² (dynes/square centimeter)
	33.45 ft of H₂O (feet of water)
	1 x 10⁵ pascals [nt/m²] (newtons/square meter)
	750.06 torr
	750.06 mm Hg (millimeters of mercury)
1 inch of water =	1.86 mm Hg (millimeters of mercury)
	249.09 pascals
	0.0361 psi (lbs/in²)
1 mm Hg (millimeter of merc.) =	1.316 x 10⁻³ atm (atmospheres)
	0.535 in H₂O (inches of water)
	1.33 mb (millibars)
	133.32 pascals
	1 torr
	0.0193 psia (pounds/square inch—absolute)
1 pascal =	9.87 x 10⁻⁶ atm (atmospheres)
	4.015 x 10⁻³ in H₂O (inches of water)
	0.01 mb (millibar)
	7.5 x 10⁻³ mm Hg (millimeters of mercury)
1 psi (pound/square inch) =	0.068 atm (atmospheres)
	27.67 in H₂O (inches of water)
	68.85 mb (millibars)
	51.71 mm Hg (millimeters of mercury)
	6,894.76 pascals
1 torr =	1.33 mb (millibars)

Units of Velocity or Speed

1 fps (feet/second) =	1.097 kmph (kilometers/hour)
	0.305 mps (meters/second)
	0.01136 mph (miles/hours)
1 mps (meter/second) =	196.9 fpm (feet/minute)
	3.6 kmph (kilometers/hour)
	2.237 mph (miles/hour)
1 mph (mile/hour) =	88 fpm (feet/minute)
	1.61 kmph (kilometers/hour)
	0.447 mps (meters/second)

Units of Density

1 gm/cc (gram/cubic cent.) =	62.43 lbs/ft³ (pounds/cubic foot)
	0.0361 lbs/in³ (pounds/cubic inch)
	8.345 lbs/gal (pounds/gallon)
1 lb/ft³ (pound/cubic foot) =	16.02 gms/ℓ (grams/liter)
1 lb/in² (pound/cubic inch) =	27.68 gms/cc (grams/cubic centimeter)
	1.728 lbs/ft³ (pounds/cubic foot)

Table 4.3   Conversion Factors by Unit Category (continued)

Units of Length

Units of Concentration

1 ppm (part/million-volume) =	1.00 ml/m^3 (milliliter/cubic meter)
1 ppm (wt) =	1.00 mg/kg (milligram/kilograms)

Radiation and Dose Related Units

1 Bq (becquerel) =	1 radioactive disintegration/second 2.7×10^{-11} Ci (curie) 2.7×10^{-8} (millicurie)
1 Ci (curie) =	3.7×10^{10} radioactive disintegration/second 3.7×10^{10} Bq (becquerel) 1,000 mCi (millicurie)
1 Gy (gray) =	1 j/kg (joule/kilogram) 100 rad 1 Sv (sievert) [unless modified through division by an appropriate factor, such as Q and/or N]
1 mCi (millicurie) =	0.001 Ci (curie) 3.7×10^{10} radioactive disintegrations/second 3.7×10^{10} Bq (becquerel)
1 rad =	100 ergs/gm (ergs/gm) 0.01 Gy (gray) 1 rem [unless modified through division by an appropriate factor, such as Q and/or N]
1 rem =	1 rad [unless modified through division by an appropriate factor, such as Q and/or N]
1 Sv (sievert) =	1 Gy (gray) [unless modified through division by an appropriate factor, such as Q and/or N]

Along with using the proper mechanics (and practice and common sense) in making conversions, probably the easiest and fastest method of converting units is to use a conversion table.

The simplest conversion requires that the measurement be multiplied or divided by a constant value. For instance, if the depth of wet cement in a form is 0.85 feet, multiplying by 12 inches per foot converts the measured depth to inches (10.2 inches). Likewise, if the depth of the cement in the form is measured as 16 inches, dividing by 12 inches per foot converts the depth measurement to feet (1.33 feet).

WEIGHT, CONCENTRATION, AND FLOW

Using table 4.4 to convert from one unit expression to another and vice versa is good practice. However, in making conversions to solve process computations in water treatment operations, for example, we must be familiar with conversion calculations based upon a relationship between weight, flow or volume, and concentration. The basic relationship is

$$\text{Weight} = \text{Concentration x Flow or Volume x Factor} \tag{4.1}$$

Table 4.5 summarizes weight, volume, and concentration calculations. With practice, many of these calculations become second nature to users.

Table 4.4   Conversion Table

To Convert	Multiply By	To Get
Feet	12	Inches
Yards	3	Feet
Yards	36	Inches
Inches	2.54	Centimeters
Meters	3.3	Feet
Meters	100	Centimeters
Meters	1,000	Millimeters
Square Yards	9	Square Feet
Square Feet	144	Square Inches
Acres	43,560	Square Feet
Cubic Yards	27	Cubic Feet
Cubic Feet	1,728	Cubic Inches
Cubic Feet (water)	7.48	Gallons
Cubic Feet (water)	62.4	Pounds
Acre-Feet	43,560	Cubic Feet
Gallons (water)	8.34	Pounds
Gallons (water)	3.785	Liters
Gallons (water)	3,785	Milliliters
Gallons (water)	3,785	Cubic Centimeters
Gallons (water)	3,785	Grams
Liters	1,000	Milliliters
Days	24	Hours
Days	1,440	Minutes
Days	86,400	Seconds
Million Gallons/Day	1,000,000	Gallons/Day
Million Gallons/Day	1.55	Cubic Feet/Second
Million Gallons/Day	3.069	Acre-Feet/Day
Million Gallons/Day	36.8	Acre-Inches/Day
Million Gallons/Day	3,785	Cubic Meters/Day
Gallons/Minute	1,440	Gallons/Day
Gallons/Minute	63.08	Liters/Minute
Pounds	454	Grams
Grams	1,000	Milligrams
Pressure, psi	2.31	Head, ft (Water)
Horsepower	33,000	Foot-Pounds/Minute
Horsepower	0.746	Kilowatts
To Get	Divide By	To Convert

The following conversion factors are used extensively in safety engineering (fire science engineering).

- 7.48 gallons per ft^3
- 3.785 liters per gallon
- 454 grams per pound
- 1,000 mL per liter
- 1,000 mg per gram
- 1 ft^3/sec (cfs) = 0.6465 MGD

Note: Density (also called specific weight) is mass per unit volume and may be registered as lb/cu ft, lb/gal, grams/mL, grams/cu meter. If we take a fixed volume container, fill it with a fluid, and weigh it, we can determine density of the fluid (after subtracting the weight of the container).

- 8.34 pounds per gallon (water)—(density = 8.34 lb/gal)
- 1 milliliter of water weighs 1 gram—(density = 1 gram/mL)
- 62.4 pounds per ft^3 (water)—(density = 8.34 lb/gal)
- 8.34 lb/gal = mg/L (converts dosage in mg/L into lb/day/MGD)

Table 4.5 Weight, Volume, and Concentration Calculations

To Calculate	Formula
Pounds	Concentration, mg/L x Tank Volume MG x 8.34 lb/MG/mg/L
Pounds/Day	Concentration, mg/L x Flow, MGD x 8.34 lb/MG/mg/L
Million Gallons/Day	$\dfrac{\text{Quantity, lb/day}}{(\text{Conc., mg/L} \times 8.34\ \text{lb/mg/L/MG})}$
Milligrams/Liter	$\dfrac{\text{Quantity, lb}}{(\text{Tank Volume, MG} \times 8.34\ \text{lb/mg/L/MG})}$
Kilograms/Liter	Conc., mg/L x Volume, MG x 3.785 lb/MG/mg/L
Kilograms/Day	Conc., mg/L x Flow, MGD x 3.785 lb/MG/mg/L
Pounds/Dry Ton	Conc., MG/kg x 0.002 lb/d.t./mg/kg

Example: 1 mg/L x 10 MGD x 8.3 = 83.4 lb/day

- 1 psi = 2.31 feet of water (head)
- 1 foot head = 0.433 psi
- °F = 9/5(°C + 32)
- °C = 5/9(°F – 32)
- average water usage: 100 gallons/capita/day (gpcd)
- persons per single family residence: 3.7

FIRE SCIENCE HYDRAULICS AND GENERAL SCIENCE CONVERSIONS

Note: Use tables 4.4 and 4.5 to make the conversions indicated in the following example problems. Other conversions are presented in appropriate sections of the text.

Example 4.4
Convert cubic feet to gallons.

$$\text{Gallons} = \text{Cubic Feet, ft}^3 \times \text{gal/ft}^3$$

Problem:
How many gallons of water can be pumped to a storage tank that has 3,600 cubic feet of volume available?

$$\text{Gallons} = 3{,}600\ \text{ft}^3 \times 7.48\ \text{gal/ft}^3 = 26{,}928\ \text{gal}$$

Example 4.5
Convert gallons to cubic feet.

$$\text{Cubic Feet} = \frac{\text{gal}}{7.48\ \text{gal/ft}^3}$$

Problem:
How many cubic feet of water are removed when 18,200 gallons are withdrawn?

$$\text{Cubic Feet} = \frac{18{,}200\ \text{gal}}{7.48\ \text{gal/ft}^3} = 2{,}433\ \text{ft}^3$$

Example 4.6
Convert pounds to gallons.

$$\text{Gallons} = \frac{\text{lb}}{8.34 \text{ lb/gal}}$$

Problem:
How many gallons of water are required to fill a tank that holds 7,540 pounds of water?

$$\text{Gallons} = \frac{7,540 \text{ lb}}{8.34 \text{ lb/gal}} = 904 \text{ gal}$$

Example 4.7
Convert million gallons per day (MGD) to gallons per minute (gpm).

$$\text{Flow} = \frac{\text{Flow, MGD x } 1,000,000 \text{ gal/MG}}{1,440 \text{ min/day}}$$

Problem:
The current flow rate is 5.55 MGD. What is the flow rate in gallons per minute?

$$\text{Flow} = \frac{5.55 \text{ MGD x } 1,000,000 \text{ gal/MG}}{1,440 \text{ min/day}} = 3,854 \text{ gpm}$$

Example 4.8
Convert million gallons per day (MGD) to gallons per day (gpd).

$$\text{Flow} = \text{Flow, MGD x } 1,000,000 \text{ gal/MG}$$

Problem:
The inflow meter reads 28.8 MGD. What is the current flow rate in gallons per day?

$$\text{Flow} = 28.8 \text{ MGD x } 1,000,000 \text{ gal/MG} = 28,800,000 \text{ gpd}$$

Example 4.9
Convert gallons per minute (gpm) to million gallons per day (MGD).

$$\text{Flow, MGD} = \frac{\text{Flow, gpm x } 1,440 \text{ min/day}}{1,000,000 \text{ gal/MG}}$$

Problem:
The flow meter indicates that the current flow rate is 1,469 gpm. What is the flow rate in MGD?

$$\text{Flow, MGD} = \frac{1,469 \text{ gpm x } 1,440 \text{ min/day}}{1,000,000 \text{ gal/MG}} = 2.12 \text{ MGD (rounded)}$$

Example 4.10
Convert flow in cubic feet per second (cfs) to million gallons per day (MGD).

$$\text{Flow, MGD} = \frac{\text{Flow, cfs}}{1.55 \text{ cfs/MG}}$$

Problem:
The flow in a channel is determined to be 3.89 cubic feet per second (cfs). What is the flow rate in million gallons per day (MGD)?

$$\text{Flow, MGD} = \frac{3.89 \text{ cfs}}{1.55 \text{ cfs/MG}} = 2.5 \text{ MGD}$$

Example 4.11
Problem:
A liquid chemical weighs 62 lb/cu ft. How much does a five-gallon can of it weigh?

Solution:
Solve for specific gravity; get lb/gal; multiply by 5.

$$\text{Specific Gravity} = \frac{\text{wt. channel}}{\text{wt. water}}$$

$$\frac{62 \text{ lb/cu ft}}{62.4 \text{ lb/cu ft}} = .99$$

$$\text{Specific Gravity} = \frac{\text{wt. chemical}}{\text{wt. water}}$$

$$.99 = \frac{\text{wt. chemical}}{8.34 \text{ lb/gal}}$$

8.26 lb/gal = wt. chemical
8.26 lb/gal x 5 gal = 41.3 lb

Example 4.12
Problem:
A wooden piling with a diameter of 16 inches and a length of 16 feet weighs 50 lb/cu ft. If it is inserted vertically into a body of water, what vertical force is required to hold it below the water surface?

Solution:
If this piling had the same weight as water, it would rest just barely submerged. Find the difference between its weight and that of the same volume of water. That is the weight needed to keep it down.

62.4 lb/cu ft (water)

-50.0 lb/cu ft (piling)

12.4 lb/cu ft difference

Volume of piling = .785 x 1.33² x 16 ft = 22.21 cu ft
                          12.4 lb/cu ft x 22.21 cu ft = 275.5 lb (needed to hold it below water surface)

Example 4.13
Problem:
A cinder block weighs 70 pounds in air. When immersed in water, it weighs 40 pounds. What is the volume and specific gravity of the cinder block?

Solution:
The cinder block displaces 30 pounds of water; solve for cu ft of water displaced (equivalent to volume of cinder block).

$$\frac{30 \text{ lb water displaced}}{62.4 \text{ lb/cu ft}} = .48 \text{ cu ft water displaced}$$

Cinder block volume = .48 cu ft; this weighs 70 lb.

$$\frac{70 \text{ lb}}{.48 \text{ cu ft}} = 145.8 \text{ lb/cu ft density of cinder block}$$

$$\text{Specific Gravity} = \frac{\text{density of cinder block}}{\text{density of water}}$$

$$= 2.34$$

TEMPERATURE CONVERSIONS

Two commonly used methods used to make temperature conversions are

°C = 5/9 (°F – 32)
°F = 9/5 (°C) + 32

Example 4.14
Problem:
At a temperature of 4°C, water is at its greatest density. What is the degree of Fahrenheit?
Solution: °F = (°C) = 9/5 + 32
                = 4 x 9/5 + 32
                = 7.2 + 32
                = 39.2

However, the difficulty arises when one tries to recall these formulas from memory. Probably the easiest way to recall these important formulas is to remember three basic steps for both Fahrenheit and Celsius conversions:

1. Add 40°.
2. Multiply by the appropriate fraction (5/9 or 9/5).
3. Subtract 40°.

Obviously, the only variable in this method is the choice of 5/9 or 9/5 in the multiplication step. To make the proper choice, you must be familiar with the two scales. The freezing point of water is 32° on the Fahrenheit scale and 0° on the Celsius scale. The boiling point of water is 212° on the Fahrenheit scale and 100° on the Celsius scale.

What does all this mean?

Note: Note, for example, that at the same temperature, higher numbers are associated with the Fahrenheit scale and lower numbers with the Celsius scale. This important relationship helps you decide whether to multiply by 5/9 or 9/5. Let's look at a few conversion problems to see how the three-step process works.

Example 4.15
Suppose that we wish to convert 240°F to Celsius. Using the three-step process, we proceed as follows:

1. Step 1: Add 40°.

    $$240° + 40° = 280°$$

2. Step 2: 280° must be multiplied by either 5/9 or 9/5. Because the conversion is to the Celsius scale, we will be moving to a number *smaller* than 280. Through reason and observation, obviously, if 280 were multiplied by 9/5, the result would be almost the same as multiplying by 2, which would double 280 rather than make it smaller. If we multiply by 5/9, the result will be about the same as multiplying by ½, which would cut 280 in half. Because in this problem we wish to move to a smaller number, we should multiply by 5/9:

    $$(5/9) (280°) = 156.0°C$$

3. Step 3: Now subtract 40°.

    $$156.0°C - 40.0°C = 116.0°C$$

Therefore, 240°F = 116.0°C

Example 4.16
Convert 22°C to Fahrenheit.

1. Step 1: Add 40°.
    $$22° + 40° = 62°$$

2. Step 2: Because we are converting from Celsius to Fahrenheit, we are moving from a smaller to a larger number, and 9/5 should be used in the multiplications.

    $$(9/5) (62°) = 112°$$

3. Step 3: Subtract 40°.
    $$112° - 40° = 72°$$

Thus, 22°C = 72°F

Obviously, knowing how to make these temperature conversion calculations is useful. However, in practical in situ or non-in situ operations, you may wish to use a temperature conversion table.

CONVERSION FACTORS: AIR POLLUTION MEASUREMENTS

Safety engineers are often tasked with monitoring environmental (air, water, and soil) conditions within the workplace or at various work sites. Although beyond the scope of this text, it is important for the safety engineer to be knowledgeable and well-trained on monitoring protocols and equipment in order to properly sample for air pollutants and ensure safe indoor air quality (IAQ). Thus, safety engineers must be familiar with types and properties of air contaminants, direct reading instruments, active and passive sampling methods, analytical methods, calibration techniques, survey and sampling strategies, recordkeeping and chain of custody, walk-through assessment, regulations, guidelines, consensus standards, and interpreting and using air sampling data. A major part of any type of environmental monitoring is mathematical calculations. Thus, we have included basic environmental sampling unit conversions in this section.

The recommended units for reporting air pollutant emissions are commonly stated in metric system whole numbers. If possible, the reported units should be the same as those that are actually being measured. For example, weight should be recorded in grams; volume of air should be recorded in cubic meters. When the analytical system is calibrated in one unit, the emissions should also be reported in the units of the calibration standard. For example, if a gas chromatograph is calibrated with a 1 ppm standard of toluene in air, then the emissions monitored by the system should also be reported in ppm. Finally, if the emission standard is defined in a specific unit, the monitoring system should be selected to monitor in that unit.

The preferred reporting units for the following types of emissions should be

- Nonmethane organic and volatile organic compound emissions   ppm, ppb
- Semivolatile organic compound emissions $\mu g/m^3$, mg/m^3
- Particulate matter (TSP/PM-10) emissions $\mu g/m^3$
- Metal compound emissions ng/m^3

Conversion from ppm to µg/m³

Often, the safety engineer must be able to convert from ppm to $\mu g/m^3$. Following is an example of how one would perform that conversion using sulfur dioxide (SO_2) as the monitored constituent.

Example 4.17

The expression *parts per million* is without dimensions, i.e., no units of weight or volume are specifically designed. Using the format of other units, the expression can be written:

$$\frac{parts}{million\ parts}$$

"Parts" are not defined. If cubic centimeters replace parts, we obtain:

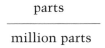

$$\frac{cubic\ centimeters}{million\ cubic\ centimeters}$$

Similarly, we might write pounds per million pounds, tons per million tons, or liters per million liters. In each expression, identical units of weight of volume appear in both the numerator and denominator and may be canceled out, leaving a dimensionless term.

An analog of parts per million is the more familiar term "percent." Percent can be written:

$$\frac{parts}{hundred\ parts}$$

To convert from part per million by volume, ppm, (µL/L), to $\mu g/m^3$ at EPA's standard temperature (25°C) and standard pressure (760 mm Hg), STP, it is necessary to know the molar volume at the given temperature and pressure and the molecular weight of the pollutant. At 25°C and 760 mm Hg, one mole of any gas occupies 24.46 liters.

Problem:

2.5 ppm by volume of sulfur dioxide (SO_2) was reported as the atmospheric concentration. What is this concentration in micrograms (µg) per cubic meter (m^3) at 25°C and 760 mmHg? What is the concentration in µg/m^3 at 37°C and 752 mm Hg?

Note: The following example problem points out the need for reporting temperature and pressure when the results are present on a weight–to-volume basis.

Solution:

Let parts per million equal µL/L then 2.5 ppm = 2.5 µL/L. The molar volume at 25°C and 760 mm Hg is 24.46 L and the molecular weight of SO_2 is 64.1 g/mole.

Step 1: 25°C and 760 mm Hg.

$$\frac{2.5\ \mu L}{L} \times \frac{1\ \mu mole}{24.46\ \mu L} \times \frac{64.1\ \mu g}{\mu mole} \times \frac{1000\ L}{m^3} = \frac{6.6 \times 10^3\ \mu g}{m^3}\ \text{at STP}$$

Step 2: 37°C and 752 mm Hg:

$$24.46\ \mu L\left(\frac{310K}{298K}\right)\left(\frac{760\ mm\ Hg}{752\ mm\ Hg}\right) = 25.72\ \mu L$$

$$\frac{2.5\ \mu L}{L} \times \frac{1\ \mu mole}{25.72\ \mu L} \times \frac{64.1\ \mu g}{\mu mole} \times \frac{1000\ L}{m^3} \times \frac{6.2 \times 10^3\ \mu g}{m^3}\ \text{at 37°, 752 mm Hg}$$

CONVERSION TABLES: AIR POLLUTION MEASUREMENTS

To assist the safety engineer in converting from one set of units to another, the following conversion factors for common air pollution measurements and other useful information are provided. The conversion tables provide factors for:

- atmospheric gases
- atmospheric pressure
- gas velocity
- concentration
- atmospheric particulate matter

Following is a list of conversions from ppm to µg/m^3 (at 25°C and 760 mm Hg) for several common air pollutants:

ppm SO_2 x 2620 = µg/m^3 SO_2 (Sulfur dioxide)
  ppm CO x 1150 = µg/m^3 CO (Carbon monoxide)
  ppm CO_x x 1.15 = mg/m^3 CO (Carbon dioxide)
  ppm CO_2 x 1.8 = mg/m^3 CO_2 (Carbon dioxide)
  ppm NO x 1230 = µg/m^3 NO (Nitrogen oxide)
  ppm NO_2 x 1880 = µg/m^3 NO_2 (Nitrogen dioxide)
  ppm O_2 x 1960 = µg/m^3 O_3 (Ozone)
  ppm CH_4 x 655 = µg/m^3 CH_4 (Methane)
  ppm CH_4 x 655 = mg/m^3 CH_4 (Methane)
  ppm CH_3SH x 2000 = µg/m^3 CH_3SH (Methyl mercaptan)
  ppm C_3H_8 x 1800 = µg/m^3 C_3H_8 (Propane)
  ppm C_3H_8 x 1.8 = mg/m^3 C_3H_8 (Propane)

ppm F- x 790 = µg/m³ F- (Fluoride)
ppm H_2S x 1400 = µg/m³ H_2S (Hydrogen sulfide)
ppm NH_3 x 696 = µg/m³ NH_3 (Ammonia)
ppm HCHO x 1230 = µg/m³ HCHO (Formaldehyde)

Table 4.6   Atmospheric Gases

To Convert From	To	Multiply By
Milligram/cu m	Micrograms/cu m	1000.0
	Micrograms/liter	1.0
	ppm by volume (20°C)	$\dfrac{24.04}{M}$
	ppm by weight	0.8347
	Pounds/cu ft	62.43×10^{-9}
Micrograms/cu ft	Milligrams/cu ft	0.001
	Micrograms/liter	0.001
	ppm by volume (20° C)	$\dfrac{0.02404}{M}$
	ppm by weight	834.7×10^{-6}
	Pounds/cu ft	62.43×10^{-12}
Micrograms/liter	Milligrams/cu m	1.0
	Micrograms/cu m	1000.0
	ppm by volume (20°C)	$\dfrac{24.04}{M}$
	ppm by weight	0.8347
	Pounds/cu ft	62.43×10^{-9}
ppm by volume (20°C)	Milligrams/cu m	$\dfrac{M}{24.04}$
	Micrograms/cu m	$\dfrac{M}{0.02404}$
	Micrograms/liter	$\dfrac{M}{24.04}$
	ppm by weight	$\dfrac{M}{28.8}$
	Pounds/cu ft	$\dfrac{M}{385.1 \times 10^{6}}$
ppm by weight	Milligrams/cu m	1.198
	Micrograms/cu m	1.198×10^{3}
	Micrograms/liter	1.198

Table 4.6 Atmospheric Gases (continued)

To Convert From	To	Multiply By
	ppm by volume (20°C)	$\dfrac{28.8}{M}$
	Pounds/cu ft	7.48×10^{-6}
Pounds/cu ft	Milligrams/cu m	16.018×10^{6}
	Micrograms/cu m	16.018×10^{9}
	Micrograms/liter	16.018×10^{6}
	ppm by volume (20°C)	$\dfrac{385.1 \times 10^{6}}{M}$
	ppm by weight	133.7×10^{3}

Table 4.7 Atmospheric Pressure

To Convert From	To	Multiply By
Atmospheres	Millimeters of mercury	760.0
	Inches of mercury	29.92
	Millibars	1013.2
Millimeters of mercury	Atmospheres	$1.316 \times 10\text{-}3$
	Inches of mercury	$39.37 \times 10\text{-}3$
	Millibars	1.333
Inches of mercury	Atmospheres	0.03333
	Millimeters of mercury	25.4005
	Millibars	33.35
Millibars	Atmospheres	0.000987
	Millimeters of mercury	0.75
	Inches of mercury	0.30
Sampling Pressures		
Millimeters of mercury (0°C)	Inches of water (60°C)	0.5358
Inches of mercury (0°C)	Inches of water (60°C)	13.609
Inches of water	Millimeters of mercury (0°C)	1.8663
	Inches of mercury (0°C)	$73.48 \times 10\text{-}2$

Table 4.8 Velocity

To Convert From	To	Multiply By
Meters/sec	Kilometers/hr	3.6
	Feet/sec	3.281
	Miles/hr	2.237
Kilometers/hr	Meters/sec	0.2778
	Feet/sec	0.9113
	Miles/hr	0.6241
Feet/hr	Meters/sec	0.3048
	Kilometers/hr	1.0973
	Miles/hr	0.6818
Miles/hr	Meters/sec	0.4470
	Kilometers/hr	1.6093
	Feet/sec	1.4667

Table 4.9 Atmospheric Particulate Matter

To Convert From	To	Multiply By
Milligrams/cu m	Grams/cu ft	283.2×10^{-6}
	Grams/cu m	0.001
	Micrograms/cu m	1000.0
	Monograms/cu ft	28.32
	Pounds/1000 cu ft	62.43×10^{-6}
Grams/cu ft	Milligrams/cu m	35.3145×10^{3}
	Grams/cu m	35.314
	Micrograms/cu m	35.314×10^{3}
	Micrograms/cu ft	1.0×10^{6}
Pounds/1000 cu ft		2.2046

Table 4.10 Concentration

To Convert From	To	Multiply By
Grams/cu m	Milligrams/cu m	1000.0
	Grams/cu ft	0.02832
	Micrograms/cu ft	1.0×10^{6}
	Pounds/1000 cu ft	0.06243
Micrograms/cu m	Milligrams/cu m	0.001
	Grams/cu ft	28.43×10^{-9}
	Grams/cu m	1.0×10^{-6}
	Micrograms/cu ft	0.02832
	Pounds/1000 cu ft	62.43×10^{-9}
Micrograms/cu ft	Milligrams/cu m	35.314×10^{-3}
	Grams/cu ft	1.0×10^{-6}
	Grams/cu m	35.314×10^{-6}
	Micrograms	35.314
	Pounds/1000 cu ft	2.2046×10^{-6}
Pounds/1000 cu ft	Milligrams/cu m	16.018×10^{3}
	Grams/cu ft	0.35314
	Micrograms/cu m	16.018×10^{6}
	Grams/cu m	16.018
	Micrograms/cu ft	353.14×10^{2}

Table 4.11 Soil Test Conversion Factors

Soil Sample Depth Inches	Multiply ppm by
3	1
6	2
7	2.33
8	2.66
9	3
10	3.33
12	4

SOIL TEST RESULTS CONVERSION FACTORS

Soil test results can be converted from parts per million (ppm) to pounds per acre by multiplying ppm by a conversion factor based on the depth to which the soil was sampled. Because a slice of soil 1 acre in area and 3 inches deep weighs approximately 1 million pounds, the conversion factors given in table 4.11 can be used.

REVIEW QUESTIONS

1. If 1,650 gallons of water are removed from the storage tank, how many pounds of water are removed?
2. The flow rate entering the storage tank is 2.89 MGD. What is the flow rate in cubic feet per second?
3. The totalizing flow meter indicates that 33,444,950 gallons of water have entered the firewater reservoir in the past twenty-four hours. What is the flow rate in MGD?
4. The water in a tank weighs 675 pounds. How many gallons does it hold?
5. A liquid chemical with a specific gravity (SG) of 1.22 is pumped at a rate of 40 gpm. How many pounds per day does the pump deliver?

5

Basic Math Operations

Most mathematical calculations required to be performed by safety engineers (as by many others) start with the basics, such as addition, subtraction, multiplication, division, and sequence of operations. Although many math operations are fundamental tools within each safety engineer's toolbox, it is important to reuse these tools on a consistent basis to remain sharp in their use. Engineers master basic math definitions and the formation of problems through aptitude, training, and experience: daily operations commonly require calculation of percentage, average, simple ratio, geometric dimensions, force, pressure, and head, and the use of dimensional analysis and advanced math operations. In this chapter we present basic math operations, provided primarily for the nonengineer who may need a quick refresher on the fundamentals.

Note: The experienced engineer may want to skip this and the other basic math review chapters and move on to the more technical areas that follow.

BASIC MATH TERMINOLOGY AND DEFINITIONS

The following basic definitions will aid in understanding the material in this chapter.

integer, or an **integral number,** is a whole number. Thus 1, 2, 3, 4, 5, 6, 7, 8, 9, 10, 11, and 12 are the first twelve positive integers.

factor, or **divisor,** of a whole number is any other whole number that exactly divides it. Thus, 2 and 5 are factors of 10.

prime number in math is a number that has no factors except itself and 1. Examples of prime numbers are 1, 3, 5, 7, and 11.

composite number is a number that has factors other than itself and 1. Examples of composite numbers are 4, 6, 8, 9, and 12.

common factor, or **common divisor,** of two or more numbers is a factor that will exactly divide each of them. If this factor is the largest factor possible, it is called the **greatest common divisor**. Thus, 3 is a common divisor of 9 and 27, but 9 is the greatest common divisor of 9 and 27.

A **multiple** of a given number is a number that is exactly divisible by the given number. If a number is exactly divisible by two or more other numbers, it is a common multiple of them. The least (smallest) such number is called the **lowest common multiple**. Thus, 36 and 72 are common multiples of 12, 9, and 4; however, 36 is the lowest common multiple.

An **even number** is a number exactly divisible by 2. Thus, 2, 4, 6, 8, 10, and 12 are even integers.

An **odd number** is an integer that is not exactly divisible by 2. Thus, 1, 3, 5, 7, 9, and 11 are odd integers.

product is the result of multiplying two or more numbers together. Thus, 25 is the product of 5 x 5. Also, 4 and 5 are factors of 20.

quotient is the result of dividing one number by another. For example, 5 is the quotient of 20 divided by 4.

dividend is a number to be divided; a **divisor** is a number that divides. For example, in $100 \div 20 = 5$, 100 is the dividend, 20 is the divisor, and 5 is the quotient.

area is the area of an object, measured in square units.

base is a term used to identify the bottom leg of a triangle, measured in linear units.

circumference is the distance around an object, measured in linear units. When determined for other than circles, it may be called the **perimeter** of the figure, object, or landscape.

cubic units are measurements used to express volume, cubic feet, cubic meters, etc.

depth is the vertical distance from the bottom of the tank to the top. This is normally measured in terms of liquid depth and given in terms of sidewall depth (SWD), measured in linear units.

diameter is the distance from one edge of a circle to the opposite edge passing through the center, measured in linear units.

height the vertical distance from the base or bottom of a unit to the top or surface.

linear units are measurements used to express distances: feet, inches, meters, yards, etc.

Pi (π) is a number in the calculations involving circles, spheres, or cones: π = 3.14.

radius is the distance from the center of a circle to the edge, measured in linear units.

sphere is a container shaped like a ball.

square units are measurements used to express area, square feet, square meters, acres, etc.

volume the capacity of the unit, how much it will hold, measured in cubic units (cubic feet, cubic meters) or in liquid volume units (gallons, liters, million gallons).

width is the distance from one side of the tank to the other, measured in linear units.

KEY WORDS

- *of:* means to multiply
- *and:* means to add
- *per:* means to divide
- *less than:* means to subtract

SEQUENCE OF OPERATIONS

Mathematical operations such as addition, subtraction, multiplication, and division are usually performed in a certain order or sequence. Typically, multiplication and division operations are done prior to addition and subtraction operations. In addition, mathematical operations are also generally performed from left to right using this hierarchy. The use of parentheses is also common to set apart operations that should be performed in a particular sequence.

Note: Again, it is assumed that the reader has a fundamental knowledge of basic arithmetic and math operations. Thus, the purpose of the following section is to provide a brief review of the mathematical concepts and applications frequently employed by safety engineers.

Sequence of Operations—Rules

Rule 1:   In a series of additions, the terms may be placed in any order and grouped in any way. Thus, 4 + 3 = 7 and 3 + 4 = 7; (4 + 3) + (6 + 4) = 17, (6 + 3) + (4 + 4) = 17, and [6 + (3 + 4) + 4]= 17.

Rule 2:   In a series of subtractions, changing the order or the grouping of the terms may change the result. Thus, 100 – 30 = 70, but 30 –100 = -70; (100 – 30) – 10 = 60, but 100 – (30 – 10) = 80.

Rule 3:   When no grouping is given, the subtractions are performed in the order written, from left to right. Thus, 100 – 30 – 15 – 4 = 51; or by steps, 100 – 30 = 70, 70 – 15 = 55, 55 – 4 = 51.

Rule 4:   In a series of multiplications, the factors may be placed in any order and in any grouping. Thus, [(2 x 3) x 5] x 6 = 180 and 5 x [2 x (6 x 3)] = 180.

Rule 5:   In a series of divisions, changing the order or the grouping may change the result. Thus, 100 ÷ 10 = 10, but 10 ÷ 100 = 0.1; (100 ÷ 10) ÷ 2 = 5, but 100 ÷ (10 + 2) = 20. Again, if no grouping is indicated, the divisions are performed in the order written, from left to right. Thus, 100 ÷ 10 ÷ 2 is understood to mean (100 ÷ 10) ÷ 2.

Rule 6:   In a series of mixed mathematical operations, the convention is as follows: whenever no grouping is given, multiplications and divisions are to be performed in the order written, then additions and subtractions in the order written.

Sequence of Operations—Examples

In a series of additions, the terms may be placed in any order and grouped in any way.

3+ 6 = 10 and 6 + 4 = 10
(4 + 5) + (3 + 7) = 19, (3 + 5) + (4 + 7) = 19, and [7 + (5 + 4)] + 3 = 19

In a series of subtractions, changing the order or the grouping of the terms may change the result.

100 − 20 = 80, but 20 − 100 = -80
(100 − 30) − 20 = 50, but 100 − (30 − 20) = 90

When no grouping is given, the subtractions are performed in the order written—from left to right.

100 − 30 − 20 − 3 = 47
or by steps, 100 − 30 = 70, 70 − 20 = 50, 50 − 3 = 47

In a series of multiplications, the factors may be placed in any order and in any grouping.

[(3 x 3) x 5] x 6 = 270 and 5 x [3 x (6 x 3)] = 270

In a series of divisions, changing the order or the grouping may change the result.

100 ÷ 10 = 10, but 10 ÷ 100 = 0.1
(100 ÷ 10) ÷ 2 = 5, but 100 ÷ (10 ÷ 2) = 20

If no grouping is indicated, the divisions are performed in the order written—from left to right.

100 ÷ 5 ÷ 2 is understood to mean (100 ÷ 5) ÷ 2

In a series of mixed mathematical operations, the rule of thumb is, whenever no grouping is given, multiplications and divisions are to be performed in the order written, then additions and subtractions in the order written.

PERCENT

The words *per cent* mean "by the hundred." Percentage is often designated by the symbol %. Thus, 15% means 15 percent or 15/100 or 0.15. These equivalents may be written in the reverse order: 0.15 = 15/100 = 15%.

When working with percent, the following key points are important:

1. **Percents** are another way of expressing a part of a whole.
2. As mentioned, the percent means **"by the hundred,"** so a percentage is the number out of 100. To determine percent, divide the quantity we wish to express as a percent by the total quantity, then multiply by 100.

$$\text{Percent } (\%) = \frac{\text{Part}}{\text{Whole}} \tag{5.1}$$

For example, 22 percent (or 22%) means 22 out of 100, or 22/100. Dividing 22 by 100 results in the decimal 0.22:

$$22\% = \frac{22}{100} = 0.22$$

3. When using percentage in calculations, the percentage must be converted to an equivalent decimal number; this is accomplished by dividing the percentage by 100.
   For example, calcium hypochlorite (HTH) contains 65% available chlorine. What is the decimal equivalent of 65%? Since 65% means 65 per hundred, divide 65 by 100: 65/100, which is 0.65.
4. Decimals and fractions can be converted to percentages. The fraction is first converted to a decimal, then the decimal is multiplied by 100 to get the percentage. For example, if a fifty-foot-high water tank has twenty-six feet of water in it, how full is the tank in terms of the percentage of its capacity?

$$\frac{26 \text{ ft}}{50 \text{ ft}} = 0.52 \text{ (decimal equivalent)}$$
$$0.52 \times 100 = 52$$

The tank is 52% full.

Example 5.1
Problem:
A foreperson removes 6,500 gal of water from a storage reservoir. The reservoir contains 135,325 gal. What is the percent of water removed?

Solution:

$$\text{Percent} = \frac{135,325 \text{ gal}}{6,500 \text{ gal}} \times 100$$

$$= 21\%$$

Example 5.2
Problem:
Convert 65% to decimal percent.

Solution:

$$\text{Decimal percent} = \frac{\text{Percent}}{100}$$

$$= \frac{65}{100}$$

$$= 0.65$$

Example 5.3
Problem:
Well water contains 5.8% minerals. What is the concentration of minerals in decimal percent?

Solution:

$$\text{Decimal percent} = \frac{5.8\%}{100} = 0.058$$

Note: Unless otherwise noted, all calculations in the text using percent values require the percent be converted to a decimal before use.

Note: To determine what quantity a percent equals, first convert the percent to a decimal then multiply by the total quantity.

$$\text{Quantity} = \text{Total} \times \text{Decimal Percent} \tag{5.2}$$

Example 5.4
Problem:
Water drawn from the storage tank contains 5% rust solids. If 2,800 gallons of water are withdrawn, how many gallons of rust solids are removed?

Solution:

$$\text{Gallons} = \frac{5\%}{100} \times 2{,}800 \text{ gallons} = 140 \text{ gal}$$

Example 5.5
Problem:
What is the percentage of 3 ppm?

Note: Because 1 liter of water weighs 1 kg (1000 g = 1,000,000 mg), milligrams per liter is parts per million (ppm).

Solution:

Since 3 parts per million (ppm) = 3 mg/L

$$3 \text{ mg/L} = \frac{3 \text{ mg}}{1 \text{ L} \times 1{,}000{,}000 \text{ mg/L}} \times 100\%$$

$$= \frac{3}{10{,}000} \%$$

$$= 0.0003\%$$

Example 5.6
Problem:
Calculate pounds per million gallons for 1 ppm (1 mg/L) of water.

Solution:
Because 1 gal of water = 8.34 lb

$$1 \text{ ppm} = \frac{1 \text{ gal}}{10^6 \text{ gal}}$$

$$= \frac{1 \text{ gal} \times 8.34 \text{ lb/gal}}{\text{mil gal}}$$

$$= 8.34 \text{ lb/mil gal}$$

Example 5.7
Problem:
How many pounds of activated carbon (AC) are needed with 42 lb of sand to make the mixture 26% AC?

Solution:
Let x be the weight of AC.

$$\frac{x}{42 + x} = 0.26$$

$$x = 0.26(42 + x)$$
$$x = 10.92 + 0.26x$$
$$(1 - 0.26)x = 10.92$$

$$x = \frac{10.92}{0.74} = 14.76 \text{ lb}$$

Example 5.8
Problem:
A motor is rated as 40 horsepower (hp). However, the output horsepower of the motor is only 26.5 hp. What is the efficiency of the motor?

Solution:

$$\text{Efficiency} = \frac{\text{hp output}}{\text{hp input}} \times 100\%$$

$$= \frac{26.5 \text{ hp}}{40 \text{ hp}} \times 100\%$$

$$= 66\%$$

SIGNIFICANT DIGITS

When rounding numbers, the following key points are important:

1. Numbers are rounded to reduce the number of digits to the right of the decimal point. This is done for convenience, not for accuracy.
2. **Rule:** A number is rounded off by dropping one or more numbers from the right and adding zeroes if necessary to place the decimal point. If the last figure dropped is 5 or more, increase the last retained figure by 1. If the last digit dropped is less than 5, do not increase the last retained figure. If the digit 5 is dropped, round off the preceding digit to the nearest *even* number.

Example 5.9
Problem:
Round off the following to one decimal.

Solution:

$$34.73 = 34.7$$

$$34.77 = 34.8$$
$$34.75 = 34.8$$
$$34.45 = 34.4$$
$$34.35 = 34.4$$

Example 5.10
Problem:
Round off 10,546 to 4, 3, 2, and 1 significant figures.

Solution:

$$10,546 = 10,550 \text{ to 4 significant figures}$$
$$10,546 = 10,500 \text{ to 3 significant figures}$$
$$10,546 = 11,000 \text{ to 2 significant figures}$$
$$10,547 = 10,000 \text{ to 1 significant figure}$$

In determining significant figures, the following key points are important:

1. The concept of significant figures is related to rounding.
2. It can be used to determine where to round off.

Note: No answer can be more accurate than the least accurate piece of data used to calculate the answer.

3. **Rule:** Significant figures are those numbers that are known to be reliable. The position of the decimal point does not determine the number of significant figures.

Example 5.11
Problem:
How many significant figures are in a measurement of 0.000135?

Solution:
Three significant figures: 1, 3, and 5. The three zeros are used only to place the decimal point.

Example 5.12
Problem:
How many significant figures are in a measurement of 103,500?

Solution:
Four significant figures: 1, 0, 3, and 5. The remaining two zeros are used to place the decimal point.

Example 5.13
Problem:
How many significant figures are in 27,000.0?

Solution:
There are six significant figures: 2, 7, 0, 0, 0, 0. In this case, the .0 means that the measurement is precise to 1/10 unit. The zeros indicate measured values and are not used solely to place the decimal point.

POWERS AND EXPONENTS

In working with powers and exponents, the following key points are important:

1. *Powers* are used to identify *area*, as in square feet, and volume as in *cubic feet*.

2. Powers can also be used to indicate that a number should be squared, cubed, etc. This later designation is the number of times a number must be multiplied times itself. For example, when several numbers are multiplied together, as 4 x 5 x 6 = 120, the numbers, 4, 5, and 6, are the *factors*; 120 is the *product*.

3. If all the factors are alike, as 4 x 4 x 4 x 4 = 256, the product is called a *power*. Thus, 256 is a power of 4, and 4 is the *base* of the power. A *power* is a *product* obtained by using a base a certain number of times as a factor.

4. Instead of writing 4 x 4 x 4 x 4, it is more convenient to use an *exponent* to indicate that the factor 4 is used as a factor four times. This exponent, a small number placed above and to the right of the base number, indicates how many times the base is to be used as a factor. Using this system of notation, the multiplication 4 x 4 x 4 x 4 is written as 4^4. The ^{4} is the *exponent*, showing that 4 is to be used as a factor 4 times.

5. These same considerations apply to letters (*a*, *b*, *x*, *y*, etc.) as well. For example:

$$z^2 = (z)(z) \quad \text{or} \quad z^4 = (z)(z)(z)(z)$$

The Powers of 1

$1^0 = 1$
$1^1 = 1$
$1^2 = 1$
$1^3 = 1$
$1^4 = 1$

The Powers of 10

$10^0 = 1$
$10^1 = 10$
$10^2 = 100$
$10^3 = 1,000$
$10^4 = 10,000$

Note: When a number or letter does not have an exponent, it is considered to have an exponent of one.

Example 5.14
Problem:
How is the term 2^3 written in expanded form?

Solution:
The power (exponent) of 3 means that the base number (2) is multiplied by itself three times.

$$2^3 = (2)(2)(2)$$

Example 5.15
Problem:
How is the term $(3/8)^2$ written in expanded form?

Note: When parentheses are used, the exponent refers to the entire term within the parentheses. Thus, in this example, $(3/8)^2$ means

Solution:
$$(3/8)^2 = (3/8)(3/8)$$

Note: When a negative exponent is used with a number or term, a number can be re-expressed using a positive exponent.

$$6^{-3} = 1/6^3$$

Another example is

$$11^{-5} = 1/11^5$$

Example 5.16
Problem:
How is the term 8^{-3} written in expanded form?

$$8^{-3} = \frac{1}{8^3} = \frac{1}{(8)(8)(8)}$$

Note: Any number or letter such as 3^0 or X^0 does not equal 3×1 or $X1$, but simply 1.

AVERAGES (ARITHMETIC MEAN)

Whether we speak of harmonic mean, geometric mean, or arithmetic mean, each is designed to find "the center" or the "middle" of the set of numbers. They capture the intuitive notion of a "central tendency" that may be present in the data. In statistical analysis an "average of data" is a number that indicates the middle of the distribution of data values.

An *average* is a way of representing several different measurements as a single number. Although averages can be useful by telling "about" how much or how many, they can also be misleading, as we demonstrate below. You find two kinds of averages in safety engineering calculations: the *arithmetic* mean (or simply *mean*) and the *median*.

Definition: The mean (what we usually refer to as an average) is the total of values of a set of observations divided by the number of observations. We simply add up all of the individual measurements and divide by the total number of measurements we took.

Example 5.17
Problem:
When working with averages, the **mean** (again, what we usually refer to as an *average*) is the total of values of a set of observations divided by the number of observations. We simply add up all of the individual measurements and divide by the total number of measurements we took. For example, a safety engineer tracks the on-the-job injuries that occur during a seven-day period, and these recordings are shown in table 5.1. Find the mean in the following table.

Table 5.1   Daily OJI Record

Day	On-the-Job Injuries (OJIs)
Monday	1
Tuesday	2
Wednesday	9
Thursday	3
Friday	1
Saturday	4
Sunday	2

Find the mean.

Solution:
Add up the seven OJI totals: 1, 2, 9, 3, 1, 4, 2 = 22. Next, divide by the number of measurements, in this case seven: 22 ÷ 7 = 3.14. The mean on-the-job injury rate for the week was 3.14 mg/l.

Example 5.18
Problem:
A water system has four wells with the following capacities: 115 gpm, 100 gpm, 125 gpm, and 90 gpm. What is the mean?

Solution:

$$\frac{115 \text{ gpm} + 100 \text{ gpm} + 125 \text{ gpm} + 90 \text{ gpm}}{4} = \frac{430}{4} = 107.5 \text{ gpm}$$

Example 5.19
Problem:
A fire water storage system has four storage tanks. Three of them have a capacity of 100,000 gallons each, while the fourth has a capacity of 1 million gallons. What is the mean capacity of the storage tanks?

Solution:
The mean capacity of the storage tanks is

$$\frac{100,000 + 100,000 + 100,000 + 1,000,000}{4} = 325,000 \text{ gal}$$

Note: Notice that no tank has a capacity anywhere close to the mean.

RATIO

A **ratio** is the established relationship between two numbers; it is simply one number divided by another number. For example, if someone says, "I'll give you four to one the Redskins over the Cowboys in the Super Bowl," what does that person mean?

Four to one, or 4:1, is a ratio. If someone gives you 4 to 1, it's his or her $4 to your $1.
When working with ratio, the following key points are important to remember.

1. One place where fractions are used in calculations is when ratios are used, such as calculating solutions.
2. A ratio is usually stated in the form A is to B as C is to D, and we can write it as two fractions that are equal to each other:

$$\frac{A}{B} = \frac{C}{D}$$

3. Cross multiplying solves ratio problems; that is, we multiply the left numerator (A) by the right denominator (D) and say that is equal to the left denominator (B) times the right numerator (C):

$$A \times D = B \times C$$
$$AD = BC$$

4. If one of the four items is unknown, dividing the two known items that are multiplied together by the known item that is multiplied by the unknown solves the ratio. For example, if $2 is needed to buy

5,000 gallons of water, how many dollars will we need to buy 100,000 gallons? We can state this as a ratio: 2 dollars is to 5,000 gallons of water as x is to 100,000 gallons.
This is set up in this manner:

$$\frac{2 \text{ dollars}}{5,000 \text{ gal water}} = \frac{x \text{ dollars}}{100,000 \text{ gal water}}$$

Cross multiplying:

$$(5,000)(x) = (2) \times (100,000)$$

Transposing:

$$x = \frac{2 \times 100,000}{5,000}$$
$$x = 40 \text{ dollars}$$

For calculating proportion, for example, five gallons of fuel costs $5.40. How much does 15 gallons cost?

$$\frac{5 \text{ gal}}{\$5.40} = \frac{15 \text{ gal}}{\$y}$$

$$5 \times y = 15 \times 5.40 = 81$$

$$y = \frac{81}{5} = \$16.20$$

Example 5.20
Problem:
Solve for x in the proportion problem given below.

Solution:

$$\frac{36}{180} = \frac{x}{4,450}$$

$$\frac{(4,450)(36)}{180} = x$$
$$= 890$$

Example 5.21
Problem:
Solve for the unknown value x in the problem given below.

Solution:

$$\frac{3.4}{2} = \frac{6}{x}$$
$$(3.4)(x) = (2)(6)$$

$$X = \frac{(2)\,(6)}{3.4}$$

$$x = 3.53$$

Example 5.22

Problem:

One pound of chemical is dissolved in 65 gallons of water. To maintain the same concentration, how many pounds of chemical would have to be dissolved in 150 gallons of water?

Solution:

$$\frac{1 \text{ lb}}{65 \text{ gal}} = \frac{x \text{ lbs}}{150 \text{ gal}}$$

$$(65)\,(x) = (1)\,(150)$$

$$x = \frac{(1)\,(150)}{65}$$

$$= 2.3 \text{ lbs}$$

Example 5.23

Problem:

It takes five workers fifty hours to complete a job. At the same rate, how many hours would it take eight workers to complete the job?

Solution:

$$\frac{5 \text{ workers}}{8 \text{ workers}} = \frac{x \text{ hours}}{50 \text{ hours}}$$

$$x = \frac{(5)\,(50)}{8}$$

$$x = 31.3 \text{ hours}$$

DIMENSIONAL ANALYSIS

Dimensional analysis is a problem-solving method that uses the fact that 1, without changing its value, can multiply any number or expression. It is a useful technique to check if a problem is set up correctly. In using dimensional analysis to check a math setup, we work with the dimensions (units of measure) only—not with numbers.

An example of dimensional analysis that is common to everyday life is the unit pricing found in many hardware stores. A shopper can purchase a one-pound box of nails for ninety-eight cents in one store, whereas a warehouse store sells a five-pound bag of the same nails for three dollars and fifty cents. The shopper will analyze this problem almost without thinking about it. The solution calls for reducing the problem to the price per pound. The pound is selected without much thought because it is the unit common to both stores. A shopper will pay seventy cents a pound for nails in the warehouse store or ninety-eight cents in the local hardware store. Implicit in the solution to this problem is knowing the unit price, which is expressed in dollars per pound ($/lb).

Note: Unit factors may be made from any two terms that describe the same or equivalent "amounts" of what we are interested in. For example, we know that

1 inch = 2.54 centimeters

In order to use the dimensional analysis method, we must know how to perform three basic operations:

1. To complete a division of units, always ensure that all units are written in the same format; it is best to express a horizontal fraction (such as gal/ft^2) as a vertical fraction.

    Horizontal to vertical

    $$\text{gal/cu ft to } \frac{\text{gal}}{\text{cu ft}}$$

    $$\text{psi to } \frac{\text{lb}}{\text{sq in.}}$$

    The same procedures are applied in the following examples.

    $$\text{ft}^3 \text{/min becomes } \frac{\text{ft}^3}{\text{min}}$$

    $$\text{s/min becomes } \frac{\text{s}}{\text{sq in.}}$$

2. We must know how to divide by a fraction. For example,

    $$\frac{\dfrac{\text{lb}}{\text{d}}}{\dfrac{\text{min}}{\text{d}}} \text{ becomes } \frac{\text{lb}}{\text{d}} \text{ x } \frac{\text{d}}{\text{min}}$$

    In the above, notice that the terms in the denominator were inverted before the fractions were multiplied. This is a standard rule that must be followed when dividing fractions.
    Another example is

    $$\frac{\dfrac{\text{mm}^2}{\text{mm}^2}}{\dfrac{}{\text{m}^2}} \text{ becomes mm}^2 \text{ x } \frac{\text{m}^2}{\text{mm}^2}$$

3. We must know how to cancel or divide terms in the numerator and denominator of a fraction. After fractions have been rewritten in the vertical form and division by the fraction has been re-expressed as multiplication as shown above, then the terms can be canceled (or divided) out.

Note: For every term that is canceled in the numerator of a fraction, a similar term must be canceled in the denominator and vice versa, as shown below:

$$\frac{Kg}{d} \times \frac{d}{min} = \frac{kg}{min}$$

$$mm^2 \times \frac{m^2}{mm^2} = m^2$$

$$\frac{gal}{min} \times \frac{ft^3}{gal} = \frac{ft^3}{min}$$

Question: How are units that include exponents calculated?

When written with exponents, such as ft^3, a unit can be left as is or put in expanded form, (ft)(ft)(ft), depending on other units in the calculation. The point is that it is important to ensure that square and cubic terms are expressed uniformly, as sq ft, cu ft, or as ft^2. For dimensional analysis, the latter system is preferred.

For example, let's say that we wish to convert 1,400 ft^3 volume to gallons, and we will use 7.48 gal/ft^3 in the conversions. The question becomes, do we multiply or divide by 7.48?

In the above instance, it is possible to use dimensional analysis to answer this question; that is, are we to multiply or divide by 7.48?

In order to determine if the math setup is correct, only the dimensions are used.

First, try dividing the dimensions:

$$\frac{ft^3}{gal/ft^3} = \frac{ft^3}{\dfrac{gal}{ft^3}}$$

Then, the numerator and denominator are multiplied to get

$$= \frac{ft^6}{gal}$$

So, by dimensional analysis we determine that if we divide the two dimensions (ft^3 and gal/ft^3), the units of the answer are ft^6/gal, not gal. It is clear that division is not the right way to go in making this conversion.

What would have happened if we had multiplied the dimensions instead of dividing?

$$(ft^3)\,(gal/ft^3) = (ft^3)\left(\frac{gal}{ft^3}\right)$$

Then, multiply the numerator and denominator to obtain

$$= \frac{(ft^3)\,(gal)}{ft^3}$$

And cancel common terms to obtain

$$= \frac{(ft^3)\,(gal)}{ft^3}$$

Obviously, by multiplying the two dimensions (ft³ and gal/ft³), the answer will be in gallons, which is what we want. Thus, because the math setup is correct, we would then multiply the numbers to obtain the number of gallons.

$$(1,400\ ft^3)\ (7.48\ gal/ft^3) = 10,472\ gal$$

Now let's try another problem with exponents. We wish to obtain an answer in square feet. If we are given the two terms—70 ft³/s and 4.5 ft/s—is the following math setup correct?

$$(70\ ft^3/s)\ (4.5\ ft/s)$$

First, only the dimensions are used to determine if the math setup is correct. By multiplying the two dimensions, we get

$$(ft^3/s)\ (ft/s) = \left(\frac{ft^3}{s}\right)\left(\frac{ft^3}{s}\right)$$

Then multiply the terms in the numerators and denominators of the fraction:

$$= \frac{(ft^3)\,(ft)}{(s)\,(s)}$$

$$= \frac{ft^4}{s^2}$$

Obviously, the math setup is incorrect because the dimensions of the answer are not square feet. Therefore, if we multiply the numbers as shown above, the answer will be wrong.

Let's try division of the two dimensions instead.

$$ft^3/s = \frac{\dfrac{ft^3}{s}}{\dfrac{ft}{s}}$$

Invert the denominator and multiply to get

$$= \left(\frac{ft^3}{(s)}\right)\left(\frac{s}{(ft)}\right)$$

$$= \frac{(ft)(ft)(ft)\,(s)}{(s)\,(ft)}$$

$$= \frac{(ft)\,(ft)\,(ft)\,(s)}{(s)\,(ft)}$$

$$= ft^2$$

Because the dimensions of the answer are square feet, this math setup is correct. Therefore, by dividing the numbers as was done with units, the answer will also be correct.

$$\frac{70\ ft^3/s}{4.5\ ft/s} = 15.56\ ft^2$$

Example 5.24
Problem:
We are given two terms 5 m/s and 7 m², and the answer to be obtained is in cubic meters per second (m³/s). Is multiplying the two terms the correct math setup?

Solution:

$$(m/s)\,(m^2) = \frac{m^2}{s} \times m^2$$

Multiply the numerators and denominator of the fraction:

$$= \frac{(m)\,(m^2)}{s}$$

$$= \frac{m^3}{s}$$

Because the dimensions of the answer are cubic meters per second (m³/s), the math setup is correct. Therefore, multiply the numbers to get the correct answer.

$$5\,(m/s)\,(7\ m^2) = 35\ m^3/s$$

Example 5.25
Problem:
A 10 gallon empty tank weighs 4.6 lbs. What is the total weight of the tank filled with 6 gal of water?

Solution:

Weight of water = 6 gal x 8.34 lb/gal
= 50.04 lb
Total weight = 50.04 + 4.6 lbs
= 54.6 lb

Example 5.26
Problem:
The depth of sand applied to the waste drying bed is 10 in. What is the depth in centimeters (2.54 cm = 1 in.)?

Solution:
10 in. = 10 x 2.54 cm
= 25.4 cm

GEOMETRICAL MEASUREMENTS

Proper design and operational control of industrial equipment and processes requires safety engineers to perform several types of calculations. Many of these calculations include parameters such as the circumference or perimeter, area, or the volume of the tank or channel as part of the information necessary to determine the result. Many process calculations require computation of surface areas. To aid in performing these calculations, the following definitions and relevant equations used to calculate areas and volumes for several geometric shapes are provided.

Definitions

- **area:** The area of an object, measured in square units.
- **base:** The term used to identity the bottom leg of a triangle, measured in linear units.
- **circumference:** The distance around an object, measured in linear units. When determined for other than circles, it may be called the perimeter of the figure, object, or landscape.
- **cubic units:** Measurements used to express volume, cubic feet, cubic meters, etc.
- **depth:** The vertical distance from the bottom of the tank to the top. Normally measured in terms of liquid depth and given in terms of sidewall depth (SWD), measured in linear units.
- **diameter:** The distance from one edge of a circle to the opposite edge passing through the center, measured in linear units.
- **height:** The vertical distance from one end of an object to the other, measured in linear units.
- **length:** The distance from one end of an object to the other, measured in linear units.
- **linear units:** Measurements used to express distances: feet, inches, meters, yards, etc.
- **pi, π:** A number in the calculations involving circles, spheres, or cones ($\pi = 3.14$).
- **radius:** The distance from the center of a circle to the edge, measured in linear units.
- **sphere:** A container shaped like a ball.
- **square units:** Measurements used to express area, square feet, square meters, acres, etc.
- **volume:** The capacity of the unit, how much it will hold, measured in cubic units (cubic feet, cubic meters) or in liquid volume units (gallons, liters, million gallons).
- **width:** The distance from one side of the tank to the other, measured in linear units.

Relevant Geometric Equations

Circumference of a circle	$C = \pi d = 2\pi r$
Perimeter of a square with side a	$P = 4a$
Perimeter of a rectangle with sides a and b	$P = 2a + 2b$
Perimeter of a triangle with sides a, b, and c	$P = a + b + c$
Area A of a circle with radius r ($d = 2r$)	$A = \pi d^2/4 = \pi r^2$
Area of duct in square feet when d is in inches	$A = 0.005454d^2$
Area of A of a triangle with base b and height h	$A = 0.5bh$
Area of A of a square with sides a	$A = a^2$
Area of A of a rectangle with sides a and b	$A = ab$
Area A of an ellipse with major axis a and minor axis b	$A = \pi ab$
Area A of a trapezoid with parallel sides a and b and height h	$A = 0.5(a + b)h$
Area A of a duct in square feet when d is in inches	$A = \pi d^2/576$
	$= 0.005454d^2$
Volume V of a sphere with a radius r ($d = 2r$)	$V = 1.33\pi r^3$
	$= 0.1667\pi d^3$
Volume V of a cube with sides a	$V = a^3$
Volume V of a rectangular solid (sides a and b and height c)	$V = abc$
Volume V of a cylinder with a radius r and height h	$V = \pi r^2 h$
	$= \pi d^2 h/4$
Volume V of a pyramid	$V = 0.33$

Geometrical Calculations

On occasion, it may be necessary to determine the distance around grounds or landscapes. In order to measure the distance around property, buildings, and basinlike structures, it is necessary to determine either perimeter or circumference. The **perimeter** is the distance around an object; a border or outer boundary. **Circumference** is the distance around a circle or circular object, such as a clarifier. Distance is linear measurement, which defines the distance (or length) along a line. Standard units of measurement like inches, feet, yards, and miles and metric units like centimeters, meters, and kilometers are used.

The **perimeter** of a rectangle (a four-sided figure with four right angles) is obtained by adding the lengths of the four sides (see figure 5.1).

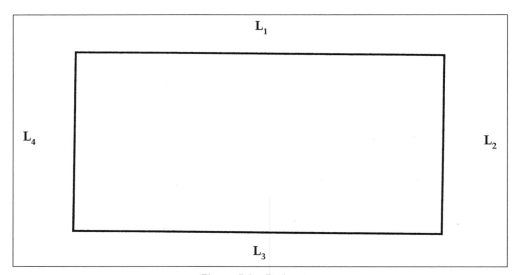

Figure 5.1   Perimeter.

$$\text{Perimeter} = L_1 + L_2 + L_3 + L_4 \tag{5.3}$$

Example 5.27
Problem:
Find the perimeter of the rectangle shown in figure 5.2.
Solution:

$$P = 35' + 8' + 35' + 8'$$
$$P = 86'$$

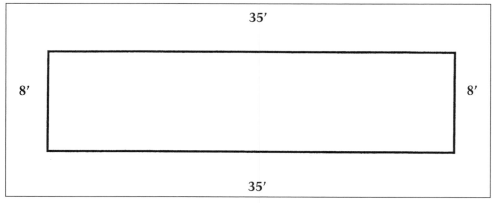

Figure 5.2   [example 5.27].

Example 5.28
Problem:
What is the perimeter of a rectangular field if its length is 100 feet and its width is 50 feet?

Solution:

$$Perimeter = 2 \text{ x length} + 2 \text{ x width}$$
$$= 2 \text{ x } 100 \text{ ft} + 2 \text{ x } 50 \text{ ft}$$
$$= 200 \text{ ft} + 100 \text{ ft}$$
$$= 300 \text{ ft}$$

The **circumference** is the length of the outer border of a circle. The circumference is found by multiplying pi (π) times the **diameter** (D) (diameter is a straight line passing through the center of a circle—the distance across the circle; see figure 5.3).

$$C = \pi D \tag{5.4}$$

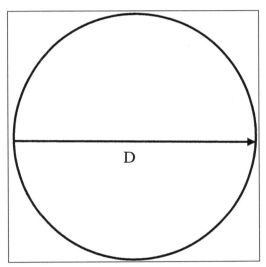

Figure 5.3   Diameter of a circle.

Where

$$C = circumference$$
$$\pi = \text{Greek letter pi} = 3.1416$$
$$D = diameter$$

Use this calculation if, for example, the circumference of a circular tank must be determined.

Example 5.29
Problem:
Find the circumference of a circle that has a diameter of 25 feet ($\pi = 3.14$)

Solution:
$$C = \pi \text{ x } 25'$$
$$C = 3.14 \text{ x } 25'$$
$$C = 78.5'$$

Example 5.30
Problem:
A circular chemical holding tank has a diameter of 18 m. What is the circumference of this tank?

Solution:

$$C = \pi\ 18\ m$$
$$C = (3.14)\ (18\ m)$$
$$C = 56.52\ m$$

Note: For **area** measurements in engineering operations, three basic shapes are particularly important, namely circles, rectangles, and triangles. Area is the amount of surface an object contains or the amount of material it takes to cover the surface. The area on top of a chemical tank is called the **surface area**. The area of the end of a ventilation duct is called the **cross-sectional area** (the area at right angles to the length of ducting). Area is usually expressed in square units, such as square inches (in.^{2}) or square feet (ft^2). Land may also be expressed in terms of square miles (sections) or acres (43,560 ft^2) or in metric system as **hectares**.

A **rectangle** is a two-dimensional box. The area of a rectangle is found by multiplying the length (L) times width (W); see figure 5.4.

$$\text{Area} = L \times W \tag{5.5}$$

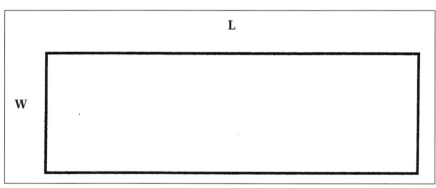

Figure 5.4   Area of a rectangle.

Example 5.31
Problem:
Find the area of the rectangle shown in figure 5.5.

Solution:

$$\text{Area} = L \times W$$
$$= 14' \times 6'$$
$$= 84\ ft^2$$

To find the **area of a circle**, we need to introduce one new term, the **radius**, which is represented by r. In figure 5.6, we have a circle with a radius of 6 feet.

The radius is any straight line that radiates from the center of the circle to some point on the circumference. By definition, all **radii** (plural of *radius*) of the same circle are equal. The surface area of a circle is determined by multiplying π times the radius squared.

$$\text{Area of Circle} = \pi r^2 \tag{5.6}$$

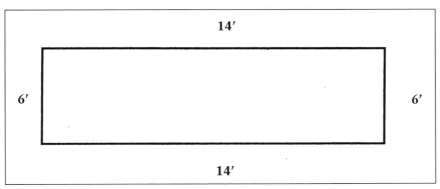

Figure 5.5   [example 5.31].

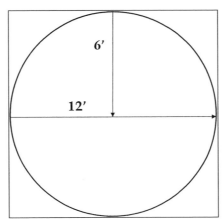

Figure 5.6   [example 5.32].

Where

A = area
π = pi (3.14)
r = radius of circle—radius is one-half the diameter

Example 5.32
Problem:
What is the area of the circle shown in figure 5.6?

$$\begin{aligned} \text{Area of circle} &= \pi r^2 \\ &= \pi 6^2 \\ &= 3.14 \times 36 \\ &= 113 \text{ ft}^2 \end{aligned}$$

If we were assigned to paint a water storage tank, we must know the surface area of the walls of the tank—we need to know how much paint is required. That is, we need to know the area of a circular or cylindrical tank. To determine the tank's surface area, we need to visualize the cylindrical walls as a rectangle wrapped around a circular base. The area of a rectangle is found by multiplying the length by the width; in this case, the width of the rectangle is the height of the wall, and the length of the rectangle is the distance around the circle, the circumference.

Thus, the area of the sidewalls of the circular tank is found by multiplying the circumference of the base (C = π x D) times the height of the wall (H):

$$A = \pi \times D \times H \qquad (5.7)$$
$$A = \pi \times 20 \text{ ft} \times 25 \text{ ft}$$
$$A = 3.14 \times 20 \text{ ft} \times 25 \text{ ft}$$
$$A = 1570 \text{ ft}^2$$

To determine the amount of paint needed, remember to add the surface area of the top of the tank, which is 314 ft². Thus, the amount of paint needed must cover 1570 ft² + 314 ft² =1884 or 1,885 ft². If the tank floor should be painted, add another 314 ft².

Note: Volume: The amount of space occupied by or contained in an object. Volume (see figure 5.7) is expressed in cubic units, such as cubic inches (in.³), cubic feet (ft³), acre feet (1 acre foot = 43,560 ft³), etc.

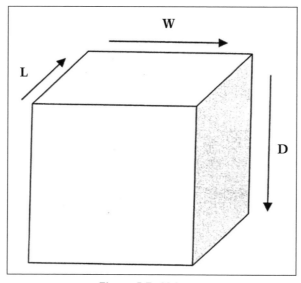

Figure 5.7   Volume.

The **volume of a rectangular object** is obtained by multiplying the length times the width times the depth or height.

$$V = L \times W \times H \qquad (5.8)$$

Where

L = length
W = width
D or H = depth or height

Example 5.33
Problem:
A unit rectangular process basin has a length of 15′, width of 7′, and a depth of 9′. What is the volume of the basin?

Solution:

$$V = L \times W \times D$$
$$= 15' \times 7' \times 9'$$
$$= 945 \text{ ft}^3$$

Representative surface areas are most often rectangles, triangles, circles, or a combination of these. Practical volume formulas used in volume calculations are given in table 5.2.

Table 5.2 Volume Formulas

Sphere volume	=	$(\pi/6)$ (diameter)^3
Cone volume	=	1/3 (volume of a cylinder)
Rectangular tank volume	=	(area of rectangle) (D or H)
	=	(LW) (D or H)
Cylinder volume	=	(area of cylinder) (D or H)
		π^2 (D or H)
Triangle volume	=	(area of triangle) (D or H)
	=	(bh/2) (D or H)

In determining the volume of round pipe and round surface areas, the following examples are helpful.

Example 5.34
Problem:
Find the volume of a 3 in. round pipe that is 300 ft long.

1. Step 1: Change the diameter of the duct from inches to feet by dividing by 12.
   D = 3 ÷ 12 = 0.25 ft

2. Step 2: Find the radius by dividing the diameter by 2.
   R = 0.25 ft ÷ 2 = 0.125

3. Step 3: Find the volume.

Solution:

$$V = L \times \pi r^2$$
$$V = 300 \text{ ft} \times 3.14 \times 0.0156$$
$$V = 14.72 \text{ ft}^2$$

To determine the volume of a cone and sphere, we use the following equations and examples.

$$\text{Volume of cone} = \frac{\pi}{12} \times \text{Diameter} \times \text{Diameter} \times \text{Height} \qquad (5.9)$$

$$\frac{\pi}{12} = \frac{3.14}{12} = 0.262$$

Note: The diameter used in the formula is the diameter of the base of the cone.

Example 5.35
Problem:
The bottom section of a circular settling tank has the shape of a cone. How many cubic feet of water are contained in this section of the tank if the tank has a diameter of 120 ft and the cone portion of the unit has a depth of 6 ft?

Solution:

$$\text{Volume, ft}^3 = 0.262 \times 120 \text{ ft} \times 120 \text{ ft} \times 6 \text{ ft} = 22,637 \text{ ft}^3$$

$$\text{Volume of sphere} = \frac{3.14}{6} \times \text{Diameter} \times \text{Diameter} \times \text{Diameter} \qquad (5.10)$$

$$\frac{\pi}{6} = \frac{3.14}{6} = 0.524$$

Example 5.36
Problem:
What is the volume of cubic feet of a gas storage container that is spherical and has a diameter of 60 ft?

Solution:

$$\text{Volume, ft}^3 = 0.524 \times 60 \text{ ft} \times 60 \text{ ft} \times 60 \text{ ft} = 113,184 \text{ ft}^3$$

Circular process and various water/chemical storage tanks are commonly found in many industrial processes. A circular tank consists of a circular floor surface with a cylinder rising above it (see figure 5.8). The **volume of a circular tank** is calculated by multiplying the surface area times the height of the tank walls.

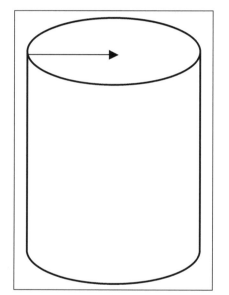

Figure 5.8   Circular or cylindrical water tank.

Example 5.37
Problem:
If a tank is 20 feet in diameter and 25 feet deep, how many gallons of water will it hold?

Hint: In this type of problem, calculate the surface area first, multiply by the height, and then convert to gallons.

Solution:

$$r = D \div 2 = 20 \text{ ft} \div 2 = 10 \text{ ft}$$
$$A = \pi \times r^2$$
$$A = \pi \times 10 \text{ ft} \times 10 \text{ ft}$$
$$A = 314 \text{ ft}^2$$
$$V = A \times H$$
$$V = 314 \text{ ft}^2 \times 25 \text{ ft}$$
$$V = 7,850 \text{ ft}^3 \times 7.5 \text{ gal/ft}^3 = 58,875 \text{ gal}$$

FORCE, PRESSURE, AND HEAD CALCULATIONS

Before we review calculations involving force, pressure, and head, we must first define these terms.

- **force:** The push exerted by water on any confined surface. Force can be expressed in pounds, tons, grams, or kilograms.
- **pressure:** The force per unit area. The most common way of expressing pressure is in pounds per square inch (psi).
- **head:** The vertical distance or height of water above a reference point. Head is usually expressed in feet. In the case of water, head and pressure are related.

Figure 5.9 helps to illustrate **force and pressure**. A cubical container measuring one foot on each side can hold one cubic foot of water. A basic fact of science states that one cubic foot of water weighs 62.4 pounds and contains 7.48 gallons. The force acting on the bottom of the container would be 62.4 pounds per square foot. The area of the bottom in square inches is

$$1 \text{ ft}^2 = 12 \text{ in. x } 12 \text{ in.} = 144 \text{ in}^2$$

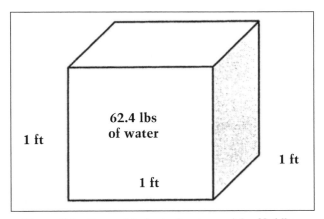

Figure 5.9   One cubic foot of water weighs 62.4 lbs.

Therefore, the pressure in pounds per square inch (psi) is:

$$\frac{62.4 \text{ lb/ft}^2}{1 \text{ ft}^2} = \frac{62.4 \text{ lb/ft}^2}{144 \text{ in}^2/\text{ft}^2} = 0.433 \text{ lb/in}^2 \text{ (psi)}$$

If we use the bottom of the container as our reference point, the head would be one foot. From this we can see that one foot of head is equal to 0.433 psi—an important parameter to remember. Figure 5.10 illustrates some other important relationships between pressure and head.

Note: Force acts in a particular direction. Water in a tank exerts force down on the bottom and out of the sides. Pressure, however, acts in all directions. A marble at a water depth of one foot would have 0.433 psi of pressure acting inward on all sides.

Using the preceding information, we can develop equations 5.11 and 5.12 for calculating pressure and head.

$$\text{Pressure (psi)} = 0.433 \text{ x Head (ft)} \tag{5.11}$$

$$\text{Head (ft)} = 2.31 \text{ x Pressure (psi)} \tag{5.12}$$

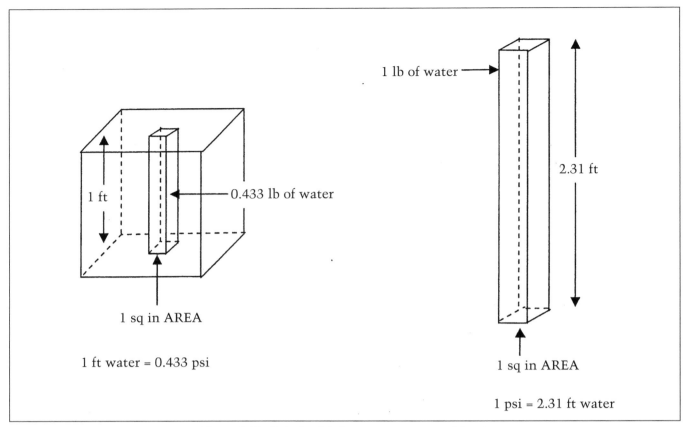

Figure 5.10   Relationship between pressure and head.

As mentioned, **head** is the vertical distance the water must be lifted from the supply tank or unit process to the discharge. The total head includes the vertical distance the liquid must be lifted (static head), the loss to friction (friction head), and the energy required to maintain the desired velocity (velocity head).

$$\text{Total Head} = \text{Static Head} + \text{Friction Head} + \text{Velocity Head} \tag{5.13}$$

Static Head is the actual vertical distance the liquid must be lifted.

$$\text{Static Head} = \text{Discharge Elevation} - \text{Supply Elevation} \tag{5.14}$$

Example 5.38
Problem:
The supply tank is located at elevation 108 ft. The discharge point is at elevation 205 ft. What is the static head in feet?

Solution:
$$\text{Static Head, ft} = 205 \text{ ft} - 108 \text{ ft} = 97 \text{ ft}$$

Friction Head is the equivalent distance of the energy that must be supplied to overcome friction. Engineering references include tables showing the equivalent vertical distance for various sizes and types of pipes, fittings, and valves. The total friction head is the sum of the equivalent vertical distances for each component.

$$\text{Friction Head, ft} = \text{Energy Losses due to Friction} \tag{5.15}$$

Velocity Head is the equivalent distance of the energy consumed in achieving and maintaining the desired velocity in the system.

$$\text{Velocity Head, ft} = \text{Energy Losses to Maintain Velocity} \qquad (5.16)$$

Total Dynamic Head (Total System Head)

$$\text{Total Head} = \text{Static Head} + \text{Friction Head} + \text{Velocity Head} \qquad (5.17)$$

Pressure/Head: The pressure exerted by water/wastewater is directly proportional to its depth or head in the pipe, tank, or channel. If the pressure is known, the equivalent head can be calculated.

$$\text{Head, ft} = \text{Pressure, psi} \times 2.31 \text{ ft/psi} \qquad (5.18)$$

Example 5.39
Problem:
The pressure gauge on the discharge line from the influent pump reads 75.3 psi. What is the equivalent head in feet?

Solution:

$$\text{Head, ft} = 75.3 \times 2.31 \text{ ft/psi} = 173.9 \text{ ft}$$

Head/Pressure: If the head is known, the equivalent pressure can be calculated by:

$$\text{Pressure, psi} = \frac{\text{Head, ft}}{2.31 \text{ ft/psi}} \qquad (5.19)$$

Example 5.40
Problem:
The tank is 15 feet deep. What is the pressure in psi at the bottom of the tank when it is filled with wastewater?

Solution:

$$\text{Pressure, psi} = \frac{15 \text{ ft}}{2.31 \text{ ft/psi}}$$

$$= 6.49 \text{ psi}$$

Before we look at a few example problems dealing with force, pressure, and head, it is important to review the key points related to force, pressure, and head.

1. By definition, water weighs 62.4 pounds per cubic foot.
2. The surface of any one side of the cube contains 144 square inches (12 in. x 12 in. = 144 in^2). Therefore, the cube contains 144 columns of water 1 foot tall and 1 inch square.
3. The weight of each of these pieces can be determined by dividing the weight of the water in the cube by the number of square inches.

$$\text{Weight} = \frac{62.4 \text{ lbs}}{144 \text{ in}^2} = 0.433 \text{ lbs/in}^2 \text{ or } 0.433 \text{ psi}$$

4. Because this is the weight of one column of water 1 foot tall, the true expression would be 0.433 pounds per square inch per foot of head or 0.433 psi/ft.

Note: 1 foot of head = 0.433 psi.

In addition to remembering the important parameter, 1 foot of head = 0.433 psi, it is important to understand the relationship between pressure and feet of head—in other words, how many feet of head 1 psi represents. This is determined by dividing 1 by 0.433.

$$\text{Feet of head} = \frac{1 \text{ ft}}{0.433 \text{ psi}} = 2.31 \text{ ft/psi}$$

If a pressure gauge were reading 12 psi, the height of the water necessary to represent this pressure would be 12 psi x 2.31 ft/psi = 27.7 feet.

Note: Both the above conversions are commonly used in water/wastewater treatment calculations. However, the most accurate conversion is 1 ft = 0.433 psi. This is the conversion we use throughout this text.

Example 5.41
Problem:
Convert 40 psi to feet head.

Solution:

$$40 \, \frac{\text{psi}}{1} \times \frac{\text{ft}}{0.433 \text{ psi}} = 92.4 \text{ feet}$$

Example 5.42
Problem:
Convert 40 feet to psi.

Solution:

$$40 \, \frac{\text{ft}}{1} \times \frac{0.433 \text{ psi}}{1 \text{ ft}} = 17.32 \text{ psi}$$

As the above examples demonstrate, when attempting to convert psi to feet, we divide by 0.433, and when attempting to convert feet to psi, we multiply by 0.433. The above process can be most helpful in clearing up the confusion on whether to multiply or divide. These is another way, however—one that may be more beneficial and easier for many operators to use. Notice that the relationship between psi and feet is almost two to one. It takes slightly more than two feet to make one psi. Therefore, when looking at a problem where the data is in pressure and the result should be in feet, the answer will be at least twice as large as the starting number. For instance, if the pressure were 25 psi, we intuitively know that the head is over 50 feet. Therefore, we must divide by 0.433 to obtain the correct answer.

Example 5.43
Problem:
Convert 15 psi to feet.

Solution:

$$15 \, \frac{\text{psi}}{1} \times \frac{1 \text{ ft}}{0.433 \text{ psi}} = 34.6 \text{ ft}$$

Example 5.44
Problem:
Between the top of a reservoir and the watering point, the elevation is 125 feet. What will the static pressure be at the watering point?

Solution:

$$125 \ \frac{psi}{1} \ x \ \frac{1 \ ft}{0.433 \ psi} = 288.7 \ ft$$

Example 5.45
Problem:
Find the pressure (psi) in a 12 foot deep tank at a point 5 feet below the water surface.

Solution:

$$\text{Pressure (psi)} = 0.433 \ x \ 5 \ ft$$
$$= 2.17 \ psi$$

Example 5.46
Problem:
A pressure gauge at the bottom of a tank reads 12.2 psi. How deep is the water in the tank?

Solution:

$$\text{Head (ft)} = 2.31 \ x \ 12.2 \ psi$$
$$= 28.2 \ ft$$

Example 5.47
Problem:
What is the pressure (static pressure) 4 miles beneath the ocean surface?

Solution:
Change miles to ft, then to psi.

$$5{,}280 \ ft/mile \ x \ 4 = 21{,}120 \ ft$$

$$\frac{21{,}120 \ ft}{2.31 \ ft/psi} = 9{,}143 \ psi$$

Example 5.48
Problem:
A 150 ft diameter cylindrical tank contains 2.0 MG water. What is the water depth? At what pressure would a gauge at the bottom read in psi?

Solution:

Step 1: Change MG to cu ft.

$$\frac{2{,}000{,}000 \ gal}{7.48} = 267{,}380 \ cu \ ft$$

Step 2: Using volume, solve for depth.

$$\text{Volume} = .785 \times D^2 \times \text{depth}$$
$$267{,}380 \text{ cu ft} = .785 \times (150)^2 \times \text{depth}$$
$$\text{Depth} = 15.1 \text{ ft}$$

Example 5.49
Problem:
The pressure in a pipe is 70 psi. What is the pressure in feet of water? What is the pressure in psf?

Solution:

Step 1: Convert pressure to feet of water.

$$70 \text{ psi} \times 2.31 \text{ ft/psi} = 161.7 \text{ ft of water}$$

Step 2: Convert psi to psf.

$$70 \text{ psi} \times 144 \text{ sq in./sq ft} = 10{,}080 \text{ psf}$$

Example 5.50
Problem:
The pressure in a pipeline is 6,476 psf. What is the head on the pipe?

$$\text{Head on pipe} = \text{ft of pressure}$$
$$\text{Pressure} = \text{Weight} \times \text{Height}$$
$$6{,}476 \text{ psf} = 62.4 \text{ lbs/cu ft} \times \text{height}$$
$$\text{Height} = 104 \text{ ft}$$

REVIEW OF ADVANCED ALGEBRA KEY TERMS AND CONCEPTS

Advanced algebraic operations (linear, linear differential, and ordinary differential equations) have in recent years become the exclusive (essential) part of the mathematical background required by safety engineers, among others. While it is not the authors' intent to provide complete coverage of the topics (engineers are normally well grounded in these critical foundational areas), it is important to review the key terms and concepts germane to the topics. Thus, in the following we provide key definitions.

algebraic multiplicity of an eigenvalue: The algebraic multiplicity of an eigenvalue c of a matrix A is the number of times the factor $(t\text{-}c)$ occurs in the characteristic polynomial of A.

basis for a subspace: A basis for a subspace W is a set of vectors $\{\mathbf{v}_1, ..., \mathbf{v}_k\}$ in which W such that:

1. $\{\mathbf{v}_1, ..., \mathbf{v}_k\}$ is linearly independent; and
2. $\{\mathbf{v}_1, ..., \mathbf{v}_k\}$ spans W.

characteristic polynomial of a matrix: The characteristic polynomial of a n by n matrix A is the polynomial in t given by the formula $\det(A - tI)$.

column space of a matrix: The subspace spanned by the columns of the matrix considered as a set of vectors (also see **row space**).

consistent linear system: A system of linear equations is consistent if it has at least one solution.

defective matrix: A matrix A is defective if A has an eigenvalue whose geometric multiplicity is less than its algebraic multiplicity.

diagonalizable matrix: A matrix is diagonalizable if it is similar to a diagonal matrix.

dimension of a subspace: The dimension of a subspace W is the number of vectors in any basis of W. (If W is the subspace $\{\mathbf{0}\}$, we say that its dimension is 0.)

echelon form of a matrix: A matrix is in row echelon form if:

1. all rows that consist entirely of zeros are grouped together at the bottom of the matrix; and
2. the first (counting left to right) nonzero entry in each nonzero row appears in a column to the right of the first nonzero entry in the preceding row (if there is a preceding row).

    eigenspace of a matrix: The eigenspace associated with the eigenvalue c of a matrix A is the null space of $A - cl$.

    eigenvalue of a matrix: An eigenvalue of a matrix A is a scalar c such that $A\mathbf{x} = c\mathbf{x}$ holds for some nonzero vector $\mathbf{x}$.

    eigenvector of a matrix: An eigenvector of a square matrix A is a nonzero vector $\mathbf{x}$ such that $A\mathbf{x} = c\mathbf{x}$ holds for some scalar c.

    elementary matrix: A matrix that is obtained by performing an elementary row operation on an identity matrix.

    equivalent linear systems: Two systems of linear equations in n unknowns are equivalent if they have the same set of solutions.

    geometric multiplicity of an eigenvalue: The geometric multiplicity of an eigenvalue c of a matrix A is the dimension of the eigenspace of c.

    homogeneous linear system: A system of linear equations $A\mathbf{x} = \mathbf{b}$ is homogeneous if $\mathbf{b} = \mathbf{0}$.

    inconsistent linear system: A system of linear equations is inconsistent if it has no solutions.

    inverse of a matrix: The matrix B is an inverse for the matrix A if $AB = BA = I$.

    invertible matrix: A matrix is invertible if it has no inverse.

    least squares solution of a linear system: A least-squares solution to a system of linear equations $A\mathbf{x} = \mathbf{b}$ is a vector $\mathbf{x}$ that minimizes the length of the vector $A\mathbf{x} - \mathbf{b}$.

    linear combination of vectors: A vector $\mathbf{v}$ is a linear combination of the vectors $\mathbf{v}_1, ..., \mathbf{v}_k$ if there exist scalars $a_1, ..., a_k$ such that $\mathbf{v} = a_1\mathbf{v}_1 + ... + a_k\mathbf{v}_k$.

    linear dependence relation for a set of vectors: A linear dependence relation for the set of vectors $\{\mathbf{v}_1, ..., \mathbf{v}_k\}$ is an equation of the form $a_1\mathbf{v}_1 + ... + a_k\mathbf{v}_k = \mathbf{0}$, where the scalars $a_1, ..., a_k$ are zero.

    linearly dependent set of vectors: The set of vectors $\{\mathbf{v}_1, ..., \mathbf{v}_k\}$ is linearly dependent if the equation $a_1\mathbf{v}_1 + ... + a_k\mathbf{v}_k = \mathbf{0}$ has a solution where not all the scalars $a_1, ..., a_k$ are zero (i.e., if $\{\mathbf{v}_1, ..., \mathbf{v}_k\}$ satisfies a linear dependence relation).

    linearly independent set of vectors: The set of vectors $\{\mathbf{v}_1, ..., \mathbf{v}_k\}$ is linearly independent if the only solution to the equation $a_1\mathbf{v}_1 + ... + a_k\mathbf{v}_k = 0$ is the solution where all the scalars $a_1, ..., a_k$ are zero. (i.e., if $\{\mathbf{v}_1, ..., \mathbf{v}_k\}$ does not satisfy any linear dependence relation).

    linear transformation: A linear transformation from V to W is a function T from V to W such that:

1. $T(\mathbf{u} + \mathbf{v}) = T(\mathbf{u}) + T(\mathbf{v})$ for all vectors $\mathbf{u}$ and $\mathbf{v}$ in V
2. $T(a\mathbf{v}) = aT(\mathbf{v})$ for all vectors $\mathbf{v}$ in V and all scalars a

    nonsingular matrix: A square matrix A is nonsingular if the only solution to the equation $A\mathbf{x} = \mathbf{0}$ is $\mathbf{x} = \mathbf{0}$.

    nullity of a linear transformation: The nullity of a linear transformation is the dimension of its null space.

    nullity of a matrix: The nullity of a matrix is the dimension of its null space.

    null space of a linear transformation: The null space of a linear transformation T is the set of vectors $\mathbf{v}$ in its domain such that $T(\mathbf{v}) = \mathbf{0}$.**null space of a matrix:** The null space of a m by n matrix A is the set of all vectors $\mathbf{x}$ in R^n such that $A\mathbf{x} = \mathbf{0}$.

    orthogonal complement of a subspace: The orthogonal complement of a subspace S of R^n is the set of all vectors $\mathbf{v}$ in R^n such that $\mathbf{v}$ is orthogonal to every vector in S.

    orthogonal linear transformation: A linear transformation T from V to W is orthogonal if $T(\mathbf{v})$ has the same length as $\mathbf{v}$ for all vectors $\mathbf{v}$ in V.

    orthogonal matrix: A matrix A is orthogonal if A is invertible and its inverse equals its transpose; i.e., $A^{-1} = A^T$.

    orthogonal set of vectors: A set of vectors in R^n is orthogonal if the dot product of any two of them is 0.

    orthonormal set of vectors: Set of vectors in R^n is orthonormal if it is an orthogonal set and each vector has length 1.

range of a linear transformation: The range of a linear transformation *T* is the set of all vectors *T*(**v**), where **v** is any vector in its domain.

rank of a linear transformation: The rank of a linear transformation (and hence of any matrix regarded as a linear transformation) is the dimension of its range. Note: A theorem tells us that the two definitions of rank of a matrix are equivalent.

rank of a matrix: The rank of a matrix A is the number of nonzero rows in the reduced row echelon form of A; i.e., the dimension of the row space of A.

reduced row echelon form of a matrix: A matrix is in reduced row echelon form if:

the matrix is in row echelon form if the first nonzero entry in each nonzero row is the number 1 and the first nonzero entry in each nonzero row is the only nonzero entry in its column

row equivalent matrices: Two matrices are row equivalent if one can be obtained from the other by a sequence of elementary row operations.

row operations: The elementary row operations performed on a matrix are

interchange two rows

multiply a row by a nonzero scalar

add a constant multiple of one row to another

row space of a matrix: The row space of a matrix is the subspace spanned by the rows of the matrix considered as a set of vectors.

similar matrices: Matrices A and B are similar if there is a square invertible matrix S such that S − 1AS = B.

singular matrix: A square matrix A is singular if the equation Ax = 0 has a nonzero solution for x.

span of a set of vectors: The span of the set of vectors {v1, …, vk} is the subspace V consisting of all linear combinations of v1, …, vk. One also says that the subspace V is spanned by the set of vectors {v1, …, vk} and that this set of vectors spans V.

subspace: A subset W of Rn is a subspace of Rn if:

the zero vector is in W

x + y is in W whenever x and y are in W

ax is in W whenever x is in W and a is any scalar.

symmetric matrix: A matrix A is symmetric if it equals its transpose; i.e., A = AT.

REVIEW QUESTIONS

1. Convert 0.55 to percent.
2. Convert 7/22 to a decimal percent to a percent.
3. How many mg/L is a 1.4% solution?
4. A pipe is laid at a rise of 140 mm in 22 m. What is the grade?
5. How many significant figures are in a measurement of 1.35 inches?
6. If a pump will fill a tank in 20 hours at 4 gpm (gallons per minute), how long will it take a 10 gpm pump to fill the same tank?
7. The flow rate in a water line is 2.3 ft^3/s. What is the flow rate expressed as gallons per minute?
8. 1 MGD equals how many cubic feet per second (cfs)?
9. What is the perimeter of a square whose side is 8 inches?
10. An influent pipe inlet opening has a diameter of 6'. What is the circumference of the inlet opening in inches?
11. Find the volume of a smokestack that is 24 in. in diameter (entire length) and 96 in. tall.
12. Convert a pressure of 45 psi to feet of head.

6

Quadratic Equations

$$X = \frac{-b \pm \sqrt{b^2 - 4ac}}{2a}$$

When studying a discipline that does not include mathematics, one thing is certain: the discipline under study has nothing or little to do with engineering.

THE QUADRATIC EQUATION AND SAFETY ENGINEERING

The logical question might be: Why is the quadratic equation important in safety engineering? The logical answer: The quadratic equation is used in safety engineering to find solutions to problems primarily dealing with length and time determinations. Stated differently: The quadratic equation is a tool—an important tool that belongs in every safety practitioner's toolbox.

To the student of mathematics, this explanation might seem somewhat strange. Math students know, for example, that there will be two solutions to a quadratic equation. In engineering, many times only one solution is meaningful. For example, if we are dealing with a length, a negative solution to the equation may be mathematically possible but it is not the solution we would use. Negative time, obviously, would also pose the same problem.

So what is the point? The point is that we often need to find "a" solution to certain mathematical problems. In safety engineering problems involving determination of length and time using a quadratic equation, we will end up with two answers. In some instances, a positive answer and a negative answer may result. One of these answers is usable; thus, we would use it. Real engineering is about modeling situations that occur naturally and using the model to understand what is happening, or maybe to predict what will happen in future. The quadratic equation is often used in modeling because it is a beautifully simple curve (Borne 2008).

Key Terms

a: Coefficient of the x^2.
b: Coefficient of the x term.
c: Number in quadratic equation (not a coefficient of any x term).
affected quadratic equation: An equation containing the first and second degree of an unknown.
pure quadratic equation: An equation in which the unknown appears only in the second degree.
simple equations: Equations in which the unknown appears only in the first degree.

QUADRATIC EQUATIONS: THEORY AND APPLICATION

The equation 6x = 12 is a form of equation most of us are familiar with. Such an equation, an equation in which the unknown appears only in the first degree, is called *simple equation* or *linear equation*.

Those experienced in mathematics know that not all equations reduce to this form. For instance, when an equation has been reduced, the result may be an equation in which the square of the unknown equals some number, as in $x^2 = 5$. Such an equation, an equation in which the unknown appears only in the second degree, is called a *pure quadratic equation*.

In some cases, when an equation is simplified and reduced, the resultant form contains the square and the first power of the unknown, which equals some number; $x^2 - 5x = 24$ is such an equation. An equation containing the first and second degree of an unknown *s* is called an *affected quadratic equation*.

Quadratic equations, and certain other forms, can be solved with the aid of factoring. However, many times factoring is either too time-consuming or not possible. The formula shown below is called the *quadratic formula*. It expresses the quadratic equation in terms of its coefficients. The quadratic formula allows us to quickly solve for *x* with no factoring.

$$X = \frac{-b \pm \sqrt{b^2 - 4ac}}{2a} \tag{6.1}$$

To use the quadratic equation, just substitute the appropriate coefficients into the equation and solve.

Derivation of the Quadratic Equation Formula

The equation $ax^2 + bx + c = 0$, where a, b, and c are any numbers, positive or negative, represents any quadratic equation in one unknown. When this general equation is solved, the solution can be used to determine the unknown value in any quadratic equation. The solution follows.

Example 6.1
Problem:
Solve $ax^2 + bc + c = 0$ for x.

Solution:
$$ax^2 + bx + x = 0$$

Subtract c from both members:
$$ax^2 + bx = -c$$

Divide both members by a:
$$x^2 = \frac{b}{a} x = \frac{c}{a}$$

Complete the square:
$$x^2 + \frac{b}{a} x + \frac{b^2}{4a^2} - \frac{b^2}{4a^2} = \frac{c}{a}$$

$$(x + \frac{b}{2a})^2 - \frac{b^2}{4a^2} - \frac{c}{a}$$

Place the right member over the lowest common denominator:
$$(x + \frac{b}{2a})^2 = \frac{b^2 - 4ac}{4a^2}$$

Extract the square root of both members:

$$x + \frac{b}{2a} = \pm \sqrt{\frac{b^2 - 4ac}{4a^2}}$$

$$x + \frac{b}{2a} = \pm \sqrt{\frac{b^2 - 4ac}{2a}}$$

Subtract b/2a from both members:

$$x = \frac{b}{2a} \pm \sqrt{\frac{b^2 - 4ac}{2a}}$$

Thus the quadratic formula:

$$x = \frac{-b \pm \sqrt{b^2 - 4ac}}{2a}$$

Using the Quadratic Equation

Example 6.2 (Use the quadratic formula to solve the following problem.)
Problem:
After conducting a study and deriving an equation representing time, we arrive at the following equation. (Note: set the equation equal to zero and all like terms combined.)

$$x^2 = 5x + 6 = 0$$

Solution:

$$x = \frac{-b \pm \sqrt{b^2 - 4ac}}{2\ a}$$

From our equation, a = 1 (the coefficient of x^2), b = -5 (the coefficient of x), and c = 6 (the constant or third term.

Substituting these coefficients in the quadratic formula:

$$x = \frac{-(-5) \pm \sqrt{(-5)^2 - 4(1)(6)}}{2(1)}$$

$$x = \frac{5 \pm \sqrt{25} - 24}{2}$$

$$x = \frac{5 \pm 1}{2}$$

$$x = 3,\ 2$$

Note: The roots may not always be rational (integers), but the procedure is the same.

REVIEW QUESTIONS

1. Use quadratic equation to solve: $4x^2 + 2x - 4 = 0$.
2. Use quadratic equation to solve: $x^2 + 10x + 25 = 0$.

REFERENCE AND RECOMMENDED READING

Borne, M. 2008. Quadratic equations. http://www.inmath.com/Quadratic-Equations/Quadratic-equations-intro.php (accessed November 27, 2008).

7

Trigonometric Ratios

$$\sin A = a/c \quad \cos A = b/c \quad \tan A = ab$$

We owe a lot to the Indians, who taught us how to count, without which no worthwhile scientific discovery could have been made.

—Albert Einstein

TRIGONOMETRIC FUNCTIONS AND THE SAFETY ENGINEER

Typically, safety engineers are called upon to make calculations involving the use of various trigonometric functions. Consider slings, for example. They are commonly used between cranes, derricks, and/or hoists and the load, so that the load may be lifted and moved to a desired location. For the safety engineer, knowledge of the properties and limitations of the slings; the type and condition of material being lifted; weight and shape of the object being lifted; angle of the lifting sling to the load being lifted; and the environment in which the lift is to be made are all important considerations to be evaluated before the safe transfer of material can take place.

In chapter 11, we put many of the following principles to work in determining sling load and working load on a ramp (inclined plane); that is, resolution of force-type problems. For now we discuss the basic trigonometric functions used to make these calculations.

TRIGONOMETRIC RATIOS

In trigonometry, all computations are based upon certain ratios (i.e., trigonometric functions). The trigonometric ratios or functions are sine, cosine, tangent, cotangent, secant, and cosecant. It is important to understand the definition of the ratios given in table 7.1 and defined in terms of the lines shown in figure 7.1.

Note: In a right triangle, the side opposite the right angle is the longest side. This side is called the *hypotenuse*. The other two sides are the *legs*.

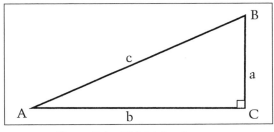

Figure 7.1   Right triangle.

Table 7.1   Definition of Trigonometric Ratios

sine of angle A = $\dfrac{\text{measure of leg opposite angle A}}{\text{measure of hypotenuse}}$	$\sin A = \dfrac{a}{c}$
cosine of angle A = $\dfrac{\text{measure of leg adjacent angle A}}{\text{measure of hypotenuse}}$	$\cos A = \dfrac{b}{c}$
tangent of angle A = $\dfrac{\text{measure of leg opposite angle A}}{\text{measure of leg adjacent angle A}}$	$\tan A = \dfrac{a}{b}$

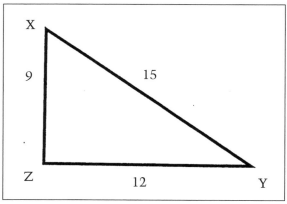

Figure 7.2   [example 7.1].

Example 7.1
Problem:
Find the sine, cosine, and tangent of angle Y in the figure.
Solution:

$$\sin Y = \frac{\text{opposite leg}}{\text{hypotenuse}} \qquad \cos Y = \frac{\text{adjacent leg}}{\text{hypotenuse}} \qquad \tan Y = \frac{\text{opposite leg}}{\text{adjacent leg}}$$

$$= \frac{9}{15} \text{ or } 0.6 \qquad = \frac{12}{15} \text{ or } 0.8 \qquad = \frac{9}{12} \text{ or } 0.75$$

Example 7.2
Problem:
Using figure 7.1, find the measure of the angle X to the nearest degree.

Solution:

$$\sin X = \frac{\text{opposite leg}}{\text{hypotenuse}} \qquad \rightarrow \qquad \sin X = \frac{12}{15} \text{ or } 0.8$$

Use a scientific calculator to find the angle measure with a sine of 0.8.
Enter: 0.8 [2nd] or [INV]

Result: 53.13010235
So, measure angle X = 53°

Example 7.3
Problem:
For the triangle shown, find sin C, cos C, and tan C.
Solution:
sin C = 2/5; cos C =4/5; tan C = 3/4

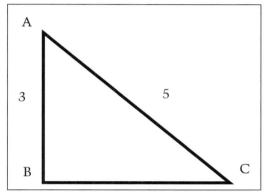

Figure 7.3   [example 7.3].

REVIEW QUESTIONS

Use a calculator to find the value of each trigonometric ratio.

1. sin 14°
2. cos 68°
3. tan 80°
4. cos 60°
5. sin 85°

REFERENCE AND RECOMMENDED READING

McKeague, C. P. 1998. *Algebra with trigonometry for college students*. Philadelphia: Saunders College Publishing.

Statistics Review

There are three kinds of lies: lies, damned lies, and statistics.

—Benjamin Disraeli

STATISTICAL CONCEPTS

Despite the protestation of Disraeli, safety engineering study is the statistical analysis of results. The principal concept of statistics is that of variation. In conducting typical environmental health functions such as toxicological or biological sampling protocol for air contamination, variation is commonly found. Variation comes from the methods that were used in the sampling process or, in this example, in the distribution of organisms. Several complex statistical tests can be used to determine the accuracy of data results. In this discussion, however, only basic calculations are reviewed. Specifically, this chapter provides the safety engineer with a survey of the basic statistical and data analysis techniques that can be used to address many of the problems that he or she will encounter on a daily basis. It covers the data analysis process from research design to data collection, analysis, reaching of conclusions, and, most importantly, presentation of findings.

Finally, it is important to point out that statistics can be used to justify the implementation of a program, identify areas that need to be addressed, or justify the impact that various safety programs have on losses and accidents. A set of safety data (or other data) is only useful if it is analyzed properly. Better decisions can be made when the nature of the data is properly characterized. For example, the importance of using statistical data in selling the safety plan or safety program, the innovation, and the winning over of those who control the purse strings cannot be overemphasized.

MEASURE OF CENTRAL TENDENCY

When talking statistics, it is usually because we are estimating something with incomplete knowledge. Maybe we can only afford to test 1 percent of the items we are interested in, and we want to say something about what the properties of the entire lot are, possibly because we destroy the sample by testing it. In that case, 100 percent sampling is not feasible if someone is supposed to get the items after we are done with them.

The questions we are usually trying to answer are "What is the central tendency of the item of interest?" and "How much dispersion about this central tendency can we expect?"

Simply, the average or averages that can be compared are measures of **central tendency** or central location of the data.

Basic Statistical Terms

Basic statistical terms include mean or average, median, mode, and range. The following is an explanation of each of these terms.

1. **Mean:** The total of the values of a set of observations divided by the number of observations.
2. **Median:** The value of the central item when the data are arrayed in size.
3. **Mode:** The observation that occurs with the greatest frequency and thus is the most "fashionable" value.
4. **Range:** The difference between the values of the highest and lowest terms.

Example 8.1
Problem:
Given the following laboratory results for the measurement of dissolved oxygen (DO) in water, find the mean, median, mode, and range.

Data:

6.5 mg/L, 6.4 mg/L, 7.0 mg/L, 6.9 mg/L, 7.0 mg/L

Solution:
   To find the mean:

$$\text{Mean} = \frac{(6.5 \text{ mg/L} + 6.4 \text{ mg/L} + 7.0 \text{ mg/L} + 6.0 \text{ mg/L} + 7.0 \text{ mg/L})}{5}$$

$$= 6.58 \text{ mg/L}$$

Mode = 7.0 mg/L (number that appears most often)
arrange in order: 6.4 mg/L, 6.5 mg/L, 6.9 mg/L, 7.0 mg/L, 7.0 mg/L
Median = 6.9 mg/L (central value)
Range = 7.0 mg/L - 6.4 mg/L = 0.6 mg/L

The importance of using statistically valid sampling methods cannot be overemphasized. Several different methodologies are available. A careful review of these methods (with the emphasis on designing appropriate sampling procedures) should be made before computing analytic results. Using appropriate sampling procedures along with careful sampling techniques will provide basic data that is accurate.

The need for statistics in safety engineering is driven by the discipline itself. As mentioned, occupational safety and health studies often deal with entities that are variable. If there were no variation in collected data, there would be no need for statistical methods.

Over a given time interval there will always be some variation in sampling analyses. Usually, the average and the range yield the most useful information. For example, in evaluating the indoor air quality (IAQ) in a factory, a monthly summary of air flow measurements, operational data, and laboratory tests for the factory would be used. For evaluation of a work center or organization, a monthly on-the-job report of accidents and illnesses, a monthly summary of reported injuries, lost-time incidents, and work-caused illnesses for the entity would be used.

In the preceding section, we used the term *sample* and the scenario sampling to illustrate the use and definition of mean, mode, median, and range. Though these terms are part of the common terminology used in statistics, the term *sample* in statistics has its own unique meaning. There is an exact delineation between both the term *sample* and the term *population*. In statistics we most often obtain data from a sample and use the results from the sample to describe an entire population. The *population* of a sample signifies that one has measured a characteristic for everyone or everything that belongs in a particular group. For example, if one wishes to measure that characteristic of the population defined as safety engineers, one would have to obtain a measure of that characteristic for every safety engineer possible. Measuring a population is difficult, if not impossible.

We use the term *subject* or *case* to refer to a member of a population or sample. There are statistical methods for determining how many cases must be selected in order to have a credible study. *Data* is another important term: it consists of the measurements taken for the purposes of statistical analysis. Data can be classified as either qualitative or quantitative. *Qualitative* data deal with characteristic of the individual or subject (e.g., gender of a person or the color of a car). *Quantitative* data describe a characteristic in terms of a number (e.g., the age of a horse or the number of lost-time injuries an organization had over the previous year).

Table 8.1   Commonly Used Statistical Symbols and Procedures

Term or Procedure	Symbol	
	Population Symbol	Sample Notation
Mean	μ	$\bar{x}$
Standard deviation	σ	s
Variance	σ^2	s^2
Number of cases	N	n
Raw number or value	X	x
Correlation coefficient	R	r

Procedure	Symbol
Sum of	Σ
Absolute value of x	$|x|$
Factorial of n	n!

Along with common terminology, the field of statistics also generally uses some common symbols. Statistical notation uses Greek letters and algebraic symbols to covey meaning about the procedures that one should follow in order to complete a particular study or test. Greek letters are used as statistical notation for a population, while English letters are used for statistical notation for a sample. Table 8.1 summarizes some of the more common statistical symbols, terms, and procedures used in statistical operations.

DISTRIBUTION

If a safety engineer wishes to conduct a research study, and the data is collected and a group of raw data is obtained, to make sense out of the data the data must be organized into a meaningful format.

The formatting begins by putting the data into some logical order, then grouping the data. Before the data can be compared to other data it must be organized. Organized data are referred to as *distributions.*

As mentioned, when confronted with masses of ungrouped data (i.e., listings of individual figures), it is difficult to generalize about the information the masses contain. However, if a frequency distribution of the figures is formed, many features become readily discernible. A frequency distribution records the number of cases that fall in each class of the data.

Example 8.2

A safety engineer gathers data on the medical costs of twenty-four on-the-job injury claims for a given year. The raw data collected was as follows.

To develop a frequency distribution, the investigator takes the values of the claims and places them in order. Then the investigator counts the frequency of occurrences for each value.

In order to develop a frequency distribution, groupings are formed using the values in the table above, ensuring that each group has an equal range. The safety engineer grouped the data into ranges of 1,000. The lowest range and highest range are determined by the data. Because it was decided to group by thousands, values will fall in the ranges of $0 to $4,999, and the distribution will end with this. The frequency distribution for this data appears in table 8.4.

Table 8.2

$60	$1,500	$85	$120
$110	$150	$110	$340
$2,000	$3,000	$550	$560
$4,500	$85	$2,300	$200
$120	$880	$1,200	$150
$650	$220	$150	$4,600

Table 8.3

Value	Frequency
$60	1
$85	2
$110	2
$120	2
$150	3
$200	1
$220	1
$340	1
$550	1
$560	1
$650	1
$880	1
$1,200	1
$1,500	1
$2,000	1
$2,300	1
$3,000	1
$4,500	1
$4,600	1
Total	24

Table 8.4 Frequency Distribution

Range	Frequency
$0 – $999	17
$1,000 – 1,999	2
$2,000 – 2,999	2
$3,000 – 3,999	1
$4,000 – 4,999	2
Total	24

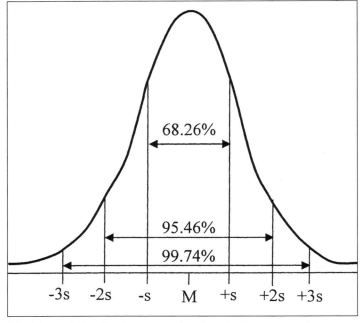

Figure 8.1 Normal distribution curve showing the frequency of a measurement.

Normal Distribution

When large amounts of data are collected on certain characteristics, the data and subsequent frequency can follow a distribution that is bell-shaped in nature—the *normal distribution.* The normal distributions are a very important class of statistical distributions. As stated, all normal distributions are symmetric and have bell-shaped curves with a single peak (see figure 8.1).

To speak specifically of any normal distribution, two quantities have to be specified: the mean μ (pronounced mu), where the peak of the density occurs, and the standard deviation σ (sigma). Different values of μ and σ yield different normal density curves and hence different normal distributions.

Although there are many normal curves, they all share an important property that allows us to treat them in a uniform fashion. All normal density curves satisfy the following property, which is often referred to as the empirical rule.

68% of the observations fall within 1 **standard deviation** of the **mean**; that is, between μ - σ and μ + σ.

95% of the observations fall within **2 standard deviations** of the **mean**, that is, between μ - 2σ and μ + 2σ.

98% of the observations fall within **3 standard deviations** of the **mean**; that is, between μ - 3σ and μ + 3σ.

Thus, for a normal distribution, almost all values lie within three standard deviations of the mean (see figure 8.1). It is important to stress that the rule applies to all normal distributions. Also remember that it applies *only* to normal distributions.

Note: Before applying the empirical rule it is a good idea to identify the data being described and the value of the mean and standard deviation. A sketch of a graph summarizing the information provided by the empirical rule should also be made.

Example 8.3

Problem:

The scores for all high school seniors taking the math section of the Scholastic Aptitude Test (SAT) in a particular year had a mean of 490 and a standard deviation of 100. The distribution of SAT scores is bell-shaped.

1. What percentage of seniors scored between 390 and 590 on this SAT?
2. One student scored 795 on this test. How did this student do compared to the rest of the scores?
3. A rather exclusive university only admits students who were among the highest 16 percent of the scores on this test. What score would a student need on this test to be qualified for admittance to this university?

The data being described are the math SAT scores for all seniors taking the test one year. Because this is describing a population, we denote the mean and standard deviation as μ = 490 and σ = 100, respectively. A bell-shaped curve summarizing the percentages given by the empirical rule is shown in figure 8.2.

1. From figure 8.2, about 68 percent of seniors scored between 390 and 590 on this SAT.
2. Because about 99.7 percent of the scores are between 190 and 790, a score of 795 is excellent. This is one of the highest scores on this test.
3. Because about 68 percent of the scores are between 390 and 590, this leaves 32 percent of the scores outside the interval. Because a bell-shaped curve is symmetric, one-half of the scores, or 16 percent, are on each end of the distribution.

STANDARD DEVIATION

The standard deviation, s or σ, is often used as an indicator of precision. The *standard deviation* is a measure of the variation (the spread in a set of observations) in the results. In order to gain better understanding and perspective of the benefits derived from using statistical methods in safety engineering, it is appropriate to consider some of the basic theory of statistics. In any set of data, the true value (mean) will lie in the middle of all the measurements taken. This is true providing the sample size is large and only random error is present in the analysis. In addition, the measurements will show a normal distribution as shown in figure 8.1.

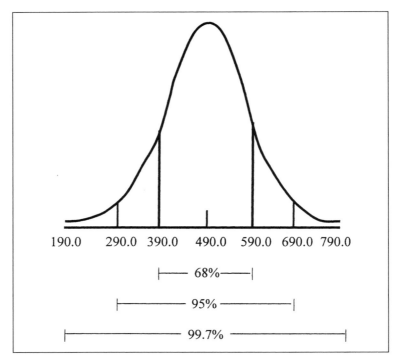

Figure 8.2   Sample Scholastic Aptitude Test (SAT) math percentages given by the empirical rule.

Figure 8.1 shows that 68.26 percent of the results fall between M + s and M - s; 95.46 percent of the results lie between M + 2s and M - 2s; and 99.74 percent of the results lie between M + 3s and M - 3s. Therefore, if precise, then 68.26 percent of all the measurements should fall between the true value estimated by the mean, plus the standard deviation and the true value minus the standard deviation. The following equation is used to calculate the sample standard deviation:

$$s = \sqrt{\frac{\sum(X - \bar{X})2}{n - 1}}$$

Where:

s = standard deviation
n = number of samples
X = the measurements from X to Xn

$\bar{X}$ = the mean
Σ = means to sum the values from X to Xn

Example 8.4
Problem:
Calculate the standard deviation, σ, of the following dissolved oxygen values:

9.5, 10.5, 10.1, 9.9, 10.6, 9.5, 11.5, 9.5, 10.0, 9.4

$\bar{X}$ = 10.0

Table 8.5

X	X - X̄	(X - X̄)2
9.5	-0.5	0.25
10.5	0.5	0.25
10.1	0.1	0.01
9.9	-0.1	0.01
10.6	0.6	0.36
9.5	-0.5	0.25
11.5	1.5	2.25
9.5	-0.5	0.25
10.0	0	0
9.4	-0.6	0.36
		3.99

$$\sigma = \sqrt{\frac{1}{(10-1)}(3.99)}$$

$$\sigma = \sqrt{\frac{3.99}{9}} = 0.67$$

SUMMARY

In this chapter we have touched only on the basics of statistics. Obviously, the practicing safety engineer needs to know much more about this valuable tool. For example, not only is it necessary for the engineer to understand both elementary probability and basic statistics—with emphasis on their application in engineering and the sciences—but also the treatment of data, sampling distributions, inferences concerning means, inferences concerning variances, inferences concerning proportions, nonparametric tests, curve fitting, analysis of variance, factorial experimentation, and much more. Because these topics are beyond the scope of this text, we highly recommend Richard A. Johnson's (1997) *Miller and Freund's Probability and Statistics for Engineers*, 6th edition, available from Prentice-Hall.

REVIEW QUESTIONS

1. Standard deviation is a good descriptor of the _____ of set of values.
2. Arithmetic mean is also referred to as the _____.
3. The occurrence of monthly accidents in an organization during a twelve-month period were 11, 11, 9, 4, 15, 6, 8, 8, 8, 12, 9, 10. Calculate the mean number of accidents per month.

REFERENCE AND RECOMMENDED READING

Wadsworth, H. M. 1990. *Handbook of statistical methods for engineers and scientists*. New York: McGraw-Hill.

9

Boolean Algebra

$$A \bullet (B + C) = (A \bullet B) + (A \bullet C)$$
$$A + B = B + A$$
$$A + (B \bullet C) = (A + B) \bullet (A + C)$$
$$A \bullet B = B \bullet A$$

For the safety practitioner, total hazard elimination is a noble goal. Yet, as long as the human element is present in the system, perfection will be impossible to attain. The recognition that hazards exist at different degrees of severity and accident causation leads to the concept of eliminating the most "important" hazards first. Thus an evaluation to determine the importance of hazards is essential if there is to be any boundary whatsoever on the problem.

—D. B. Brown (1976)

WHY BOOLEAN ALGEBRA?

No rational person wants to cause pain or injury either to himself or herself or to their fellow humans. The continuing occurrence of accidents despite the obvious lack of intent indicates that there is a logical flaw in some part of the reasoning process. No doubt you have heard the old saying, "the best laid plans of ...," and so forth and so on. Simply, when an accident occurs, a breakdown in the cause-effect reasoning process is apparent. In this chapter, the goal is to describe a procedure for evaluating the logical reasoning process (system safety analysis).

The methodology we briefly describe is based on the concepts of Boolean algebra. One might ask, "Why describe the basic concepts of Boolean algebra?" The simple answer is safety practitioners are expected to use the tools of logic to mitigate hazardous situations; Boolean algebra is one of these tools. The compound answer is safety practitioners are often expected to know the "basics" of Boolean algebra to score on certification examinations for licensure and/or are taught Boolean algebra concepts in formalized training programs or on-the-job training. Moreover, Boolean algebra plays an instrumental role in probability theory and reliability and various safety studies.

Having evolved in the 1950s, Boolean algebra is a branch of mathematics (a variant of algebra) that was developed systematically, because of its applications to logic, by the English mathematician George Boole. Closely related are its applications to sets and probability. A *set* is any well-defined list or collection of objects (elements). Usually, sets are denoted by capital letters such as A, B, and C and their elements by the lowercase letters such as e, f, and g. Boolean algebra also underlies the theory of relations. Currently the most prominent use of Boolean algebra is in the design of electronic switching circuits used in digital computers (Marcus 1967). Its original application, however, was in the context of logical reasoning (Boole 1951; 2003). Common practice today is to use Boolean algebra to solve the logical portion of problems involving probability calculations (Boole 1952; 2003). All these applications can be of some benefit to the safety practitioner in his or her goal of reducing accidents. Of particular current interest is the use of Boolean techniques in fault tree analysis

(FTA), which is used in reliability analysis of engineering systems (Kolodner 1971; Clemens and Simmons 1986). Fault tree analysis is not only an effective combination of probability and backward reasoning but also Boolean algebra. The probability of some high-level event is the result of a combination of lower-level events. Estimating (or knowing) the probabilities of failure of events in the fault tree allows one to estimate the probability of success or failure.

Key Term

Boolean variable: usually represented by a capital letter, represents a distinct event or fact.

> A, B, and C = Boolean variables
> (+) = the logic OR operator
> (•) = the logic AND operator

Safety practitioners and students of safety are familiar with algebra; it is a required core subject. Algebra is a logical outgrowth of arithmetic, and many of the methods of arithmetic are used in algebra, although in modified, expanded, or original form. To a limited extent, the same relationship can be used in describing Boolean algebra. That is, many of the laws for Boolean variables are not much different than the laws of numeric algebra. This relationship is readily seen in the commutative law, distributive law, and less so in addition (identity and inverse variables) used for real algebra operations.

TECHNICAL OVERVIEW

Boolean algebra variables are usually depicted by capital letters, representing distinct events or facts. For example, we may let A represent the event that the belt on a certain machine pulley system breaks. If this occurs, we say that $A = T$ or A is true. If the event fails to occur, we say $A = F$ or A is false. Of course, there must be some finite time during which the system is under consideration, and there is a probability associated with event A, although it is often unknown.

The most obvious way to simplify Boolean expressions is to manipulate them in the same way as normal algebraic expressions are manipulated. The manipulations, expressed as true or false (i.e., occurrence or non-occurrence), are two modes of a Boolean algebra variable. These modes can be formed into functions, also into modes, as a combination of Boolean algebra variables. With regard to logic relations in digital forms, a set of rules for symbolic manipulation is needed in order to solve for the unknowns.

A set of rules formulated by the originator of Boolean expressions, George Boole, described certain propositions whose outcomes correspond to the either true or false relationship described above. Again, with regard to digital logic, these rules are used to describe circuits or possibilities whose state can be either 1 (true) or 0 (false). In order to fully understand this, the AND, OR, and NOT operators should be appreciated. A number of rules can be derived from these relations, as table 9.1 demonstrates.

Table 9.2 shows the basic Boolean laws we are concerned with in this text. Note that every law has two expressions, (a) or (b). This is known as duality. These are obtained by changing every AND (•) to OR (+) to AND (•) and all 1s to 0s and vice versa.

Table 9.1   Boolean Postulates

Postulate 1	X = 0 or X = 1
Postulate 2	0 • 0 = 0
Postulate 3	1 + 1 = 1
Postulate 4	0 + 0 = 0
Postulate 5	1 • 1 = 1
Postulate 6	1 • 0 = 0 • 1 = 0
Postulate 7	1 + 0 = 0 + 1 = 1

Did You Know?

In most cases it has become conventional to drop the • (AND symbol); that is, A • B is written as AB.

Table 9.2 Boolean Laws

Commutative Law	A + B = B + A
	AB = BA
Distributive Law	A (B + C) = AB + AC
	A + (BC) = (A + B) (A + C)
Identity and Inverse Laws	A + A = A
	AA = A
	A + not (A) = I
	A • not (A) = 0

Commutative Law

The *commutative law* for Boolean algebra is not much different than the commutative law for numeric algebra. For both the AND or OR logic operations, the order in which a variable is presented will not affect the outcome. So the first equation, A + B = B + A, may be read as "A OR B equals B OR A." It follows that AB + BA also reads as "A AND B equals B AND A."

Note: In algebra, the commutative laws for addition and multiplication are similar to the commutative law for Boolean variable. Commutative law for addition (algebra) states: The sum of two quantities is the same in whatever order they are added. The commutative law for multiplication (algebra) states: The product of two quantities is the same whatever the order of multiplication.

Distributive Law

The equation A (B + C) = (AB) + (AC) is the first distributive law for Boolean variables, which, again, is not that different from the distributive law for numeric algebra. This simply states that A AND with the results of B OR C is the same as the results of A AND B and A AND C undergoing OR. On the other hand, the equation A + (BC) = (A + B) (A + C) is the second distributive law for Boolean logic, which is not so logical from real algebra. It states that A OR with the results of B AND C is the same as the results of A OR B and A OR C undergoing AND.

Note: In algebra, the distributive law states: The product of an expression of two or more terms multiplied by a single factor is equal to the sum of the products of each term of the expression multiplied by the single factor.

Identity and Inverse Variables

Equations A + A = A, AA= A, A + not (A) = 1, and A not (A) = 0 are statements for the existence of an identity and inverse element for a given Boolean variable. Similar to addition for real algebra, the identity for the OR operator is 0. For the AND operator it is 1, similar to multiplication in real algebra.

Note: Because the *bit* of a Boolean variable (similar to the digit of a real number variable) is either 0 or 1 by definition, then the inverse is found by simply switching all the bits to the other Boolean value (e.g., 0 becomes 1 and 1 becomes 0). Unlike numeric algebra, the inverse of Boolean variable AND with the Boolean variable is 0. If the Boolean variable undergoes the OR operation with its inverse, then the result is 1. Also, if a Boolean variable undergoes either the OR or AND operation with itself, the result is simply the original Boolean variable.

BOOLEAN SYNTHESIS

To this point we have focused briefly on those Boolean laws and variables important to the safety practitioner. There is, however, much more to Boolean algebra, but the additional concepts are beyond the scope of this text.

Notwithstanding the brevity of the overview of Boolean principles provided to this point, we can now provide a basic illustration of how these Boolean principles can be applied in the real world. Note: The following example is taken from *Multidisciplinary Accident Investigation Summaries* (1998), U.S. Department of Transportation, DOT-HS-600, Vol. 2, No. 4, page 143.

Example 9.1
The following facts are to be considered in this example:

1. The time was 8:40 a.m. on a Sunday.
2. A single car, occupied by the driver only, struck the concrete uprights of a bridge pier.
3. The driver was pronounced dead on arrival at the hospital fifteen minutes after the crash. He suffered a compound comminuted fracture of the skull after being thrown from the vehicle and striking his head on a concrete pillar.
4. The roadway was wet, and it was possibly raining hard during the accident; it was daylight.
5. The roadway was a divided, limited-access highway with three lanes each way.
6. The accident took place 200 feet before an exit ramp to the right. The uprights that were struck were off the road to the right.
7. Traffic controls consisted of speed-limit signs (45 mph), broken white-lane-divider lines, a solid-white-line-dividing roadway from a ten-foot shoulder, and exit signs.
8. The vehicle was a 1964 Pontiac Catalina, recently inspected, that had traveled 72,843 miles. Mismatched tire sizes and tread types may have contributed to the accident.
9. At impact the left front door detached as a result of broken hinges. The driver was thrown from the vehicle. Lap belts were not being used at the time of the accident. The car spun around 180 degrees clockwise.
10. The driver, nineteen years of age, was alleged to suffer blackouts and nystagmus. He was familiar with the area. Alcohol tests proved negative.
11. According to a witness, the vehicle was traveling at 90 to 100 mph when the vehicle veered to the right beneath the overpass and struck one of the concrete uprights.
12. After impact, the left rear tire was seen to be deflated.

The "pertinent" facts for Boolean analysis are as follows (with the applicable item number in parentheses):

A. wet roadway contributed to loss of control (4)
B. raining hard, limited visibility (4)
C. confusion between bridge uprights and exit ramp (6)
D. improper traffic controls (7)
E. mismatched tires caused loss of control (8)
F. weak hinges (9)
G. lap belts not used (9)
H. shoulder belts not installed
I. driver illness caused accident (10)
J. speed contributed to loss of control (11)
K. left rear tire blew out and caused accident (12)
L. there were no guardrails to prevent direct impact with the bridge piers

Although the Boolean operators are stated as facts, there is no implication that they are all true. In fact, the following "contradictions" are significant.

Not (C): driver was familiar with the area, no confusion.
Not (D): traffic controls were standard for this type of highway; speed limit (45 mph) was adequate.

Not (E): mismatched tires would only cause loss of control in sudden braking; no evidence of this was available.

Not (K): deflation of left rear tire would have caused a pull to the left, which was not observed.

Also, if the tire blew out, this would make the effect of mismatched tires insignificant, further strengthening the probability of note (E). The deflation of the left rear tire could have been an effect of the car spinning around after impact.

The objective now is to synthesize a Boolean expression for the accident in terms of each contributing factor. First, consider the hazard caused by the weather expressed as W, where

$$W = A + B$$

A second hazard could have been caused by the roadway, given by R, where

$$R = C + D$$

A third hazard could have been caused by the vehicle, given by V, where

$$V = E + K$$

A fourth factor will be charged to shortcomings (not necessarily intentionally) of the driver. These could be in terms of errors in judgment, illness, or intentional violation of the law. Calling this factor P,

$$P = I + J$$

Finally, there are factors that did not cause the accident but did add to the severity. Calling these severity factors S,

$$S = F + G + H + L$$

In the absence of other evidence, it will be assumed that the concurrence of events W, R, V, P, and S lead to the particular accident. Thus, letting X be the occurrence of an accident of this type and severity:

$$X = WRVPS$$
$$= (A + B) (C + D) (E + K) (I + J) (F + G + H + L)$$

Note: This Boolean expression is in its simplest form and would be best left as a product of sums.

Before leaving this example, consider the reduction that would result if only the presence of an accident was considered and not the total event occurrence X. Letting Y be the accident only:

$$Y = WRVP$$

Most importantly, those who are to perform modification on the roadway, vehicle, or driver might restrict their considerations to those pertinent subexpressions.

REVIEW QUESTIONS

1. The inverse of a Boolean variable AND with the Boolean variable is _____.
2. The inverse of the Boolean variable 0 becomes _____ and 1 becomes _____.
3. For both the OR and AND logic operations, the order in which a variable is presented will _____ affect the outcome.

4. (+) is the logic _____ operator.

5. • A(B + C) = _____.

REFERENCES AND RECOMMENDED READING

Boole, G. [1854] 2003. *An investigation of the laws of thought*. New York: Prometheus Books.

Boole, G. [1951] 2003. *The mathematical analysis of logic*. Oxford: Basil Blackwell.

Boole, G. [1952] 2003. *Studies in logic and probability*. LaSalle, IL: Open Court Publishing Company.

Brown, D. B. 1976. *Systems analysis and design for safety*. Englewood Cliffs, NJ: Prentice-Hall.

Clemens, P., and R. Simmons. 1986. *System safety and risk management*. Cincinnati, OH: U.S. Deptartment of Health and Human Services, NIOSH.

Kolodner, H. H. 1971. The fault tree techniques of system safety analysis as applied to the occupational safety situation. *ASSE Monograph 1*.

Marcus, M. P. 1967. *Switching circuits for engineers*. Englewood Cliffs, NJ: Prentice-Hall.

10

Engineering Economy

$$A = P \left(\frac{i(1 + i)^n}{(1 + i)^n - 1} \right) \qquad P = A \left(\frac{(1 + i)^n - 1}{I(1 + i)^n} \right)$$

Engineering activities of analysis and design are not an end in themselves, but are a means for satisfying human wants. Thus, engineering has two aspects. One aspect concerns itself with the materials and forces of nature; the other is concerned with the needs of mankind. Because of this, engineering must be closely associated with economics

—H. G. Thuesen et al. (1971)

SAFETY ENGINEERING AND ECONOMICS

In regard to economics, safety practitioners don't necessarily have to be economists but should have a "general" footing in a variety of economic principles. This makes sense when we consider that most safety engineering decisions are based on economic considerations—a situation that is unlikely to change in the years ahead. Moreover, it is also important to consider that the prevention of hazards or correction of hazardous situations is not effected without cost. Unfortunately, even the conscientious safety practitioner often forgets or overlooks the financial implications in providing additional design measures for safety or corrective applications for mitigation, and, by doing so, ends up with a plan that does not get funded.

The person tasked with protecting people and property through safety engineering practices often feels that nothing is more important than accomplishing this goal. When told that his or her plan must be justified through cost-benefit analysis and must add value to the business, safety engineers sometimes balk at the idea that anyone can, could, or should put a price on human life and/or safety. The fact is, in the real world, we are required do this every day—the safety person must justify his or her existence within the organization. While safety professionals feel that safety compliance is the sine qua non—the indispensable—element of any business success, the average business manager views safety as a cost that does not add to the bottom line. Safety in this type of shortsighted business manager's mind is simply persona non grata.

For those of us who have worked in the safety profession for any length of time, we are accustomed to this type of dysfunctional thinking. It must be pointed out, however, that we also learn (sooner rather than later) that most of us work in the real world where we have to deal with or within the constraints of an economic bottom line. It does not take a rocket-science mentality to understand the implausibility of recommending a very costly "fix" to a present or potential hazard situation, when, at the same time, bringing about such a fix would bankrupt the company.

The implementation of safety engineering practices (fixes) must be tempered not only by common sense but also by the economic bottom line.

Having stated the obvious (hopefully, it is obvious) that cost-benefit analysis has its place in the practice of safety engineering, in this chapter we present a few of the economic principles the safety engineering practitioner should be familiar with. That is, we present mathematical techniques and practical advice for evaluating decisions in the design and preparation of engineering systems. These procedures support selection and justification of design alternatives, operating policies, and capital expenditure. Thus, what follows is a brief introduction to economic equations and formulae commonly used in the practice of safety engineering. Also keep in mind that historically, many of the math operations discussed in the following sections have been presented as the types of engineering economy questions in professional certification examinations.

Note: For those desiring a more in-depth treatment of engineering economics, we recommend Leland Blank and Anthony Tarquin's text, *Engineering Economy*, available from McGraw-Hill.

Key Terms

i: Interest rate per interest period. In the equations, the interest rate is stated as a decimal (i.e., 6% interest is 0.06).
n: Number of interest periods.
P: A present sum of money.
F: A future sum of money. The future sum F is an amount, n interest periods from the present, that is equivalent to P with interest rate i.
A: An end-of-period cash receipt or disbursement in a uniform series, continuing for n periods, the entire series equivalent to P or F at interest rate i.
sinking fund: A separate fund into which one makes a uniform series of money deposits (A) with the goal of accumulating some desired future sum (F) at a given future point in time.

CAPITAL-RECOVERY FACTOR (EQUAL-PAYMENT SERIES)

Annual amounts of money ($) to be received or paid are the equivalent of either a single amount in the future or a single amount in the present, when the annual amounts are compounded over a period of years at a given interest rate (i). The value of the annual amounts can be calculated from a single present amount (P), or a single future amount (F).

We can use the *capital-recovery factor* (sometimes called the uniform series capital recovery factor or annual payment from a present value) to determine the annual payments (A) from an investment. This is accomplished using equation 10.1. The equation is based on present value (P), the interest rate (i) at which that present value is invested, and the period (term) over which it is invested (n).

$$A = P \left(\frac{i (1 + i)^n}{(1 + i)^n - 1} \right) \tag{10.1}$$

Where
   A = annual investment or payment ($)
   P = present value ($)
   i = interest rate (%)
   n = number of years

Example 10.1
Problem:
How much will an investment of $5,000 yield annually over eight years at an interest rate of 5%?

Solution:

$$A = \$5,000 \left(\frac{0.05 \, (1 + 0.05)^8}{(1 + 0.05)^8 - 1} \right)$$

$$= 0.1547$$
$$= \$5,000 \, (0.1547) = \$773.50$$

$5,000 invested at 5% for eight years will yield an annual payment $773.50.

Note: The higher the interest rate (i) earned by the investment, the higher the annual amount will be, because the annual amounts compound at a higher rate. On the other hand, the longer the term of the investment (n), the lower the annual amount will be because there are more annual payments being made that compound for a longer time.

UNIFORM SERIES PRESENT WORTH (VALUE) FACTOR

The present worth of an amount of money is the equivalent of either a single amount in the future (the future amount) of a series of amounts to be received or paid annually over a period of years as compounded at an interest rate over a period of years. Stated differently, what is the investment that needs to be made now so that the future series of money can be received? The present worth can be calculated from a single future amount (F), or an annual amount (A).

Here, the present worth (P) of a series of equal annual amounts (A) can be calculated by using equation 10.2, which compound the interest (%) at which the annual amounts are invested over the term of the investment in years (n).

$$P = A \left(\frac{(1 + i)^n - 1}{i \, (1 + i)^n} \right) \tag{10.2}$$

Example 10.2
Problem:
The present worth of a series of eight equal annual payments of $154.72 at an interest rate of 5% compounded annually will be what?

Solution:

$$P = \$154.72 \left(\frac{(1 + 0.05)^8 - 1}{0.05 \, (1 + 0.05)^8} \right)$$

$$= \$154.72 \, (6.4632)$$
$$= \$1,000$$

FUTURE VALUE

The future value (or uniform series compound amount factor) of an amount of money is the equivalent of either a single amount today (the present amount) or a series of amounts to be received or paid annually over a period of years as compounded at an interest rate over a period of years. The future value can be calculated from either a single present amount (P) or an annual amount (A).

The future (F) value of a series of equal annual amounts can be calculated by using equation 10.3. The equation compounds the interest (i) at which the annual amounts (A) are invested over the term of the investment in years (n).

$$F = A \left(\frac{(1 + i)^n - 1}{i} \right) \tag{10.3}$$

Example 10.3
Problem:
A woman deposits $500 in a bank at the end of each year for five years. The bank pays 5% interest, compounded annually. At the end of five years, immediately following her fifth deposit, how much will she have in her account?

Solution:

$$F = A \left(\frac{(1 + i)^n - 1}{i} \right) = A \, (F/A, I, n)$$

where A = $500, n =5, I = 0.05, F = unknown. Filling in the know variables:

$$F = \$500 \, (F/A \, 5\%, \, 5) = \$500 \, (5.526) = \$2763$$

She will have $2,763 in her account following the fifth deposit.

Note: The higher the interest rate (i) earned by the investment, the higher the future value will be because the investment compounds at a higher rate. The longer the term of the investment (n), the higher the future value will be because there are more annual payments being made that compound for a longer time.

ANNUAL PAYMENT (UNIFORM SERIES SINKING FUND)

Annual amounts of money to be received or paid are the equivalent of either a single amount in the future or a single amount in the present, when the annual amounts are compounded over a period of years at a given interest rate. The value of the annual amounts can be calculated from a single present amount (P), or a single future amount (F).
 The annual payments into an investment can be calculated by using equation 10.4.

$$A = F \left(\frac{i}{(1 + i)^n - 1} \right) \tag{10.4}$$

Example 10.4
Problem:
A man read that in the western United States, a ten-acre parcel of land could be purchased for $1,000 cash. The man decided to save a uniform amount at the end of each month so that he would have the required $1,000 at the end of one year. The local bank pays 1/2% (0.005) interest, compounded monthly. How much would the man have to deposit each month?

Solution:

$$A = F \left(\frac{i}{(1 + i)^n - 1} \right)$$

Where
 F = $1000

n = 12
 i = 0.005
 A = ?

$$A = 1000 \ (A/F, \ \tfrac{1}{2}\%, \ 12) = 1000 \ (0.0811) = \$81.10$$

The man would have to deposit $81.10 each month.

Note: The higher the interest rate (i) earned by the investment, the lower the annual amount will be, because the annual amounts can compound at a lower rate to reach the same future amount. The longer the term of the investment (n), the higher the annual amount will be, because there are more annual payments being made that compound for a longer time.

PRESENT VALUE OF FUTURE DOLLAR AMOUNT

The present value of an amount of money is the equivalent of either a single amount in the future (the future amount) or a period of years as compounded at an interest rate over a period of years. The present value can be calculated from a single amount (F), or an annual amount (A).

The present value (P) of a future dollar amount (F) can be calculated by using equation 10.5. The equation compounds the interest in % (i) at which the present value (P) is invested over the term of the investment in years (n).

$$P = F \ (1 + i)^{-n} \tag{10.5}$$

Example 10.5
Problem:
What is the present value of $6,000 to be received in five years, if it is invested at 6%?

Solution:

$$P = F \ (1 + i)^{-n}$$
$$P = \$6,000 \ (1 + .06)^{-5}$$
$$P = \$6,000 \ (1/.06)^{5}$$
$$P = \$6,000 \ (.747)$$
$$P = \$4,482$$

The present value of $6,000 to be received in five years, if it is invested at 6%, is $4,482.

Note: The higher the interest rate (i) earned by the investment, the lower the present value will be, because the investment compounds at a higher rate. The longer the term of the investment (n), the lower the present value will be because the investment compounds over a longer time.

FUTURE VALUE OF A PRESENT AMOUNT

The future value of an amount of money is the equivalent of either a single amount today (the present amount) or a series of amounts to be received or paid annually over a period of years as compounded at an interest rate over a period of years. The future value can be calculated from either a single present amount (P), or an annual amount (A).

The future value (F) of a present dollar amount can be calculated by using equation 10.6. The equation compounds the interest in $ (i) at which the present value (P) is invested over the term of the investment in years (n).

$$F = P \ (1 + i)^{n} \tag{10.6}$$

Example 10.6
Problem:
If $6,000 is invested for 5 years at 6% interest per year, what will the investment be worth in five years?

$$F = P (1 + i)^n$$
$$F = \$6,000 (1 + .06)^5$$
$$F = \$6,000 (1.06)^5$$
$$F = \$6,000 (1.338)$$
$$F = \$8,029$$

The $6,000 investment will be worth $8,029 in five years.

Note: The higher the interest rate (i) earned by the investment, the higher the future value will be because the investment compounds at a higher rate. The longer the term of the investment (n), the higher the future value will be because the investment compounds for a longer time.

REVIEW QUESTIONS

1. The present worth of a series of ten equal annual payments of $200.10 at an interest rate of 3.5% compounded annually will be what?
2. The present worth of a series of four equal annual payments of $312.50 at an interest rate of 6% compounded annually will be what?
3. A man deposits $6,000 in a bank at the end of each year for six years. The bank pays 6% interest, compounded annually. At the end of five years, immediately following his fifth deposit, how much will he have in his account?
4. What is the present value of $8,000 to be received in eight years, if it is invested at 5%?
5. If $8,000 is invested for six years at 6% interest per year, what will the investment be worth in five years?

REFERENCES AND RECOMMENDED READING

Blank, L. T., and A. J. Tarquin. 1997. *Engineering economy*. New York: McGraw-Hill.
Thuesen, H. G., W. J. Fabrycky, and G. J. Thuesen. 1971. *Engineering economy*. 4th ed. Englewood Cliffs, NJ: Prentice-Hall.

III

SAFETY ENGINEERING CONCEPTS

Fundamental Engineering Concepts

A life without adventure is likely to be unsatisfying, but a life in which adventure is allowed to take whatever form it will, is likely to be short.

—Bertrand Russell

Education can only go so far in preparing the safety engineer for on-the-job experience. A person who wishes to become a safety engineer is greatly assisted by two personal factors.

First, a well-rounded, broad development of experience in many areas is required, and results in the production of the classic generalist. Second, although safety engineers cannot possibly attain great depth in all areas, they must have the desire and the aptitude to do so. They must be interested in—and well-informed about—many widely differing fields of study. The necessity for this in the safety application is readily apparent. Why? Simply because the range of problems encountered is so immense that a narrow education will not suffice; safety engineers must handle situations that call upon skills as widely diverse as the ability to solve psychological and sociological problems with people to the ability to perform calculations required for mechanics and structures. We have pointed out that the would-be practicing safety engineer can come from just about any background and that a narrow education does not preclude students and others from broadening themselves later; however, quite often those who are very specialized lack appreciation for other disciplines as well as the adaptability necessary for safety engineering.

INTRODUCTION

Though individual learning style is important in your determination of a career choice as a safety engineer, again we stress generalized education as the key ingredient in the mix that produces the safety engineer. That is, along with basic and applied sciences (mathematics, natural and behavioral sciences—which are applied to the solution of technological, biological, and behavioral problems), education in engineering and technology are a must. Topics including applied mechanics, properties of materials, electrical circuits and machines, fire science hydraulics, principles of engineering design, and computer science fall into this category.

In this chapter, we concentrate on applied mechanics and in particular, forces and the resolution of forces. Why? Because many accidents and resulting injuries are caused by forces of too great a magnitude for a machine, material, or structure. To inspect systems, devices, or products to ensure their safety, safety engineers must account for the forces that act or might act on them. Safety engineers must also account for forces from objects that may act on the human body (an area of focus that is often overlooked).

Important areas that are part of or that interface with applied mechanics are the properties of materials and engineering design considerations (fire science hydraulics, electrical circuits, and safety concerns associated with both these important areas are addressed separately later in this text). We cannot discuss all of the engineering aspects related to these areas in this text. Instead, our goal is to look at some fundamental concepts and their relationships to safety engineering.

RESOLUTION OF FORCES

In safety engineering, we tend to focus our attention on those forces that are likely to cause failure or damage to some device or system, resulting in an occurrence that is likely to produce secondary and tertiary damage to other devices or systems and harm to individuals. Typically, large forces are more likely to cause failure or damage than small ones.

Safety engineers must understand force. He or she must understand how a force acts on a body: 1. the direction of force, 2. the point of application (location) of force, 3. the area over which force acts, 4. the distribution or concentration of forces that act on bodies, and 5. how essential these elements are in evaluating the strength of materials. For example, a 40 lb force applied to the edge of a sheet of plastic and parallel to it probably will not break it. If a sledgehammer strikes the center of the sheet with the same force, the plastic will probably break. A sheet metal panel of the same size undergoing the same force will not break.

Practice tells us that different materials have different strength properties. Striking the plastic panel will probably cause it to break, whereas striking a sheet metal panel will cause a dent. The strength of a material and its ability to deform are directly related to the force applied.

Important physical, mechanical, and other properties of materials are

- crystal structure
- strength
- melting point
- density
- hardness
- brittleness
- ductility
- modulus of elasticity
- wear properties
- coefficient of expansion
- contraction
- conductivity
- shape
- exposure to environmental conditions
- exposure to chemicals
- fracture toughness
- and many others

Note: All these properties can vary, depending on whether forces are crushing, corroding, cutting, pulling, or twisting.

The forces an object can encounter are often different from the forces that an object would be able to withstand. The object may be designed to withstand only minimal force before it fails (a toy doll may be designed of very soft, pliable materials or designed to break or give way in certain places when a child falls on it, preventing injury). Other devices may be designed to withstand the greatest possible load and shock (e.g., a building constructed to withstand an earthquake).

When working with any material to go in an area with a concern for safety, a safety factor (or factor of safety) is often introduced. Safety factor (SF) (as defined by ASSE 1988) is the ratio allowed for in design, between the ultimate breaking strength of a member, material, structure, or equipment and the actual working stress or safe permissible load placed on it during ordinary use. Simply put, including a factor of safety—into the design of a machine, for example—makes an allowance for many unknowns (inaccurate estimates of real loads or irregularities in materials, for example) related to the materials used to make the machine, related to the machine's assembly, and related to the use of the machine. Safety factor can be determined in several ways. One of the most commonly used ways is

$$SF = \frac{failure - producing\ load}{allowable\ stress} \qquad (11.1)$$

Forces on a material or object are classified by the way they act on the material. For example, if the force pulls a material apart, it is called tensile force. Forces that squeeze a material or object are called compression forces. Shear forces cut a material or object. Forces that twist a material or object are called torsional forces. Forces that cause a material or object to bend are called bending forces. A bearing force occurs when one material or object presses against or bears on another material or body.

So, what is force?

Force is typically defined as any influence that tends to change the state of rest or the uniform motion in a straight line of a body. The action of an unbalanced or resultant force results in the acceleration of a body in the direction of action of the force, or it may (if the body is unable to move freely) result in its deformation. Force is a vector quantity, possessing both magnitude and direction (see figure 11.1.A–B); its SI unit is the newton (equal to 3.6 ounces, or 0.225 lb).

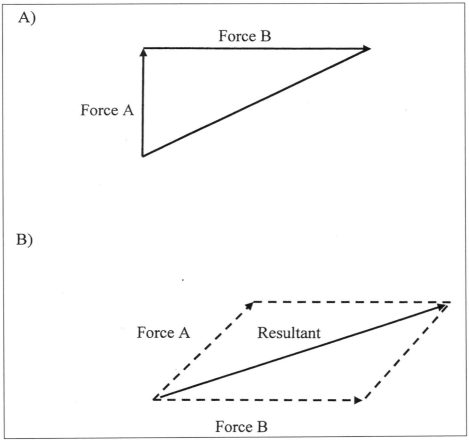

Figure 11.1 a, b   Force vector quantities.

According to Newton's second law of motion, the magnitude of a resultant force is equal to the rate of change of momentum of the body on which it acts. The unit of force is the pound force in the English or engineering system and is the newton in the SI system. The pound force is defined as the force required to give one slug of mass an acceleration of one foot per second per second. The newton is defined as the force required to give one kilogram of mass acceleration of one meter per second per second. The unit of mass in the English system is the slug that is the mass of 32.2 standard pounds. Since the mass of an object is commonly expressed as its weight in pounds, we can make the conversion to slugs by dividing this weight by the gravitational constant, $g_c = 32.2$ ft/s^2, for use in the formula

$$F = ma \tag{11.2}$$

Example 11.1
Problem:
How much force must a seat belt be capable of withstanding to safely restrain a 180 lb woman when her car comes to a sudden (i.e., 1.0 sec) stop if initially traveling at 60 mph? (Assume a safety factor of 4.)

Solution:

$$F = ?$$
$$m = 180 \text{ lb}/32.2 \text{ ft/s}^2 = 5.59 \text{ slugs}$$
$$a = v/t = (60 \text{ m/hr})(5280 \text{ ft/m})(\text{hr}/3600 \text{ sec})/1.0 \text{ sec} = 88 \text{ ft/s}^2$$
$$F = ma \text{ x S.F. of } 4$$
$$F = (5.59)(88) \text{ lbs x } 4 = 1968 \text{ lbs}$$

Another important relationship whereby force is a key player is in the concept of work. *Work* is the product of the force and the effective displacement of its application point. The equation for calculating force is

$$W = Fs \tag{11.3}$$

Where
W = work in foot-pounds (ft-lbs)
F = force in pounds (lbs)
s = distance in feet (ft)

If a force is applied to an object and no movement occurs, no effective work is done. The energy possessed by a body determines the amount of work it can do. Newton's Third Law states that for every action there is an equal and opposite reaction.

Example 11.2
Problem:
A system of pulleys having a mechanical advantage of 5 is used to move a 1 ton weight up an inclined plane. The plane has an angle of 45° from horizontal. The weight has to be moved a vertical distance of 30 ft. Ignoring friction, how much work is required to be done to move the weight up to the top (vertical lift work)?

Solution:

$$W = ?$$
$$F = 2000 \text{ lb}$$
$$S = 30 \text{ ft}$$
$$W = Fs$$
$$W = (2000 \text{ lb})(30 \text{ ft}) = 60,000 \text{ ft-lb}$$

In regard to safety engineering, a key relationship between a force F and a body on which it acts is

$$F = sA \tag{11.4}$$

Where
s = force or stress per unit area (e.g., pounds per square inch)
A = area (square inches, square feet, etc.) over which a force acts

---

Did You Know?

The stress a material can withstand is a function of the material and the type of loading.

Frequently, two or more forces act together to produce the effect of a single force, called a resultant. This resolution of forces can be accomplished in two ways: triangle and/or parallelogram law.

The *triangle law* provides that if two concurrent forces are laid out vectorially with the beginning of the second force at the end of the first, the vector connecting the beginning and the end of the forces represents the resultant of the two forces (see figure 11.1.A).

The *parallelogram law* provides that if two concurrent forces are laid out vectorially, with either forces pointing toward or both away from their point of intersection, a parallelogram represents the resultant of the force. The concurrent forces must have both direction and magnitude if their resultant is to be determined (see figure 11.1.B).

After the triangle or parallelogram has been completed, and if the individual forces are known or one of the individual forces and the resultant are known, the resultant force may be simply calculated by either the trigonometric method (sines, cosines, and tangents), or by the graphic method (which involves laying out the known force, or forces, to an exact scale and exact direction in either a parallelogram or triangle, and then measuring the unknown to the same scale).

SLINGS

The United States Department of Energy (2009) points out that slings must be used in accordance with recommendations of the sling manufacturer. Slings manufactured from conventional three-strand natural or synthetic fiber rope are not recommended for use in lifting service. Natural or synthetic fiber rope slings may be used only if other sling types are suitable for the unique application. For natural or synthetic rope slings, the requirements of ASME B30.9 and OSHA 1910.184(h) must be followed. All types of slings must have, as a minimum, the rated capacity clearly and permanently marked on each sling. Each sling must receive a documented inspection at least annually, more frequently if recommended by the manufacturer or made necessary by service conditions.

Note: Slings are commonly used between cranes, derricks, and/or hoists and the load, so that the load may be lifted and moved to a desired location. For the safety engineer, knowledge of the properties and limitations of the slings, the type and condition of material being lifted, the weight and shape of the object being lifted, the angle of the lifting sling to the load being lifted, and the environment in which the lift is to be made are all important considerations to be evaluated—before the transfer of material can take place safely.

Let's take a look at a few example problems involving forces that the safety engineer might be called upon to calculate. In our examples, we use lifting slings under different conditions of loading.

Example 11.3
Let us assume a load of 2,000 pounds supported by a two-leg sling; the legs of the sling make an angle of 60° with the load. What force is exerted on each leg of the sling? (Note: In solving this type of problem, you should always draw a rough diagram as shown by figure 11.2.) A resolution of forces provides the answer. We use the trigonometric method to solve this problem, but remember that it may also be solved using the graphic method. Using the trigonometric method with the parallelogram law, the problem could be solved as follows: (Again, make a drawing showing a resolution of forces similar to figure 11.3; it should look like figures 11.2 and 11.3):

We could consider the load (2,000 pounds) as being concentrated and acting vertically, which can be indicated by a vertical line. The legs of the slings are at a 60° angle, which can be shown as ab and ac. The parallelogram can now be constructed by drawing lines parallel to ab and ac, intersecting at d. The point where cb and ad intersect can be indicated as e. The force on each leg of the sling (ab for example) is the resultant of two forces, one acting vertically (ae), the other horizontally (be), as shown in the force diagram. Force ae is equal to one half of ad (the total force acting vertically, 2,000 pounds, so ae = 1,000). This value remains constant regardless of the angle ab makes with bd, because as the angle increases or decreases, ae also increases or decreases. But ae is always ad/2. The force ab can be calculated by trigonometry using the right triangle abe:

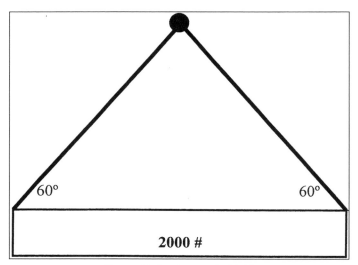

Figure 11.2   [example 11.3].

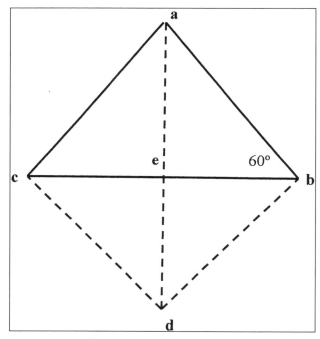

Figure 11.3   [example 11.3].

Sine of angle $= \dfrac{\text{opposite}}{\text{hypotenuse}}$

therefore, sine 60 degree $= \dfrac{ae}{ab}$

transposing,   ab $= \dfrac{ae}{\text{sine 60 degree}}$

substituting known value, ab $= \dfrac{1000}{0.866} = 1155$

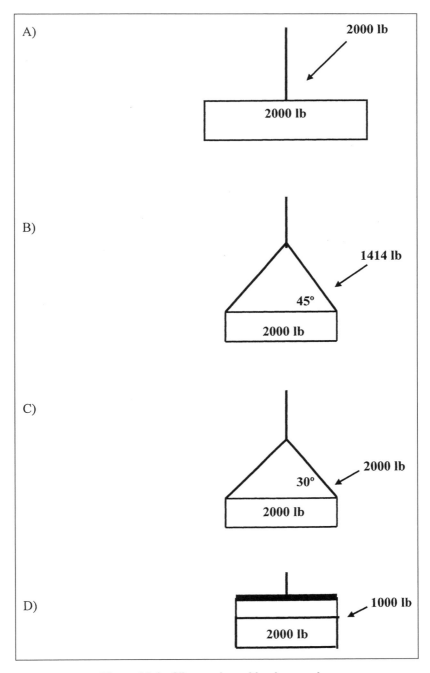

Figure 11.4   Sling angle and load examples.

The total weight on each leg of the sling at a 60° angle from the load is 1,155 pounds. Note that the weight is more than half the load, because the load is made up of two forces—one acting vertically, the other horizontally. An important point to remember is—the smaller the angle, the greater the load (force) on the sling. For example, at a 15° angle, the force on each leg of a 2,000 pound load increases to 3,864 pounds.

Note: Sling angles less than 30° are *not* recommended.

Let's take a look (see figure 11.4) at what the force would be on each leg of a 2,000 pound load at different angles (angles that are common for lifting slings).

Now let's work a couple of example problems.

Example 11.4
Problem: We have a 3,000 lb load to be lifted with a two-leg sling whose legs are at a 30° angle with the load. The load (force) on each leg of the sling is as follows:

Solution:

$$\text{Sine A} = \frac{a}{c}$$

$$\text{Sine 30} = 0.500$$

$$a = \frac{3,000}{2} \quad 1500$$

$$c = \frac{a}{\text{Sine A}}$$

$$c = \frac{1,500}{0.5}$$

$$c = 3,000 \text{ lb}$$

Example 11.5
Problem: Given a two-rope sling supporting 10,000 pounds, what is the load (force) on the left sling? Sling angle to load is 60°.

Solution:

$$\text{Sine A} = \frac{a}{c}$$

$$\text{Sine A} = \frac{60}{0.866}$$

$$a = \frac{10,000}{2}$$

$$c = \frac{a}{\text{Sine A}}$$

$$c = \frac{5,000}{0.866}$$

$$c = 5,774 \text{ lb}$$

Rated Sling Loads

In the preceding section we demonstrated simple math operations used to determine the rated sling load that a particular sling can safely bear. In the field and on the job, knowing how to use simple math to make sling angle-load determinations is important. It is also important to point out, however, that many tables showing rated loads on slings are available. In table 11.1, for example, we present a rated load for alloy steel chain slings chart provided by U.S. DOE (2009).

Table 11.1   Alloy Steel Chain Sling Load Angle Factors

Tension in Each Sling Leg = Load/2 = Load Angle Factor	
Horizontal Sling Angle	Load Angle Factor
90°	1.000
85°	1.004
80°	1.015
75°	1.035
70°	1.064
65°	1.104
60°	1.155
55°	1.221
50°	1.305
45°	1.414
40°	1.555
35°	1.742
30°	2.000
25°	2.364*
20°	2.924*
15°	3.861*
10°	5.747*
5°	11.490*

* Not recommended.

INCLINED PLANE

Another common problem encountered by the safety engineer involving the resolution of forces occurs in material handling operations in moving a load (a cart, for example) up (or down) an *inclined plane* (a ramp or tilted surface, in our example). The safety implications in this type of work activity should be obvious. Objects are known to accelerate down inclined planes because of an unbalanced force (any time we deal with unbalanced forces, safety issues are present and must be addressed). To understand this type of motion, it is important to analyze the forces acting upon an object on an inclined plane. Figure 11.5 depicts the two forces acting upon a load that is positioned on an inclined plane (assuming no friction). As shown in figure 11.5, there are always at least two forces acting upon any load that is positioned on an inclined plane—the force of gravity (also known as weight) and the normal (perpendicular) force. The force of gravity acts in a downward direction, yet the normal force acts in a direction perpendicular to the surface.

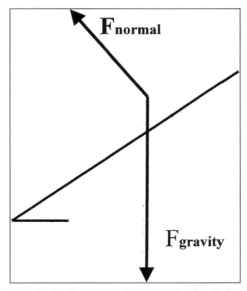

Figure 11.5   Forces acting on an inclined plane.

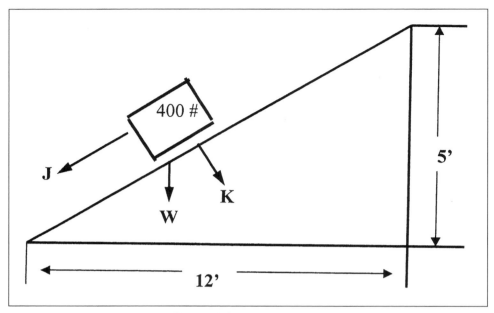

Figure 11.6 [example 11.6].

Let's take a look at a typical example of how to determine the force needed to pull a fully loaded cart up a ramp (an inclined plane).

Example 11.6
We assume that a fully loaded cart weighing 400 pounds is to be pulled up a ramp that has a five-foot rise for each 12 feet, measured along the horizontal (again, make a rough drawing; see figure 11.6).
What force is required to pull it up the ramp?

Note: For illustrative purposes, we assume no friction. Without friction, of course, the work done in moving the cart in a horizontal direction would be zero (once the cart was started, it would move with constant velocity—the only work required is that necessary to get it started). However, a force equal to J is necessary to pull the cart up the ramp or to maintain the car at rest (in equilibrium). As the angle (slope) of the ramp is increased, greater force is required to move it, because the load is being raised as it moves along the ramp, thus doing work (remember, this is not the case when the cart is moved along a horizontal plane without friction—in actual practice, however, friction can never be ignored and some "work" is accomplished in moving the cart).

To determine the actual force involved, we can again use a resolution of forces. The first step is to determine the angle of the ramp. This can be calculated by the following formula:

$$\text{Tangent (angle of ramp)} = \frac{\text{opposite side}}{\text{adjacent side}} \quad 5/12 = 0.42$$

Then, arctan 0.42 = 22.8 degree

Now you need to draw a force parallelogram (see figure 11.7) and apply the trigonometric method. The weight of the cart W (shown as force acting vertically) can be resolved into two components: one a force J parallel to the ramp, the other a force K perpendicular to the ramp. The component K, being perpendicular to the inclined ramp (plane) does not hinder movement up the ramp. The component J represents a force that would accelerate the cart down the ramp. To pull the cart up the ramp, a force equal to or greater than J is necessary.
Applying the trigonometric method, the angle WOK is the same as the angle of the ramp.

OJ = WK & OW = 400 lb

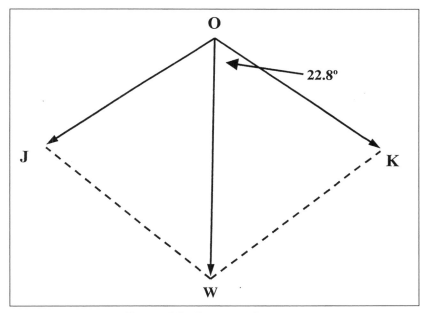

Figure 11.7   Force parallelogram.

$$\text{Sine of angle WOK (22.8°)} = \frac{\text{opposite side (WK)}}{\text{hypotenuse (OW)}}$$

Transposing,   WK = OW x sine (22.8°)

WK = 400 x .388

WK = 155.2

Thus, a force of 155.2 pounds is necessary to pull the cart up the 22.8° angle of the ramp (friction ignored). Note that the total amount of work is the same, whether the cart is lifted vertically (400 pounds x 5 feet = 2000 foot-pounds), or pulled up the ramp (155.2 pounds x 13 feet = 2000 foot-pounds). The advantage gained in using a ramp instead of a vertical lift is that less force is required—but through a greater distance.

Materials and Principles of Mechanics

To be able to recognize hazards and to select and implement appropriate controls, safety engineers must have a good understanding of the properties of materials and principles of mechanics. In this section, we start with the properties of materials and then cover the wide spectrum that is mechanics and soil mechanics. Our intent is to clearly illustrate the wide scope of knowledge required in areas germane to the properties of materials and the principles of mechanics and those topics on the periphery—all of which are blended in to the safety knowledge mix—the mix that helps to produce the well-rounded, knowledgeable safety engineer.

PROPERTIES OF MATERIALS

When we speak of the properties of materials or materials properties, what are we referring to, and why should we concern ourselves with this topic? The best way to answer this question is to use an example where the safety engineer, working with design engineers in a preliminary design conference, might typically be exposed to (should be exposed to) engineering design data, parameters, and specifications related to the properties of a particular construction material to be used in the fabrication of, for example, a large mezzanine in a warehouse. In constructing this particular mezzanine, consideration was given to the fact that it would be used to store large, heavy equipment components. The demands placed on the finished mezzanine create the need for the mezzanine to be built using materials that can safely support a heavy load.

For illustration, let's say that the design engineers plan to use an aluminum alloy type structural—No. 17ST. Before they decide to include No. 17ST and determine the required quantity needed to build the mezzanine, they are concerned with determining its mechanical properties to ensure that it will be able to handle the intended load (they will also factor in, many times over, for safety—selecting a type of material that will handle a load much greater than expected).

Using a table on the Mechanical Properties of Engineering Materials in Urquhart's *Civil Engineering Handbook*, 4th ed. (1959), they check the following for No. 17ST:

1. Ultimate strength, psi (defined as the ultimate strength in compression for ductile materials which is usually taken as the yield point) includes tension: 58,000 psi, compression: 35,000 psi, shear: 35,000 psi
2. Yield point tension, psi: 35,000
3. Modulus of elasticity, tension, or compression, psi: 10,000,000
4. Modulus of elasticity, shear, psi: 3,750,000
5. Weight per cu in., lb: 0.10

The question becomes "Is this information important to the safety engineer?" Well, in a specific sense, no—not exactly; however, in a general sense, yes. What is important to the safety engineer is: 1. that a procedure such as the one just described was actually specified; that is, professional engineers actually took the time to determine the correct materials to use in constructing the mezzanine; and 2. when exposed to this type of information, to specific terms, the safety engineer must know enough about the "language" used to know what the design engineers are talking about—and to understand its significance. A materials property is an intensive, often quantitative property of a material. Thus, it is important for safety engineers to understand the material property descriptor and also the units used as metrics of value. In short, we must know and understand the meaning of nomenclature of the materials to be used and their limits. This is important in order to compare the benefits of one material versus another to assist in proper materials selection. Remember Voltaire: "If you wish to converse with me, define your terms."

Let's take a look at a few other engineering terms and their definitions so that we will be able to converse. Many of the following engineering terms are from Heisler's *The Wiley Engineer's Desk Reference: A Concise Guide for the Professional Engineer* (1984), Tapley's *Eshbach's Handbook of Engineering Fundamentals*, 4th ed. (1990), and Giachino and Weeks's *Welding Skills* (1985)—which should be standard reference texts for safety engineers.

mechanics: That branch of science that deals with forces and motion.

rupture: Describes the ultimate failure of tough ductile materials loaded in tension.

stress: The internal resistance a material offers to being deformed. Measured in terms of the applied load over the area (see figure 11.8.A).

buckling: A failure mode characterized by a sudden failure of a structural member subjected to high compressive stresses, where the actual compressive stress at the point of failure is less than the ultimate compressive stresses that the material is capable of withstanding.

strain: The deformation that results from a stress. Expressed in terms of the amount of deformation per inch (see figure 11.8.B).

corrosion: The breaking down of essential properties in a material due to chemical reactions with its surroundings.

intensity of stress: The stress per unit area, usually expressed in pounds per square inch. It is due to a force of P pounds producing tension, compression, or shear on an area of A square inches, over which it is uniformly distributed. The simple term, *stress*, is normally used to indicate intensity of stress.

creep: The tendency of a solid material to slowly move or deform permanently under the influence of stresses.

ultimate stress: The greatest stress that can be produced in a body before rupture occurs.

allowable stress or working stress: The intensity of stress that the material of a structure or a machine is designed to resist.

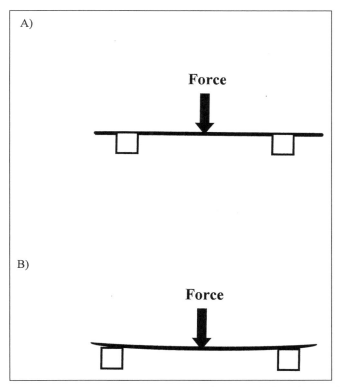

**Figure 11.8 a,b   Stress—measured in terms of the applied load over the area
Strain—expressed in terms of amount per square inch.**

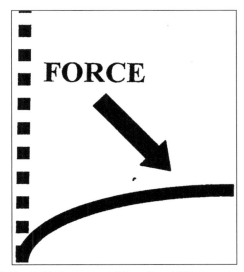

**Figure 11.9   Elasticity and elastic limit—a metal has the ability to return to its original shape after
being elongated or distorted, unless it reaches maximum stress point.**

elastic limit: The maximum intensity of stress to which a material may be subjected and return to its original shape upon the removal of stress (see figure 11.9).

yield point: The intensity of stress beyond which the change in length increases rapidly with little (if any) increase in stress.

fracture: The local separation of an object or material into two, or more, pieces under the action of stress.

modulus of elasticity: The ratio of stress to strain, for stresses below the elastic limit. By checking the modulus of elasticity, the comparative stiffness of different materials can readily be ascertained. Rigidity and stiffness are very important for many machine and structural applications.

Poisson's ratio: The ratio of the relative change of diameter of a bar to its unit change of length under an axial load that does not stress it beyond the elastic limit.

intensity of stress: The stress per unit area, usually expressed in pounds per square inch. Due to a force of P pounds producing tension, compression, or shear on an area of A square inches, over which it is uniformly distributed. The simple term, *stress*, is normally used to indicate intensity of stress.

tensile strength: The property that resists forces acting to pull the metal apart—a very important factor in the evaluation of a metal (see figure 11.10).

Figure 11.10   A metal with tensile strength resists pulling forces.

Figure 11.11   Compressive strength—a metal's ability to resist crushing forces.

compressive strength: The ability of a material to resist being crushed (see figure 11.11).

bending strength: That quality that resists forces from causing a member to bend or deflect in the direction in which the load is applied—actually a combination of tensile and compressive stresses (see figure 11.12).

torsional strength: The ability of a metal to withstand forces that causes a member to twist (see figure 11.13).

shear strength: How well a member can withstand two equal forces acting in opposite directions (see figure 11.14).

fatigue strength: The property of a material to resist various kinds of rapidly alternating stresses.

impact strength: The ability of a metal to resist loads that are applied suddenly and often at high velocity.

ductility: The ability of a metal to stretch, bend, or twist without breaking or cracking.

hardness: The property in steel that resists indentation or penetration.

brittleness: A condition whereby a metal will easily fracture under low stress.

toughness: May be considered as strength, together with ductility. A tough material can absorb large amounts of energy without breaking.

malleability: The ability of a metal to be deformed by compression forces without developing defects, such as encountered in rolling, pressing, or forging.

mechanical overload: The failure or fracture of a product or component in a single event.

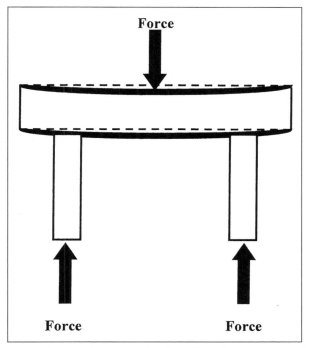

Figure 11.12   Bending strength (stress)—a combination of tensile strength and compressive strength.

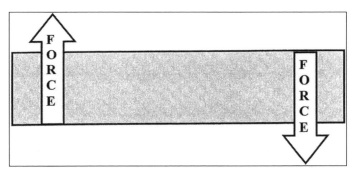

Figure 11.13   Torsional strength—a metal's ability to withstand twisting forces.

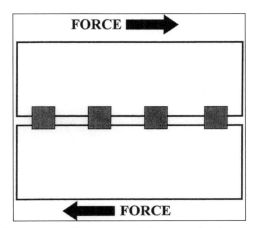

Figure 11.14   Sheer strength—two equal forces acting in opposite directions.

Friction

Earlier, in discussing the principle of the inclined plane, we ignored the effect of friction. In actual use, friction cannot be ignored and you must have some understanding of its characteristics and applications.

Friction results when one body (object) on the verge of sliding, rotating, rolling, spinning, or in the process of any of these, is in contact with another body. Friction allows us to walk, ski, drive vehicles, and power machines (among other things). Whenever one body slides over another, frictional forces opposing the motion are developed between them. Friction force is the force tangent to the contact surface that resists motion. If motion occurs, the resistance is due to kinetic friction, which is normally lower than the value for static friction. Contrary to common perception, the degree of smoothness of a surface area is not responsible for these frictional forces; instead, the molecular structure of the materials is responsible. The coefficient of friction M (which differs among different materials) is the ratio of the frictional force (F) to the normal force (N) between two bodies.

$$M = \frac{F}{N} \tag{11.5}$$

For dry surfaces, the coefficient of friction remains constant, even if the weight of an object (i.e., the force N) is changed. The force of friction (F) required for moving the block changes proportionally. Note that the coefficient of friction is independent of the area of contact, which means that pushing a brick across the floor requires the same amount of force, whether it is on end, on edge, or flat. The coefficient of friction is useful in determining the force necessary to do a certain amount of work. Temperature changes only slightly affect friction. Friction causes wear. To overcome this wear problem (to reduce friction), lubricants are used.

Example 11.7
Problem:
How much force is required to move a 300 lb box if the static coefficient of friction between it and the horizontal surface upon which it is resting is 0.66?

Solution:
$$F = (0.66) (300 \text{ lbs}) = 396 \text{ lbs}$$

Example 11.8
Problem:
A 200 lb box is placed on a plane inclined at a 30° angle from the horizontal. The plane has a static coefficient of friction of 0.66. What is the minimum push or pull force required to move the box down the plane?

Solution:
$$\begin{aligned} \text{Force} &= (0.66) (200 \text{ lbs} \times \cos 30) \\ &= (0.66) (200 \text{ lbs} \times 0.866) \\ &= (0.66) (173.2 \text{ lbs}) = 114 \text{ lbs} \end{aligned}$$

Note, however, that a component of the weight of the box is causing a force (due to gravity) to already be acting in the downward direction of incline. This force is equal to Wsinθ or 200 x 0.50 = 50 lbs. Therefore, the only additional push/push force needed to get the box moving is

$$114 \text{ lbs} - 50 \text{ lbs} = 64$$

Specific Gravity

Specific gravity is the ratio of the weight of a liquid or solid substance to the weight of an equal volume of water, a number that can be determined by dividing the weight of a body by the weight of an equal volume of water. Since the weight of any body per unit of volume is called density, then:

$$\text{Specific gravity} = \frac{\text{density of body}}{\text{density of water}} \qquad\qquad (11.6)$$

Example 11.9
Problem:
The density of a particular material is 0.24 pounds per cubic inch, and the density of water per cubic inch is 0.0361 pounds per cubic inch. What is the specific gravity of the material?

Solution:

$$\text{Specific gravity of the material} = \frac{0.24}{0.0361} = 6.6$$

The material is 6.6 times as heavy as water. This ratio does not change, regardless of the units that may be used, which is an advantage for two reasons: 1. it will always be the same for the same material, and 2. it is less confusing than the term *density*, which changes as the units change.

Force, Mass, and Acceleration

Let's review what we learned earlier about force, mass, and acceleration. According to Newton's Second Law of Motion:

> The acceleration produced by an unbalanced force acting on a mass is directly proportional to the unbalanced force, in the direction of the unbalanced force, and inversely proportional to the total mass being accelerated by the unbalanced force.

If we express Newton's Second Law mathematically, it is greatly simplified and becomes

$$F = ma \qquad\qquad (11.7)$$

This equation is extremely important in physics and engineering. It simply relates acceleration to force and mass. Acceleration is defined as the change in velocity divided by the time taken. This definition tells us how to measure acceleration. $F = ma$ tells us what causes the acceleration—an unbalanced force. Mass may be defined as the quotient obtained by dividing the weight of a body by the acceleration caused by gravity. Since gravity is always present, we can, for practical purposes, think of mass in terms of weight, making the necessary allowance for gravitational acceleration.

Centrifugal and Centripetal Forces

Two terms the safety engineer should be familiar with are centrifugal and centripetal forces. Centrifugal force is a concept based on an apparent (but not real) force. It may be regarded as a force that acts radially outward from a spinning or orbiting object (a ball tied to a string whirling about), thus balancing a real force, the centripetal force (the force that acts radially inward).

This concept is important in safety engineering because many of the machines encountered on the job may involve rapidly revolving wheels or flywheels. If the wheel is revolving fast enough, and if the molecular structure of the wheel is not strong enough to overcome the centrifugal force, it may fracture. Pieces (shrapnel) of the wheel would fly off tangent to the arc described by the wheel. The safety implications are obvious. Any worker using such a device or near it may be severely injured when the rotating member ruptures. This is what happens when a grinding wheel on a pedestal grinder "bursts." Rim speed determines the centrifugal force, and rim speed involves both the speed (rpm) of the wheel and the diameter of the wheel.

Stress and Strain

In materials, stress is a measure of the deforming force applied to a body. Strain (which is often erroneously used as a synonym for stress) is really the resulting change in its shape (deformation). For perfectly elastic material, stress is proportional to strain. This relationship is explained by Hooke's Law, which states that the deformation of a body is proportional to the magnitude of the deforming force, provided that the body's elastic limit is not exceeded. If the elastic limit is not reached, the body will return to its original size once the force is removed. For example, if a spring is stretched by 2 cm by a weight of 1 N, it will be stretched by 4 cm by a weight of 2 N, and so on; however, once the load exceeds the elastic limit for the spring, Hooke's Law will no longer be obeyed, and each successive increase in weight will result in a greater extension until the spring finally breaks.

Stress forces are categorized in three ways:

1. Tension (or tensile stress), in which equal and opposite forces that act away from each other are applied to a body, tends to elongate a body.
2. Compression stress, in which equal and opposite forces that act toward each other are applied to a body, tends to shorten it.
3. Shear stress, in which equal and opposite forces that do not act along the same line of action or plane are applied to a body, tends to change its shape without changing its volume.

PRINCIPLES OF MECHANICS

In this section, we discuss mechanical principles: statics, dynamics, fluid mechanics, welds, moments, beams, columns, bending moments, floors, industrial noise, and radiation. The field safety engineer should have at least some familiarity with all of these. Note: The safety engineer whose function is to verify design specifications (with safety in mind) should have more than just a familiarity with these topics.

Statics

Statics is the branch of mechanics concerned with the behavior of bodies at rest and forces in equilibrium and is distinguished from dynamics (concerned with the behavior of bodies in motion). Forces acting on statics do not create motion. Static applications are bolts, welds, rivets, load-carrying components (ropes and chains), and other structural elements. A common example of a static situation is shown in the bolt and plate assembly. The bolt is loaded in tension and holds two elements together. One force acting on it is the load on the lower element (160 lb load plus 15 lb suspending elements). Another force is that caused by the tightened nut (25 lb). The total effective load on the bolt is 200 lb (160 + 15 + 25). The plate will fail in shear if the head of the bolt pulls through the plate.

Dynamics

Dynamics (kinetics in mechanics) is the mathematical and physical study of the behavior of bodies under the action of forces that produce changes of motion in them. In dynamics, certain properties are important: displacement, velocity, acceleration, momentum, kinetic energy, potential energy, work, and power. Safety engineers work with these properties to determine, for example, if rotating equipment will fly apart and cause injury to workers, or to determine the distance needed to stop a vehicle in motion.

Hydraulics and Pneumatics—Fluid Mechanics

Hydraulics (liquids only) and pneumatics (gases only) make up the study of fluid mechanics, which in turn is the study of forces acting on fluids (liquids and gases are considered fluids). Safety engineers encounter many fluid mechanics problems and applications of fluid mechanics. In particular, safety engineers working in chemical industries or in or around processes using or producing chemicals need an understanding of flowing liquids or gases to be able to predict and control their behavior.

Welds

Welding is a method of joining metals to achieve a more efficient use of the materials and faster fabrication and erection. Welding also permits the designer to develop and use new and aesthetically appealing designs and saves weight because connecting plates are not needed and allowances need not be made for reduced load-carrying ability due to holes for rivets, bolts, and so on (Heisler 1984).

Simply put, the welding process joins two pieces of metal together by establishing a metallurgical bond between them. Most processes use a fusion technique; the two most widely used are arc welding and gas welding.

In the welding process, where two pieces of metal are joined together, the mechanical properties of metals are important, of course. The mechanical properties of metals primarily measure how materials behave under applied loads—in other words, how strong a metal is when it comes in contact with one or more forces. The important point is that if you know and use the strength properties of a metal, you can build a structure that is both safe and sound.

In welding, the welder must know the strength of his weld as compared with the base metal to produce a weldment that is strong enough to do the job. Thus, the welder is just as concerned with the mechanical properties of metals as is the engineer.

Moment

Moment is a synonym for *torque*. However, in many engineering applications, the two terms are not interchangeable. Torque is usually used to describe a rotational force down a shaft, for example, a turning of a pump shaft, whereas *moment* is more often used to describe a bending force on a beam. Moment is the product of the force magnitude (F) and the distance from the point to its action line. The perpendicular distance (d) is called the arm of the force, and the point is the origin or center of the moment. The product is the measure of the tendency of the force to cause rotation (e.g., bending, twisting). The name of the unit of moment is a combination of the names of force and distance units, as pound-foot (lb-ft), to distinguish it from ft-lb, which is the unit for work or energy.

$$F_1 d_1 = F_2 d_2 \tag{11.8}$$

$$\sum M_o = 0$$

Beams and Columns

A *beam* is a structural member whose length is large compared to its transverse dimensions and is subjected to forces acting transverse to its longitudinal axis (Tapley 1990).

Safety engineers are not only concerned with structural members (beams and flooring support members, for example) but also with columns. *Columns* are structural members with an unsupported length ten times greater than the smallest lateral dimension and are loaded in compression. When a column is subjected to small compressive loads, the column axially shortens. If continually larger loads are applied, a load is reached at which the column suddenly bows out sideways. This load is referred to as the column's critical or buckling load. These sideways deformations are normally too large to be acceptable; consequently, the column is considered to have failed. For slender columns, the axial stress corresponding to the critical load is generally below the yield strength of the material. Since the stresses in the column just prior to buckling are within the elastic range, the failure is referred to as *elastic buckling*. The term *elastic stability* is commonly used to designate the study of elastic buckling problems. For short columns, yielding or rupture of the column may govern failure while it is still axially straight. Failure of short columns may also be caused by inelastic buckling; that is, a large sideways deformation that occurs when the nominal axial stress is greater than the yield strength.

Beams, floors, and columns are all critical elements for safe loading. As an example, in late October 2003, the top five floors of a parking garage under construction in Atlantic City, New Jersey, collapsed while workers were pouring concrete on the structure's top floor. The incident killed four workers and critically injured six others. OSHA's investigation of the collapse involved close examination of the blueprints for the garage and evaluation of the cure rate for the concrete. No matter who or what element of construction, design, or engineering was ultimately to blame in this collapse, because obviously the load limits were exceeded for that moment. Perhaps a longer concrete curing interval would have prevented the collapse. Regardless of the cause, the cost in lives and dollars both is too high.

Safety engineers are primarily interested in beams because the load on a beam induces stresses in the material—stresses that could be dangerous. The structural aspect of beams is most important to the safety engineer because the strength of the beam material and the kind of loading determine the size of load that it can safely carry. For example, in the construction of the storage mezzanine discussed earlier, and in the construction of other load-bearing structures, the beams used to support the load are an important (critical) consideration.

Refer to figure 11.15. The neutral axis is the plane that undergoes no change in length from bending and along which the direct stress is zero. The fibers on one side of the neutral axis are stressed in tension, and on the other side in compression, and the intensities of these stresses in homogeneous beams are directly proportional to the distances of the fibers from the neutral axis (Heisler 1984).

In determining the load that can be carried, two properties of the beam are important: the moment of inertia (I) and the section modulus (Z). The moment of inertia (I) is the sum of differential areas multiplied by the square of the distance from a reference plane (usually the neutral axis) to each differential area. Note that the strength of a beam increases rapidly as its cross section is moved farther from the neutral axis because the distance is squared. This is why a rectangular beam is much stronger when it is loaded along its thin dimension than along its flat dimension. The section modulus (Z) is the moment of inertia divided by the distance from the neutral axis to the outside of the beam cross-section.

Of special interest to safety engineers is the type of beams and allowable loads on beams. Allowable loads differ from maximum loads that produce failure by some appropriate factor of safety (Brauer 1994). The following types of beams are approximations of actual beams used in practice (Tapley 1990).

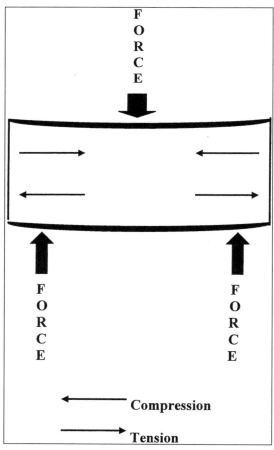

Figure 11.15 Distribution of stress in a beam cross-section during bending.

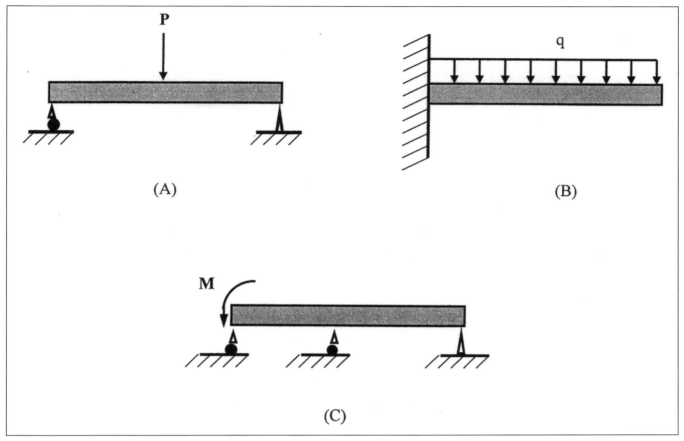

Figure 11.16 Classification of beams and bending: (A) simple beam, concentrated load; (B) cantilever beam, distributed load; (C) continuous beam, concentrated moment.

- **simple beams:** [see figure 11.16 (A); shows concentrated load] Supported beams that have a roller support at one end and pin support at the other. The ends of a simple beam cannot support a bending moment but can support upward and downward vertical loads. Stated differently, the ends are free to rotate but cannot translate in the vertical direction. The end with the roller support is free to translate in the axial direction.
- **cantilever beams:** [see figure 11.16 (B); shows distributed load] Beams that are rigidly supported on only one end. The beam carries the load to the support where it is resisted by moment and shear stress.
- **continuously supported beam:** [see figure 11.16 (C); shows concentrated load] A beam resting on more than two supports.
- **fixed beam:** Rigidly fixed at both ends.
- **restrained beam:** Rigidly fixed at one end and simply supported at the other.
- **overhanging beam:** Projects beyond one or both ends of its supports.

Bending Moment

A *bending moment* (or internal torque) exists in a structural element when a moment is applied to the element so that the element bends. In engineering design (and safety engineering), it is important to determine the point(s) along a beam where the bending moment (and shear force) is maximum since it is at these points that the bending stresses reach their maximum values. When a beam is in equilibrium, the sum of all moments about a particular point is zero.

In determining the maximum bending moment (uniform loading), we use the following equation:

$$M = wl^2/8$$

(11.9)

Where

M = the maximum bending moment (ft-lb)
w = the uniform weight loading per foot of the beam (lb/ft)
l = the length of the beam (ft)

Example 11.10
Problem:
An 11 ft long beam supported at two points, one at each end, is uniformly loaded at a rate of 250 lb/ft. What is the maximum bending moment in this beam?

Solution:

$M = wl^2/8$
$M = ?$
$w = 250$ lb/ft
$l = 11$ ft
$M = (250$ lb/ft$)(11$ ft$)^2/8$
    $= 3781$ ft-lb

To determine the bending moment (concentrated load at center) the following equation can be used.

$$M = Pl/4 \tag{11.10}$$

Where

M = maximum bending moment (ft-lbs)
P = the concentrated load applied at the center of the beam (lb)
l = the length of the beam (ft)

Example 11.11
Problem:
A ten-ton hoist is suspended at the midpoint of a 10 ft long beam that is supported at each end. What is the maximum bending moment in this beam? (Neglect the weight of the beam.)

Solution:

$M = ?$
$P = 10$ tons $= 20,000$ lbs
$l = 10$ ft
$M = Pl/4$
$M = (20,000$ lbs$) (10$ ft$) / 4$
    $= 50,000$ ft-lb

To determine the bending moment (concentrated load off-center) the following equation can be used.

$$M = Pab/l \tag{11.11}$$

Where

M = the maximum bending moment (ft-lb)
P = a concentrated load on the beam (lb)
$a + b$ = the respective distances from the left and right supports of the beam (ft)
l = the length of the beam (ft)

Example 11.12
Problem:
A 10 ft long beam supported at each end is loaded with a weight of 1400 lb at a point 2 ft left of its center. What is the maximum bending moment in this beam? (Note: Neglect weight of the beam.)

Solution:

M = ?
P = 1400 lb
a = 2 ft
b = 8 ft
l = 10 ft
M = Pab/l
M = (1400 lb) (2 ft) (8 ft) / 10 ft
  = (1400 lb)(2 ft)(8 ft) / 10 ft
  = 2240 ft-lb

Radius of Curvature (Beam)

The *radius of curvature* provides one of several measures of the deformation of a beam. The radius of curvature of a beam experiencing a bending moment (in., feet, meters) can be determined by using the following equation:

$$p = EI/M \qquad (11.12)$$

Where

p = the radius of curvature (in., feet, meters)
E = the modulus of elasticity of the beam material (psi, KPa)
I = the cross-sectional area moment of inertia about its centroid (m^4, in^4, ft^4)
M = the moment or torque applied at the cross-section (ft-lb)

Column Stress

Stress is defined as force per unit area and is expressed in pound per square inch (psi). The amount of stress is a real indicator of how severely the member is loaded. *Tensile stress* occurs when a member is in tension (force acts to stretch it); *compressive stress* is when a member is in compression (force acts to shorten or flatten it); and *shear stress* is when a member is in shear (force acts to cause one part of the material to slide over another part).

In determining the stress (loading) of a column, we use the following equation:

$$P/A = \sigma \qquad (11.13)$$

Where

P = the loading on the column (lb)
A = the cross-sectional area of the member (in^2)
σ = the stress (lb/in^2 or psi)

Example 11.13
Problem:
A column 4 inches in diameter supports a load of 6,000 lbs. What is the stress on it?

Solution:

8 = ?
P = 6,000 lbs.
A = πr^2 = (3.14)(2 in)^2 = 12.56 in^2
σ = P/A
σ = 6,000 lb / 12.56 in
  = 478 psi

Beam Flexure

In mechanics, *flexure* (also known as bending) characterizes the behavior of a structural element (beam) subjected to an external load applied perpendicular to the axis of the element.

A closet rod sagging under the weight of clothes on clothes hangers is an example of a beam experiencing flexure (bending). The equation used to calculate flexure is known as the *flexure formula* (see below), which shows the relationship between the maximum bending stress, σ, and the maximum bending moment, M.

$$\sigma = Mc/l \qquad\qquad (11.14)$$

Where

σ = maximum stress (lbs/in^2 or psi)
M = the moments in in-lbs
c/l = the inverse of the section modulus (in^3) (taken from an applicable table)

Example 11.14
Problem:
A rectangular beam that is 12 ft long by 6 inches high by 3 inches wide has a two-ton concentrated load applied at its center. What is the maximum bending stress on this beam?

Solution:

σ = Mc/l
σ = ?
M = Pl/4
    = (4,000 lb)(12 ft x 12 in/ft) / 4
    = 144,000 in-lb
c/l = 6/bd^2 (from applicable table)
    = 6 /(3 in)(6 in)^2
    = .05555 in^{-3}
σ = (144,000 lb-in)(.05555in^{-3})
    = 7,999 psi

Reaction Force (Beams)

Uniform Loading—Unknown forces exerted upward by the supports at each end of a beam to hold it up are called *reactions*. Known forces that act on beams are called *loads*. Reactions balance the load to keep the beam in a state of equilibrium. To determine the reactions on a beam's supports, we use the following equation.

$$R_L = R_R = wl/2 \qquad\qquad (11.15)$$

Where

$R_L = R_R$ = the left and right reactions (lb)
w = the uniform weight loading per foot of the bean (lb/ft)
l = the length of the beam (ft)

Example 11.15
Problem:
A 10 ft long beam supported at two points, one at each end, is uniformly loaded at a rate of 200 lb/ft. What are the reactions on the supports?

Solution:

$R_L = R_R = wl/2$
$R_L = R_R = ?$
w = 200 lb/ft

l = 10 ft
$R_L = R_R = (200 \text{ lb/ft})(10 \text{ ft})/2$
= 1000 lb

Concentrated Load at Center—To calculate the reactions, $R_{L\&R}$, in simple beams (those beams supported by two points, one at each end) of any length that are loaded at their center with a concentrated load, we use the following equation.

$$R_L = R_R = P/2 \qquad\qquad (11.16)$$

Where

$R_L = R_R$ = are the left and right reactions in pounds (lb)
P = the concentrated load at the center of the beam (lb/ft)

Example 11.16
Problem:
A ten-ton hoist is suspended at the midpoint of a beam that is supported at each end. What would the design basis need to be for these supports if we specified a factor of safety of four?

Solution:

$R_L = R_R = P/2$
$R_L = R_R = ?$
P = 10 tons = 20,000 lb
$R_L = R_R$ = 20,000 lb/2 = 10,000 lb

Applying a safety factor of four requires that each support be designed to handle at least 32,000 lbs, or 16 tons of force.

Load at any Point—The equation below is used to calculate the reactions, $R_{L\&R}$, in simple beams of length *l* that are loaded off-center with a weight of *P*.

$$R_R = Pa/l \quad\quad R_L = Pb/l \qquad\qquad (11.17)$$

Where

R_R & R_L = the right and left reactions (lb)
P = a concentrated load on the beam (lb)
a & b = the respective distances from the left and right supports of the beam (ft)
l = the length of the beam (ft)

Example 11.17
A 10 ft long beam supported at each end is loaded with a weight of 1,200 lb at a point 2 ft left of its center. What are the reactions on the supports? (Note: Neglect weight of the beam.)

Solution:

P = 1200 lb
a = 2 ft
b = 8 ft
l = 10 ft
$R_R = Pa/l$
R_R = (1200 lb)(2 ft) / 10 ft
= 240 lb
$R_L = Pb/l$
R_L = (1200 lb)(8 ft) / 10 ft
= 960 lb

Buckling Stress (Wood Columns)

Columns, which support a load, have a tendency to bend. If the bending becomes too great the column will become unstable and fail by buckling under the pressure from the load.

The equation below is for columns made of wood, with intermediate length.

$$P/A = \sigma[1 - 1/3\ (l/Kd)^4 \qquad\qquad (11.18)$$

Where

P = maximum acceptable applied load to the column (lb, N)
A = cross-sectional area of the column (in^2, m^2)
σ = allowable unit compressive stress parallel to the grain (psi, kPa)
l = the free unsupported length of the column (in, m)
K = is the minimum value of l/d for which Euler column mechanics many be used.
d = the dimension (width) of the column in the expected direction of buckling (in, m).

Floors

Safety engineers commonly spend a considerable amount of time and attention on ensuring the proper maintenance of floors and flooring in general, in ways that range from the mundane (housekeeping) to the structurally essential (calculating floor load). Housekeeping is always a focal point; passageways, storerooms, and service rooms need to be scrutinized daily to ensure they are kept clean and orderly, in a sanitary condition, and free from fire hazards. Typically the responsibility for housekeeping (in most places of employment) falls to the supervisor and to the employees themselves; however, safety engineers cannot avoid the responsibility of ensuring that workplace housekeeping is kept to the highest standards. Housekeeping also includes maintaining the floors of every workroom in a clean and dry condition. Where wet processes are used, drainage must be maintained, and platforms, mats, or other dry, standing places must be provided where practicable. All floors must be kept free from protruding nails, splinters, holes, or loose boards.

Along with housekeeping, safety engineers are concerned with floor load protection. In fact, one of the safety engineer's commonly requested services is to determine the safe loads on a floor. To determine the safe floor load for any floor, safety engineers have to take into consideration two load components: 1. *dead loads* (the weight of the building and its components) and 2. *live loads* (loads placed on the floor). The safety engineer's main concern is to ensure that no load exceeds that for which a floor (or roof) is approved by the design engineer and/or approving official. Safety engineers should ensure that the approved floor rating is properly posted in a conspicuous place in each space to which it relates.

NOISE

High noise levels in the workplace are a hazard to employees. High noise levels are a physical stress that may produce psychological effects by annoying, startling, or disrupting the worker's concentration, which can lead to accidents. High levels can also result in damage to worker's hearing, resulting in hearing loss. In this section, we discuss the basics of noise, including those elements safety engineers need to know to ensure that their organization's hearing conservation program is in compliance with OSHA.

Key Terms

There are many specialized terms used to express concepts in noise, noise control, and hearing-loss prevention. Safety engineering practitioners responsible for ensuring proper engineering design to reduce workplace noise levels and the in-house safety engineers responsible for compliance with OSHA's Hearing Conservation Program requirements must be familiar with these terms. The NIOSH (2005) definitions below were written in as nontechnical a fashion as possible.

acoustic trauma: Single incident which produces an abrupt hearing loss. Welding sparks (to the eardrum), blows to the head, and blast noise are examples of events capable of providing acoustic trauma.

action level: The sound level which when reached or exceeded necessitates implementation of activities to reduce the risk of noise-induced hearing loss. OSHA currently uses an eight-hour time-weighted average of 85 dBA as the criterion for implementing an effective hearing conservation program.

attenuate/attenuation: To reduce the amplitude of sound pressure (noise).

real ear attenuation at threshold (REAT): A standardized procedure for conducting psychoacoustic tests on human subjects designed to measure sound protection features of hearing protective devices. Typically, these measures are obtained in a calibrated sound field and represent the difference between subjects' hearing thresholds when wearing a hearing protector vs. not wearing the protector.

real-world: Estimated sound protection provided by hearing protective devices as worn in "real-world" environments.

audible range: The frequency range over which normal ears hear: approximately 20 Hz through 20,000 Hz.

audiogram: A chart, graph, or table resulting from an audiometric test showing an individual's hearing threshold levels as a function of frequency.

audiologist: A professional, specializing in the study and rehabilitation of hearing, who is certified by the American Speech-Language-Hearing Association or licensed by a state board of examiners.

background noise: Noise coming from sources other than the particular noise sources being monitored.

baseline audiogram: A valid audiogram against which subsequent audiograms are compared to determine if hearing thresholds have changed. The baseline audiogram is preceded by a quiet period so as to obtain the best estimate of the person's hearing at that time.

continuous noise: Noise of a constant level as measured over at least one second using the "slow" setting on a sound level meter. Note that a noise which is intermittent, e.g., on for over a second and then off for a period, would be both variable and continuous.

controls:

administrative: Efforts, usually by management, to limit workers' noise exposure by modifying workers' schedules or locations or by modifying the operating schedule of noisy machinery.

engineering: Any use of engineering methods to reduce or control the sound level of a noise source by modifying or replacing equipment or by making any physical changes at the noise source or along the transmission path (with the exception of hearing protectors).

criterion sound level: A sound level of 90 decibels.

dB (decibel): The unit used to express the intensity of sound. The decibel was named after Alexander Graham Bell. The decibel scale is a logarithmic scale in which 0 dB approximates the threshold of hearing in the mid-frequencies for young adults and in which the threshold of discomfort is between 85 and 95 dB SPL and the threshold for pain is between 120 and 140 dB SPL (sound pressure level).

dosimeter: When applied to noise, refers to an instrument that measures sound levels over a specified interval, stores the measures, and calculates the sound as a function of sound level and sound duration and describes the results in terms of dose, time-weighted average, and (perhaps) other parameters such as peak level, equivalent sound level, sound exposure level, etc.

double hearing protection: A combination of both ear plug and ear muff type hearing protection devices is required for employees who have demonstrated temporary threshold shift during audiometric examination and for those who have been advised to wear double protection by a medical doctor in work areas that exceed 104 dBA.

equal-energy rule: The relationship between sound level and sound duration based upon a 3 dB exchange rate, i.e., the sound energy resulting from doubling or halving a noise exposure's duration is equivalent to increasing or decreasing the sound level by 3 dB, respectively.

exchange rate: The relationship between intensity and dose. OSHA uses a 5 dB exchange rate. Thus, if the intensity of an exposure increases by 5 dB, the dose doubles. Sometimes, this isalso referred to as the doubling rate. The U.S. Navy uses a 4 dB exchange rate; the U.S. Army and Air Force use a 3 dB exchange rate. NIOSH recommends a 3 dB exchange rate. Note that the equal-energy rule is based on a 3 dB exchange rate.

frequency: Rate in which pressure oscillations are produced. Measured in hertz (Hz).

hazardous noise: Any sound for which any combination of frequency, intensity, or duration is capable of causing permanent hearing loss in a specified population.

hazardous task inventory: A concept based on using work tasks as the central organizing principle for collecting descriptive information on a given work hazard. It consists of a list(s) of specific tasks linked to a database containing the prominent characteristics relevant to the hazard(s) of interest that are associated with each task.

hearing conservation record: Employee's audiometric record. Includes name, age, job classification, time-weighted average (TWA) exposure, date of audiogram, and name of audiometric technician. The records are to be retained for the duration of employment for OSHA and kept indefinitely for workers' compensation.

hearing damage risk criteria: A standard that defines the percentage of a given population expected to incur a specified hearing loss as a function of exposure to a given noise exposure.

hearing handicap: A specified amount of permanent hearing loss usually averaged across several frequencies that negatively impacts employment and/or social activities. Handicap is often related to an impaired ability to communicate. The degree of handicap will also be related to whether the hearing loss is in one or both ears and whether the better ear has normal or partial hearing.

hearing loss: Hearing loss is often characterized by the area of the auditory system responsible for the loss. For example, when injury or a medical condition affects the outer ear or middle ear (i.e., from the pinna, ear canal, and eardrum to the cavity behind the eardrum—which includes the ossicles) the resulting hearing loss is referred to as a *conductive* loss. When an injury or medical condition affects the inner ear or the auditory nerve that connects the inner ear to the brain (i.e., the cochlea and the VIIIth cranial nerve) the resulting hearing loss is referred to as a *sensorineural* loss. Thus, a welder's spark that damaged the eardrum would cause a conductive hearing loss. Because noise can damage the tiny hair cells located in the cochlea, it causes a sensorineural hearing loss.

hearing loss prevention program audit: An assessment performed prior to putting a hearing-loss prevention program into place or before changing an existing program. The audit should be a top-down analysis of the strengths and weaknesses of each aspect of the program.

HTL (hearing threshold level): The hearing level, above a reference value, at which a specified sound or tone is heard by an ear in a specified fraction of the trials. Hearing threshold levels have been established so that 0 dB HTL reflects the best hearing of a group of persons.

Hz (Hertz): The unit of measurement for audio frequencies. The frequency range for human hearing lies between 20 Hz and approximately 20,000 Hz. The sensitivity of the human ear drops off sharply below about 500 Hz and above 4,000 Hz.

impulsive noise: Used to generally characterize impact or impulse noise which is typified by a sound that rapidly rises to a sharp peak and then quickly fades. The sound may or may not have a "ringing" quality (such as striking a hammer on a metal plate or a gunshot in a reverberant room). Impulsive noise may be repetitive or may be a single event (as with a sonic boom). Note: If impulses occur in very rapid succession (such as with some jackhammers), the noise would not be described as impulsive.

loudness: The subjective attribute of a sound by which it would be characterized along a continuum from "soft" to "loud." Although this is a subjective attribute, it depends primarily upon sound pressure level, and to a lesser extent, the frequency characteristics and duration of the sound.

material hearing impairment: As defined by OSHA, a material hearing impairment is an average hearing threshold level of 25 dB HTL at the frequencies of 1000, 2000, and 3000 Hz.

medical pathology: A disorder or disease. For purposes of this program, a condition or disease affecting the ear, which a physician specialist should treat.

noise: Any unwanted sound.

noise dose: The noise exposure expressed as a percentage of the allowable daily exposure. For OSHA, a 100 percent dose would equal an eight-hour exposure to a continuous 90 dBA noise; a 50 percent dose would equal an eight-hour exposure to an 85 dBA noise; or a four-hour exposure to a 90 dBA noise. If 85 dBA is the maximum permissible level, then an eight-hour exposure to a continuous 85 dBA noise would equal a 100 percent dose. If a 3 dB exchange rate is used in conjunction with an 85 dBA maximum permissible level, a 50 percent dose would equal a two-hour exposure to 88 dBA, or an eight-hour exposure to 82 dBA.

noise dosimeter: An instrument that integrates a function of sound pressure over a period of time to directly indicate a noise dose.

noise hazard area: Any area where noise levels are equal to or exceed 85 dBA. OSHA requires employers to designate work areas, post warning signs, and warn employees when work practices exceed 90 dBA as a "noise hazard area." Hearing protection must be worn whenever 90 dBA is reached or exceeded.

noise hazard work practice: Performing or observing work where 90 dBA is equaled or exceeded. Some work practices will be specified, however, as a "rule of thumb." Whenever attempting to hold normal conversation with someone who is one foot away and one must shout to be heard, one can assume that a

90 dBA noise level or greater exists and hearing protection is required. Typical examples of work practices where hearing protection is required are jackhammering, heavy grinding, heavy equipment operations, and similar activities.

noise-induced hearing loss: A sensorineural hearing loss that is attributed to noise and for which no other etiology can be determined.

noise level measurement: Total sound level within an area. Includes workplace measurements indicating the combined sound levels of tool noise (from ventilation systems, cooling compressors, circulation pumps, etc.).

NRR (noise reduction rating): The NRR is a single-number rating method that attempts to describe a hearing protector based on how much the overall noise level is reduced by the hearing protector. When estimating A-weighted noise exposures, it is important to remember to first subtract 7 dB from the NRR and then subtract the remainder from the A-weighted noise level. The NRR theoretically provides an estimate of the protection that should be met or exceeded by 98 percent of the wearers of a given device. In practice, this does not prove to be the case, so a variety of methods for "de-rating" the NRR have been discussed.

ototoxic: A term typically associated with the sensorineural hearing loss resulting from therapeutic administration of certain prescription drugs.

ototraumatic: A broader term than ototoxic. As used in hearing-loss prevention, refers to any agent (e.g., noise, drugs, or industrial chemicals) that has the potential to cause permanent hearing loss subsequent to acute or prolong exposure.

presbycusis: The gradual increase in hearing loss that is attributable to the effects of aging and is not related to medical causes or noise exposure.

sensorineural hearing loss: A hearing loss resulting from damage to the inner ear (from any source).

sociacusis: A hearing loss related to nonoccupational noise exposure.

sound intensity (I): Sound intensity at a specific location is the average rate at which sound energy is transmitted through a unit area normal to the direction of sound propagation.

sound level meter (SLM): A device that measures sound and provides a readout of the resulting measurement. Some provide only A-weighted measurements, others provide A- and C-weighted measurements, and some can provide weighted, linear, and octave (or narrower) ban measurements. Some SLMs are also capable of providing time-integrated measurements.

sound power: The total sound energy radiated by a source per unit time. Sound power cannot be measured directly.

SPL (sound pressure level): A measure of the ratio of the pressure of a sound wave relative to a reference sound pressure. Sound pressure level in decibels is typically referenced to 20 mPa. When used alone (e.g., 90dBA), a given decibel level implies an unweighted sound pressure level.

significant threshold shift: NIOSH uses this term to describe a change of 15 dB or more at any frequency, 5000 through 6000 Hz, from baseline levels that is present on an immediate retest in the same ear and at the same frequency. NIOSH recommends a confirmation audiogram within thirty days with the confirmation audiogram preceded by a quiet period of at least fourteen hours.

standard threshold shift: OSHA uses the term to describe a change in hearing threshold relative to the baseline audiogram of an average of 10 dB or more at 2000, 3000, and 4000 Hz in either ear. Used by OSHA to trigger additional audiometric testing and related followup.

threshold shift: Audiometric monitoring programs will encounter two types of changes in hearing sensitivity, i.e., threshold shifts: permanent threshold shift (PTS) and temporary threshold shift (TTS). As the names imply, any change in hearing sensitivity that is persistent is considered a PTS. Persistence may be assumed if the change is observed on a thirty-day follow-up exam. Exposure to loud noise may cause a temporary worsening in hearing sensitivity (i.e., a TTS) that may persist for fourteen hours (or even longer in cases where the exposure duration exceeded twelve to sixteen hours). Hearing health professionals need to recognize that not all threshold shifts represent decreased sensitivity and not all temporary or permanent threshold shifts are due to noise exposure. When a permanent threshold shift can be attributable to noise exposure, it may be referred to as a noise-induced permanent threshold shift (NIPTS).

velocity (c): The speed at which the regions of sound-producing pressure changes move away from the sound source.

wavelength (λ): This term refers to the distance required for one complete pressure cycle to be completed (1 wavelength) and is measured in feet or meters.

weighted measurements: Two weighting curves are commonly applied to measures of sound levels to account for the way the ear perceives the "loudness" of sounds.

A-weighting: A measurement scale that approximates the "loudness" of tones relative to a 40 db SPL 1000 Hz reference tone. A-weighting has the added advantage of being correlated with annoyance measures and is most responsive to the midfrequencies, 500 to 4000 Hz.

C-weighting: A measurement scale that approximates the "loudness" of tones relative to a 90 dB SPL 1000 Hz reference tone. C-weighting has the added advantage of providing a relatively "flat" measurement scale that includes very low frequencies.

Occupational Noise Exposure

As mentioned above, *noise* is commonly defined as any unwanted sound. Noise literally surrounds us every day and is with us just about everywhere we go. However, the noise we are concerned with here is that produced by industrial processes. Excessive amounts of noise in the work environment (and outside it) cause many problems for workers, including increased stress levels, interference with communication, disrupted concentration, and most importantly, varying degrees of hearing loss. Exposure to high noise levels also adversely affects job performance and increases accident rates.

One of the major problems with attempting to protect workers' hearing acuity is the tendency of many workers to ignore the dangers of noise. Because hearing loss, like cancer, is insidious, it's easy to ignore. It sneaks up slowly and is not apparent (in many cases) until after the damage is done. Alarmingly, hearing loss from occupational noise exposure has been well documented since the eighteenth century, yet since the advent of the Industrial Revolution, the number of exposed workers has greatly increased (Mansdorf 1993). However, today the picture of hearing loss is not as bleak as it has been in the past, as a direct result of OSHA's requirements. Now that noise exposure must be controlled in all industrial environments, that well-written and well-managed hearing conservation programs must be put in place, and that employee awareness must be raised to the dangers of exposure to excessive levels of noise, job-related hearing loss is coming under control.

Determining Workplace Noise Levels

The unit of measurement for sound is the decibel. *Decibels* are the preferred unit for measuring sound, derived from the *bel*, a unit of measure in electrical communications engineering. The decibel is a dimensionless unit used to express the logarithm of the ratio of a measured quantity to a reference quantity.

In regard to noise control in the workplace, the safety engineer's primary concern is first to determine if any "noise-makers" in the facility exceed the OSHA limits for worker exposure—exactly which machines or processes produce noise at unacceptable levels. Making this determination is accomplished by conducting a noise level survey of the plant or facility. Sound measuring instruments are used to make this determination. These include noise dosimeters, sound level meters, and octave-band analyzers. The uses and limitations of each kind of instrument are discussed below.

1. **Noise Dosimeter**. The noise dosimeters used by OSHA meet the American National Standards Institute (ANSI) Standard S1.25-1978, "Specifications for Personal Noise Dosimeter," which set performance and accuracy tolerances. For OSHA use, the dosimeter must have a 5 dB exchange rate, use a 90 dBA criterion level, be set at slow response, and use either an 80 dBA or 90 dBA threshold gate or a dosimeter that has both capabilities, whichever is appropriate for evaluation.

2. **Sound Level Meter (SLM)**. When conducting the noise level survey, the industrial hygienist or survey technician should use an ANSI-approved sound level meter (SLM)—a device used most commonly to measure sound pressure. The SLM measures in decibels. One decibel is one-tenth of a bel and is the minimum difference in loudness that is usually perceptible. The sound level meter (SLM) consists of a microphone, an amplifier, and an indicating meter that responds to noise in the audible frequency range of about 20 to 20,000 Hz. Sound level meters usually contain "weighting" networks designated "A," "B," or "C." Some meters have only one weighting network; others are equipped with all three. The A-network approximates the equal loudness curves at low sound pressure levels, the B-network is used for medium sound pressure levels, and the C-network is used for high levels. In

conducting a routine workplace sound level survey, using the A-weighted network (referenced dBA) in the assessment of the overall noise hazard has become common practice. The A-weighted network is the preferred choice because it is thought to provide a rating of industrial noise that indicates the injurious effects such noise has on the human ear (gives a frequency response similar to that of the human ear at relatively low sound pressure levels). With an approved and freshly calibrated (always calibrate test equipment prior to use) sound level meter in hand, the industrial hygienist is ready to begin the sound level survey. In doing so, the industrial hygienist is primarily interested in answering the following questions:

1. What is the noise level in each work area?
2. What equipment or process is generating the noise?
3. Which employees are exposed to the noise?
4. How long are they exposed to the noise?

In answering these questions, industrial hygienists record their findings as they move from workstation to workstation, following a logical step-by-step procedure. The first step involves using the sound level meter set for A-scale slow response mode to measure an entire work area. When making such measurements, restrict the size of the space being measured to less than 1,000 square feet. If the maximum sound level does not exceed 80 dBA, it can be assumed that all workers in this work area are working in an environment with a satisfactory noise level. However, a note of caution is advised: The key words in the preceding statement are "maximum sound level." To assure an accurate measurement, the industrial hygienist must ensure that all "noise makers" are actually in operation when measurements are taken. Measuring an entire work area does little good when only a small percentage of the noise makers are actually in operation.

The next step depends on the readings recorded when the entire work area was measured. For example, if the measurements indicate sound levels greater than 80 dBA, then another set of measurements need to be taken at each worker's workstation. The purpose here, of course, is to determine two things: which machine or process is making noise above acceptable levels (i.e., >80 dBA), and which workers are exposed to these levels. Remember that the worker who operates the machine or process might not be the only worker exposed to the noise maker. You need to inquire about other workers who might, from time to time, spend time working in or around the machine or process. Our experience in conducting workstation measurements has shown us noise levels usually fluctuate. If this is the case, you must record the minimum and maximum noise levels. If you discover that the noise level is above 90 (dBA) (and it remains above this level), you have found a noise maker that exceeds the legal limit (90 dBA). However, if your measurements indicate that the noise level is never greater than 85 (dBA) (OSHA's action level), the noise exposure can be regarded as satisfactory.

If workstation measurements indicate readings that exceed the 85 dBA level, you must perform another step. This step involves determining the length of time of exposure for workers. The easiest, most practical way to make this determination is to have the worker wear a noise dosimeter, which records the noise energy to which the worker was exposed during the work shift.

Note: This parameter assumes that the worker has good hearing acuity with no loss. If the worker has documented hearing loss, exposure to 95 dBA or higher, without proper hearing protection, may be unacceptable under any circumstances.

3. **Octave-Band Noise Analyzers.** Several Type 1 sound level meters (such as the GenRad 1982 and 1983 and the Quest 155) used by OSHA have built-in octave band analysis capability. These devices can be used to determine the feasibility of controls for individual noise sources for abatement purposes and to evaluate hearing protectors.

Octave-band analyzers segment noise into its component parts. The octave-band filter sets provide filters with the following center frequencies: 31.5; 63; 125; 250; 500; 1,000; 2,000; 4,000; 8,000; and 16,000 Hz.

The special signature of a given noise can be obtained by taking sound level meter readings at each of these settings (assuming that the noise is fairly constant over time). The results may indicate those octave-bands that contain the majority of the total radiated sound power.

Octave-band noise analyzers can assist industrial hygienists in determining the adequacy of various types of frequency-dependent noise controls. They also can be used to select hearing protectors because they can measure the amount of attenuation offered by the protectors in the octave-bands responsible for most of the sound energy in a given situation.

Engineering Control for Industrial Noise

When safety engineers investigate the possibility of using engineering controls to control noise, the first thing they recognize is that reducing and/or eliminating all noise is virtually impossible. And this should not be the focus in the first place—eliminating or reducing the "hazard" is the goal. While the primary hazard may be the possibility of hearing loss, the distractive effect (or its interference with communication) must also be considered. The distractive effect or excessive noise can certainly be classified as hazardous whenever the distraction might affect the attention of the worker. The obvious implication of noise levels that interfere with communications is emergency response. If ambient noise is at such a high level that workers can't hear fire or other emergency alarms, this is obviously an unacceptable situation.

So what does all this mean? Safety engineers must determine the "acceptable" level of noise. Then he or she can look into applying the appropriate noise control measures. These include making alterations in engineering design (obviously this can only be accomplished in the design phase) or making modifications after installation. Unfortunately, this latter method is the one the safety engineer is usually forced to apply—and also the most difficult, depending upon circumstances.

Let's assume that the safety engineer is trying to reduce noise levels generated by an installed air compressor to a safe level. Obviously, the first place to start is at the *source*: the air compressor. Several options are available to employ at the source. First, the safety engineer would look at the possibility of modifying the air compressor to reduce its noise output. One option might be to install resilient vibration mounting devices. Another might be to change the coupling between the motor and the compressor—install an insulator-cushioning device between the couplings to dampen noise and vibration.

If the options described for use at the source of the noise are not feasible or are only partially effective, the next component the safety engineer would look at is the *path* along which the sound energy travels. Increasing the distance between the air compressor and the workers could be a possibility. (Note: Sound levels decrease with distance.) Another option might be to install acoustical treatments on ceilings, floors, and walls. The best option available (in this case) probably is to enclose the air compressor, so that the dangerous noise levels are contained within the enclosure and the sound leaving the space is attenuated to a lower, safe level. If total enclosure of the air compressor is not practicable, then erecting a barrier or baffle system between the compressor and the open work area might be an option.

The final engineering control component the safety engineer might incorporate to reduce the air compressor's noise problem is to consider the *receiver* (the worker/operator). An attempt should be made to isolate the operator by providing a noise reduction or soundproof enclosure or booth for the operator.

Noise Units, Relationships, and Equations

A number of noise units, relationships, and equations that are important to safety engineers involved with controlling noise hazards in the workplace are discussed below.

1. **Sound Power (w)** of a source is the total sound energy radiated by the source per unit time. It is expressed in terms of the sound power level (L_w) in decibels referenced to 10^{-12} watts (w_0). The relationship to decibels is shown below:

   $$L_w = 10 \log w/w_0$$

   Where:

   L_w = sound power level (decibels)
   w = sound power (watts)
   w_0 = reference power (10^{-12} watts)
   $\log$ = a logarithm to the base 10

2. Units used to describe **sound pressures** are

$$1 \ \mu bar = 1 \ dyne/cm^2 = 0.1 \ N/cm^2 = 0.1 \ Pa$$

3. **Sound pressure level** or SPL = $10 \log p^2/p_0$

Where:

SPL = sound pressure level (decibels)

p = measured root-mean-square (rms) sound pressure (N/m^2, μbars). Root-mean-square (rms) value of a changing quantity, such as sound pressure, is the square root of the mean of the squares of the instantaneous values of the quantity.

p_0 = reference rms sound pressure 20 μPa (N/m^2, μbars)

4. **Speed of sound** (c) = $c = f\lambda$
5. **Wavelength** (λ) = c/f
6. Calculation of **frequency of octave bands** can be calculated using the following formulas:

Upper frequency band: $f_2 = 2f_1$

Where:

f_2 = upper frequency band
f_1 = lower frequency band

One-half octave band: $f_2 = \sqrt{2}(f_1)$

Where:

f_2 = ½ octave band
f_1 = lower frequency band

One-third octave band: $f_2 = \sqrt[3]{2}(f_1)$

Where:

f_2 = 1/3 octave band
f_1 = lower frequency band

7. Formula for **adding noise sources**, when sound power is known:

$$L_w = 10 \log (w_1 + w_2)/(w_0 + w_0)$$

Where:

L_w = sound power in watts
w_1 = sound power of noise source 1 in watts
w_2 = sound power of noise source 2 in watts
w_0 = reference sound power (reference 10^{-12}) watts

8. Formula for **sound pressure additions**, when sound pressure is known:

$$SPL = 10 \log p^2/p_0^2$$

Where:

$$p^2/p_0^2 = 10^{SPL/10}$$

and:

SPL = sound pressure level (decibels)
p = measured root-mean-square (rms) sound pressure (N/m^2, μbars)

p_0 = reference rms sound pressure (20 µPa, N/m², µbars)

For three sources, the equation becomes:

$$SPL = 10 \log \left(10^{SPL_1/10}\right) + \left(10^{SPL_2/10}\right) + \left(10^{SPL_3/10}\right)$$

When adding any number of sources, whether the sources are identical or not, the equation becomes

$$SPL = 10 \log \left(10^{SPL_1/10} \ldots + {}^{10SPL_n/10}\right)$$

Determining the sound pressure level from multiple identical sources:

$$SPL_f = SPL_i + 10 \log n$$

Where:

SPL_f = total sound pressure level (dB)
SPL_i = individual sound pressure level (dB)
n   = number of identical sources

9. The equation for determining **noise levels in a free field** is expressed as:

$$SPL = L_w - 20 \log r - 0.5$$

Where:

SPL = sound pressure (reference 0.00002 N/m²)
L_w = sound power (reference 10^{-12} watts)
r   = distance in feet

10. Calculation for **noise levels with directional characteristics** is expressed as:

$$SPL = L_w - 20 \log r - 0.5 + \log Q$$

Where:

SPL = sound pressure (reference 0.00002 N/m²)
L_w = sound power (reference 10^{-12} watts)
r   = distance in feet
Q   = directivity factor
       Q = 2 for one reflecting plane
       Q = 4 for two reflecting planes
       Q = 8 for three reflecting planes

11. Calculating the **noise level at a new distance** from the noise source can be computed as follows:

$$SPL = SPL_1 + 20 \log (d_1)/(d_2)$$

Where:

SPL = sound pressure level at new distance (d_2)
SPL_1 = sound pressure level at d_1
d_n = distance from source

12. **Calculating daily noise dose** can be accomplished using the following formula, which combines the effects of different sound pressure levels and allowable exposure times.

$$\text{Daily Noise Dose} = \frac{C_1 + C_2 + C_3 \ldots + C_n}{T_1 \quad T_2 \quad T_3 \quad T_n}$$

Where:

$$C_i = \text{number of hours exposed at given } SPL_i$$
$$T_i = \text{number of hours exposed at given } SPL_i$$

13. **Calculating OSHA permissible noise levels** using the formula below:

$$T_{SPL} = 8/2^{(SPL - 90)/5}$$

Where:

$$T_{SPL} = \text{time in hours at given SPL}$$
$$SPL = \text{sound pressure level (dBA)}$$

14. Formula for converting noise dose measurements to the **equivalent eight-hour TWA:**

$$TWA_{eq} = 90 + 16.61 \log (D)/(100)$$

Where:

$$TWA_{eq} = \text{eight-hour equivalent TWA in dBA}$$
$$D = \text{noise dosimeter reading in \%}$$

15. **Noise reduction in duct system**

$$NR = 12.6\, P\alpha^{1.4}/A$$

Where:

$$NR = \text{noise reduction (dB/ft)}$$
$$P = \text{perimeter of duct (in)}$$
$$\alpha = \text{absorption coefficient of the lining material at frequency of interest}$$
$$A = \text{cross-sectional area of duct (in}^2\text{)}$$

16. **Sound intensity level** is the power passing through a unit area as the sound power radiates in free space.

$$L_1 = 10 \log I/I_o \text{ dB}$$

Where

$$L_1 = \text{sound pressure level (dB)}$$
$$I = \text{sound intensity (watts/m}^2\text{)}$$
$$I_o = \text{reference sound intensity (watts/m}^2\text{)}$$

17. **Noise reduction by absorption**
The amount of noise absorption from room surfaces is measured in sabins.

$$dB = 10\log_{10}(A_2/A_1)$$

Where

$$dB = \text{noise reduction (dB)}$$
$$A_1 = \text{total amount of absorption in room before treatment (sabins)}$$
$$A_2 = \text{total amount of absorption in room after treatment (sabins)}$$

RADIATION

Key Terms

absorbed dose: The energy imparted by ionizing radiation per unit mass of irradiated material. The units of absorbed dose are the rad and the gray (Gy).

activity: The rate of disintegration (transformation) or decay of radioactive material. The units of activity are the curie (Ci) and the becquerel (Bq).

alpha particle: A strongly ionizing particle emitted from the nucleus of an atom during radioactive decay, containing two protons and neutrons and having a double positive charge.

alternate authorized user: Serves in the absence of the authorized user and can assume any duties as assigned.

Authorized user: An employee who is approved by the RSO and RSC and is ultimately responsible for the safety of those who use radioisotopes under his/her supervision.

beta particle: An ionizing charge particle emitted from the nucleus of an atom during radioactive decay, equal in mass and charge to an electron.

bioassay: The determination of kinds, quantities, or concentrations, and, in some cases, the locations of radioactive material in the human body, whether by direct measurement (in vivo counting) or by analysis and evaluation of materials excreted or removed from the human body.

biological half-life: The length of time required for one-half of a radioactive substance to be biologically eliminated from the body.

Bremsstrahlung: Electromagnetic (x-ray) radiation associated with the deceleration of charged particles passing through matter.

contamination: The deposition of radioactive material in any place where it is not wanted.

controlled area: An area, outside of a restricted area but inside the site boundary, access to which can be limited by the licensee for any reason.

counts per minute (cpm): The number of nuclear transformations from radioactive decay able to be detected by a counting instrument in a one-minute time interval.

curie (Ci): A unit of activity equal to 37 billion disintegrations per second.

declared pregnant woman: A woman who has voluntary informed her employer, in writing, of her pregnancy and the estimated date of childbirth.

disintegrations per minute (dpm): The number of nuclear transformation from radioactive decay in a one-minute time interval.

dose equivalent: A quantity of radiation dose expressing all radiation on a common scale for calculating the effective absorbed dose. The units of dose equivalent are the rem and sievert (SV).

dosimeter: A device used to determine the external radiation dose a person has received.

effective half-life: The length of time required for a radioactive substance in the body to lose one-half of its activity present through a combination of biological elimination and radioactive decay.

exposure: The amount of ionization in air from x-rays and gamma rays.

extremity: Hand, elbow, and arm below the elbow, foot, knee, or leg below the knee.

gamma rays: Very penetrating electromagnetic radiations emitted from a nucleus and an atom during radioactive decay.

half-life: The length of time required for a radioactive substance to lose one-half of its activity by radioactive decay.

limits (dose limits): The permissible upper bounds of radiation doses.

permitted worker: A laboratory worker who does not work with radioactive materials but works in a radiation laboratory.

photon: A type of radiation in the form of an electromagnetic wave.

rad: A unit of radiation-absorbed dose. One rad is equal to 100 ergs per gram.

radiation (ionizing radiation): Alpha particles, beta particles, gamma rays, x-rays, neutrons, high-speed electrons, high-speed protons, and other particles capable of producing ions.

radiation workers: Those personnel listed on the Authorized User Form of the supervisor to conduct work with radioactive materials.

radioactive decay: The spontaneous process of unstable nuclei in an atom disintegrating into stable nuclei, releasing radiation in the process.

radioisotope: A radioactive nuclide of a particular element.

rem: A unit of dose equivalent. One rem is approximately equal to one rad of beta, gamma, or x-ray radiation, or 1/20 of alpha radiation.

restricted area: An area, access to which is limited by the licensee for the purpose of protecting individuals against undue risks from exposure to radiation and radioactive materials.

roentgen: A unit of radiation exposure. One roentgen is equal to 0.00025 coulombs of electrical charge per kilogram of air.

thermoluminescent dosimeter (TLD): A dosimeter worn by radiation workers to measure their radiation dose. The TLD contains crystalline material that stores a fraction of the absorbed ionizing radiation and releases this energy in the form of light photons when heated.

total effective dose equivalent (TEDE): The sum of the deep-dose equivalent (for external exposures) and the committed effective dose equivalent (for internal exposures).

unrestricted area: An area, access to which is neither limited nor controlled by the licensee.

x-rays: A penetrating type of photon radiation emitted from outside the nucleus of a target atom during bombardment of a metal with fast electrons.

Ionizing Radiation

Ionization is the process by which atoms are made into ions by the removal or addition of one or more electrons; they produce this effect by the high kinetic energies of the quanta (discrete pulses) they emit. Simply, ionizing radiation is any radiation capable of producing ions by interaction with matter. Direct ionizing particles are charged particles (e.g., electrons, protons, alpha particles, etc.) having sufficient kinetic energy to produce ionization by collision. Indirect ionizing particles are uncharged particles (e.g., photons, neutrons, etc.) that can liberate direct ionizing particles. Ionizing radiation sources can be found in a wide range of occupational settings, including health care facilities, research institutions, nuclear reactors and their support facilities, nuclear weapon production facilities, and other various manufacturing settings, just to name a few.

These ionizing radiation sources can pose a considerable health risk to affected workers if not properly controlled. Ionization of cellular components can lead to functional changes in the tissues of the body. Alpha, beta, neutral particles, x-rays, gamma rays, and cosmic rays are ionizing radiations.

Three mechanisms for external radiation protection include time, distance, and shielding. A shorter time in a radiation field means less dose. From a point source, dose rate is reduced by the square of the distance and expressed by the inverse square law:

$$I_1(d_1)^2 = I_2(d_2)^2 \tag{11.19}$$

Where:

I_1 = dose rate or radiation intensity at distance d_1
I_2 = dose rate or radiation intensity at distance d_2

Radiation is reduced exponentially by thickness of shielding material.

Effective Half-Life

The half-life is the length of time required for one-half of a radioactive substance to disintegrate. The formula depicted below is used when the industrial hygienist is interested in determining how much radiation is left in a worker's stomach after a period of time. Effective half-life is a combination of radiological and biological half-lives and is expressed as:

$$T_{eff} = \frac{(T_b)(T_r)}{T_b + T_r} \tag{11.20}$$

Where:

T_b = biological half-life
T_r = radiological half-life

It is important to point out that T_{eff} will always be shorter than either T_b or T_r. T_b may be modified by diet and physical activity.

Alpha Radiation

Alpha radiation is used for air ionization—elimination of static electricity (Po-210), clean room applications, and smoke detectors (Am-241). It is also used in air density measurement, moisture meters, nondestructive testing, and oil well logging. Naturally occurring alpha particles are also used for physical and chemical properties, including uranium (coloring of ceramic glaze, shielding) and thorium (high temperature materials).

The characteristics of alpha radiation are listed below.

- Alpha (α) radiation is a particle composed of two protons and neutrons with source: Ra-226 → Rn 222 → Accelerators.
- Alpha radiation is not able to penetrate skin.
- Alpha-emitting materials can be harmful to humans if the materials are inhaled, swallowed, or absorbed through open wounds.
- A variety of instruments have been designed to measure alpha radiation. Special training in the use of these instruments is essential for making accurate measurements.
- A civil defense instrument (CD V-700) cannot detect the presence of radioactive materials that produce alpha radiation unless the radioactive materials also produce beta and/or gamma radiation.
- Instruments cannot detect alpha radiation through even a thin layer of water, blood, dust, paper, or other material because alpha radiation is not penetrating.
- Alpha radiation travels a very short distance through air.
- Alpha radiation is not able to penetrate turnout gear, clothing, or a cover on a probe. Turnout gear and dry clothing can keep alpha emitters off the skin.

Alpha Radiation Detectors

The types of high-sensitivity portable equipment used to evaluate alpha radiation in the workplace include:

- Geiger-Mueller counter
- scintillators
- solid-state analysis
- gas proportional devices

Beta Radiation

Beta radiation is used for thickness measurements for coating operations; radioluminous signs; tracers for research; and air ionization (gas chromatograph, nebulizers).

The characteristics of beta radiation are listed below.

- Beta (β) is a high-energy electron particle with source: Sr-90 → Y-90 → electron beam machine.
- Beta radiation may travel meters in air and is moderately penetrating.
- Beta radiation can penetrate human skin to the "germinal layer," where new skin cells are produced. If beta-emitting contaminants are allowed to remain on the skin for a prolonged period of time, they may cause skin injury.
- Beta-emitting contaminants may be harmful if deposited internally.
- Most beta emitters can be detected with a survey instrument (such as a DC V-700, provided the metal probe cover is open). Some beta emitters, however, produce very low energy, poorly penetrating radiation that may be difficult or impossible to detect. Examples of these are carbon-14, tritium, and sulfur-35.
- Beta radiation cannot be detected with an ionization chamber such as CD V-715.
- Clothing and turnout gear provide some protection against most beta radiation. Turnout gear and dry clothing can keep beta emitters off the skin.
- Beta radiation presents two potential exposure methods, external and internal. External beta radiation hazards are primarily skin burns. Internal beta radiation hazards are similar to alpha emitters.

Beta Detection Instrumentation

The types of equipment used to evaluate beta radiation in the workplace include:

- Geiger-Mueller counter
- Gas proportional devices
- Scintillators
- Ion chambers
- Dosimeters

Shielding for Beta Radiation

Shielding for beta radiation is best accomplished by using materials with a low atomic number (low z materials) to reduce Bremsstrahlung radiation (i.e., secondary x-radiation produced when a beta particle is slowed down or stopped by a high-density surface). The thickness is critical to stop maximum energy range and varies with the type of material used. Typical shielding material includes lead, water, wood, plastics, cement, plexiglas, and wax.

Gamma Radiation and X-rays

Gamma radiation and x-rays are used for sterilization of food and medical products; radiography of welds, castings, and assemblies; gauging of liquid levels and material density; and oil well logging and material analysis.

The characteristics of gamma radiation and x-rays are listed below.

- Gamma (γ) is not a particle (electromagnetic wave) composed of high-energy electron with source: Tc-99.
- X-rays are composed of photons (generated by electrons leaving an orbit) with a source, such as most radioactive materials and x-ray machines that are secondary to β.
- Gamma radiation and x-rays are electromagnetic radiation like visible light, radio waves, and ultraviolet light. These electromagnetic radiations differ only in the amount of energy they have. Gamma rays and x-rays are the most energetic of these.
- Gamma radiation is able to travel many meters in air and many centimeters in human tissue. It readily penetrates most materials and is sometimes called "penetrating radiation."
- X-rays are like gamma rays. They, too, are penetrating radiation.
- Radioactive materials that emit gamma radiation and x-rays constitute both an external and internal hazard to humans.
- Dense materials are needed for shielding from gamma radiation. Clothing and turnout gear provide little shielding from penetrating radiation but will prevent contamination of the skin by radioactive materials.
- Gamma radiation is detected with survey instruments, including civil defense instruments. Low levels can be measured with a standard Geiger counter, such as the CD V-700. High levels can be measured with an ionization chamber, such as a CD V-715.
- Gamma radiation or x-rays frequently accompany the emission of alpha and beta radiation.
- Instruments designed solely for alpha detection (such as an alpha scintillation counter) will not detect gamma radiation.
- Pocket chamber (pencil) dosimeters, film badges, thermoluminescent, and other types of dosimeters can be used to measure accumulated exposure to gamma radiation.
- The principal health concern associated with gamma radiation is external exposure by penetrating radiation and physically strong source housing. Sensitive organs include the lens of the eye, the gonads, and damage to the bone marrow.

Gamma Detection Instrumentation

The types of equipment used to evaluate gamma radiation in the workplace include:

- ion chamber
- gas proportional
- Geiger-Mueller

Shielding for Gamma and X-rays

Shielding gamma and x-radiation depends on energy level. Protection follows an exponential function of shield thickness. At low energies, absorption can be achieved with millimeters of lead. At high energies, shielding can attenuate gamma radiation.

Radioactive Decay Equations

Radioactive materials emit alpha particles, beta particles, and photon energy and lose a proportion of their radioactivity with a characteristic half-life. This is known as radioactive decay. To calculate the amount of radioactivity remaining after a given period of time, use the following basic formulae for decay calculations:

$$\text{Later activity} = (\text{earlier activity}) \ e^{-\lambda} (\text{elapsed time})$$
$$A = A_i e^{-\lambda t}\sim \tag{11.21}$$

Where:

$$\lambda = LN2/T$$

and:

λ = lambda decay constant (probability of an atom decaying in a unit time)
t   = time
LN2 = 0.693
 T = radioactive half-life (time period in which half of a radioactive isotope decays)
 A = new or later radioactivity level
 A_i = initial radioactivity level

In determining time required for a radioactive material to decay (A_o to A), use:

$$t = (- LN \ A/A_i) \ (T/LN2) \tag{11.22}$$

Where:

λ = lambda decay constant (probability of an atom decaying in a unit time)
t   = time
LN2 = 0.693
 T = radioactive half-life (time period in which half of a radioactive isotope decays)
 A = new or later radioactivity level
 A_i = initial radioactivity level

Basic rule of thumb: In seven half-lives, reduced to <1%; 10 half-lives <0.1%.

In determining the rate of radioactive decay, keep in mind that radioactive disintegration is directly proportional to the number of nuclei present. Thus, the radioactive decay rate is expressed in nuclei disintegrated per unit time.

$$A_i = (0.693/T) \ (N_i) \tag{11.23}$$

Where:

 A_i = initial rate of decay

N_i = initial number of radionuclei

T = half-life

As mentioned earlier, half-life is defined as the time it takes for a material to lose 50 percent of its radio-activity. The following equation can be used to determine half-life.

$$A = A_i \, (0.5)^{t/T} \tag{11.24}$$

Where:

A = activity at time t

A_i = initial activity

t = time

T = half-life

Radiation Dose

In the United States, radiation *absorbed dose*, *dose equivalent*, and *exposure* are often measured and stated in the traditional units called *rad*, *rem*, or *roentgen (R)*. For practical purposes with gamma and x-rays, these units of measure for exposure or dose are considered equal. This exposure can be from an external source irradiating the whole body, an extremity, or other organ or tissue resulting in an *external radiation dose*. Alternately, internally deposited radioactive material may cause an *internal radiation dose* to the whole body or other organ or tissue.

A prefix is often used for smaller measured fractional quantities, such as milli (m) means 1/1,000. For example, 1 rad = 1,000 mrad. Micro (μ) means 1/1,000,000. So, 1,000,000 μrad = 1 rad, or 10 μR = 0.000010 R.

The SI system (System International) for radiation measurement is now the official system of measurement and uses the "gray" (Gy) and "sievert" (Sv) for absorbed dose and equivalent dose respectively. Conversions are as follows:

- 1 Gy = 100 rad
- 1 mGy = 100 mrad
- 1 Sv = 100 rem
- 1 mSv = 100 mrem

Radioactive transformation event (radiation counting systems) can be measured in units of "disintegrations per minute" (dpm) and, because instruments are not 100 percent efficient, "counts per minute" (cpm). Background radiation levels are typically less than 10 μR per hour, but due to differences in detector size and efficiency, the cpm reading on fixed monitors and various handheld survey meters will vary considerably.

SOIL MECHANICS

When dealing with nature's building material, soil, the engineer should keep the following statement in mind:

> Observe always that everything is the result of change, and get used to thinking that there is nothing
> Nature loves so well as to change existing forms and to make new ones like them.

—*Meditations*, Marcus Aurelius

Soil for Construction

By the time a student reaches the third or fourth year of elementary school, he or she is familiar with the Leaning Tower of Pisa. Many are also familiar with Galileo's experiments with gravity and the speed of falling objects dropped from the top of the tower. The Leaning Tower of Pisa, this twelfth-century bell tower, has been a curiosity for literally millions of people from the time it was first built to the present. Eight stories high and 180 feet tall, with a base diameter of fifty-two feet, the tower began to lean by the time the third story was completed, and leans about 1/25 inches more each year.

How many people know why the tower is leaning in the first place—and who would be more than ordinarily curious about why the Leaning Tower leans? If you are a soil scientist or an engineer, this question has real significance—the question requires an answer. And, in fact, the Leaning Tower of Pisa should never have acquired the distinction of being a "leaning" tower in the first place.

So, why does the Leaning Tower of Pisa lean?—Because it rests on a nonuniform consolidation of clay beneath the structure. This ongoing process may eventually lead to failure of the tower.

As you might have guessed, the mechanics of why the Leaning Tower of Pisa leans is what this section is all about. More specifically, it is about the mechanics and physics of the soil—important factors in making the determination as to whether a particular building site is viable for building. Simply put, these two factors are essential in answering the question, "Will the soils present support buildings?"

Soil Characteristics

When we refer to the characteristics of soils, we are referring to its mechanical characteristics, physical factors important to safety engineers. Safety engineers focus on the characteristics of the soil related to its suitability as a construction material and as a substance to be excavated (can the soil in question be safely excavated without caving in on workers?). Simply put, the safety engineer must understand the response of a particular volume of soil to internal and external mechanical forces. Obviously, to be able to determine the soil's ability to withstand the load applied by structures of various types and its ability to remain stable when excavated is important.

From a purely engineering point of view, soil is any surficial material of the earth that is unconsolidated enough to be excavated with tools (from bulldozers to shovels). The engineer takes into consideration both the advantages and disadvantages of using soil for engineering purposes. The obvious key advantage of using soil for engineering is that there is (in many places) no shortage of it—it may already be on the construction site, thus avoiding the expense of hauling it from afar. Another advantage of using soil for construction is its ease of manipulation; it may be easily shaped into almost any desired form. Soil also allows for the passage of moisture, or, as needed, it can be made impermeable.

We said that the engineer looks at both the advantages and disadvantages of using soil for construction projects. The most obvious disadvantage of using soil is its variability from place to place and from time to time. Soil is not a uniform material for which reliable data related to strength can be compiled or computed. Cycles of wetting and drying and freezing and thawing affect the engineering properties of soil. A particular soil may be suitable for one purpose, but not for another. Stamford clay in Texas, for example, is rated as "very good" for sealing of farm ponds—and "very poor" for use as base for roads and buildings (Buol et al. 1980).

How does the engineer know if a particular soil is suitable for use as a base for roads or buildings? The engineer determines the suitability of a particular soil (for whatever purpose) by studying soil survey maps and reports. The engineer also checks with soil scientists and other engineers familiar with the region and the soil types of that region. Any good engineer will also want to conduct field sampling—to ensure that the soil product he or she will be working with possesses the soil characteristics required for its intended purpose.

What are the soil characteristics (the kinds of information) that the engineer is interested in? Important characteristics of soils for engineering purposes include:

- Soil texture
- Kinds of clay present
- Depth to bedrock
- Soil density
- Erodibility
- Corrosivity
- Surface geology
- Plasticity
- Content of organic matter
- Salinity
- Depth to seasonal water table

The engineer will also want to know the soil's density, space, and volume or weight-volume relationships, stress-strain, slope stability, and compaction. Because these concepts are of paramount importance to the engineer (and others), these concepts are presented in the following sections.

Soil Weight-Volume or Space and Volume Relationships

All natural soil consists of at least three primary components or phases, solid (mineral) particles, water, and/or air (within void spaces between the solid particles). The physical relationships (for soils in particular) between these phases must be examined.

The volume of the soil mass is the sum of the volumes of three components, or

$$V_T = V_a + V_W + V_S \qquad (11.25)$$

Where:

V_T = Volume total
V_a = Volume air
V_w = Volume water
V_s = Volume solids

The volume of the voids is the sum of V_a and V_W. However, regarding weight (because the weighing of air in the soil voids would be done within the earth's atmosphere as with other weighings), the weight of the solids is determined on a different basis. We consider the weight of air in the soil to be zero, and the total weight is expressed as the sum of the weights of the soil solids and the water:

$$W_T = W_s + W_w \qquad (11.26)$$

Where:

W_T = weight total
W_s = weight solids
W_w = weight water

The relationship between weight and volume can be expressed as

$$W_m = V_m G_m\,_{-w} \qquad (11.27)$$

Where:

W_m = weight of the material (solid, liquid, or gas)
V_m = volume of the material
G_m = specific gravity of the material (dimensionless)
-w = unit weight of water

With the relationships described above, a few useful problems can be solved. When an engineer determines that within a given soil, the proportions of the three major components need to be mechanically adjusted, this can be accomplished by reorienting the mineral grains by compaction or tilling. The engineer, probably for a specific purpose, may want to blend soil types to alter the proportions, such as increasing or decreasing the percentage of void space.

How do we go about doing this? Relationships between volumes of soil and voids are described by the void ratio (e) and porosity (η). To accomplish this, we must first determine the void ratio (the ratio of the void volume to the volume of solids):

$$e = \frac{Vv}{Vs} \qquad (11.28)$$

We must also determine the ratio of the volume of void spaces to the total volume. This can be accomplished by determining the porosity (η) of the soil, which is the ratio of void volume to total volume. Porosity is usually expressed as a percentage.

$$\eta = Vv/Vt \times 100\% \tag{11.29}$$

Two additional relationships, known as the moisture content, (w), and degree of saturation, (S), relate the water content of the soil and the volume of the water in the void space to the total void volume:

$$w = \frac{Wv}{Ws} \times 100\% \tag{11.30}$$

And

$$S = \frac{Vw}{Vv} \times 100\% \tag{11.31}$$

Soil Particle Characteristics

The size and shape of particles in the soil, as well as density and other characteristics, relate to sheer strength, compressibility, and other aspects of soil behavior. Engineers use these index properties to form engineering classifications of soil. Simple classification tests are used to measure index properties (see table 11.2) in the lab or the field.

From table 11.2 we see that an important division of soils (from the engineering point of view) is the separation of the cohesive (fine-grained) from the incohesive (coarse-grained) soils. Let's take a closer look at these two important terms.

Cohesion indicates the tendency of soil particles to stick together. Cohesive soils contain silt and clay. The clay and water content makes these soils cohesive through the attractive forces between individual clay and water particles. The influence of the clay particles makes the index properties of cohesive soils somewhat more complicated than the index properties of cohesionless soils. The resistance of a soil at various moisture contents to mechanical stresses or manipulations is called the soil's consistency—the arrangement of clay particles, and is the most important characteristic of cohesive soils.

Another important index property of cohesive soils is its *sensitivity*. Simply defined, sensitivity is the ratio of unconfined compressive strength in the undisturbed state to strength in the remolded state (see equation 11.32). Soils with high sensitivity are highly unstable.

Table 11.2 Index Property of Soils

SOIL TYPE	INDEX PROPERTY
Cohesive (fine-grained)	Water content Sensitivity Type and amount of clay Consistency Atterberg limits
Incohesive (coarse-grained)	Relative density In-place density Particle-size distribution Clay content Shape of particles

Source: Adapted from A. E. Kehew, *Geology for Engineers and Environmental Scientists*, 2nd ed., 1995, 284.

$$\text{Sensitivity} = \frac{\text{Strength in undisturbed condition}}{\text{Strength in remolded condition}} \qquad (11.32)$$

Soil water content (described earlier) is an important factor that influences the behavior of the soil. The water content values of soil are known as the Atterburg limits, a collective designation of so-called limits of consistency of fine-grained soils, which are determined with simple laboratory tests. They are usually presented as the liquid limit (LL) and plastic limit (PL).

Note: Plasticity is exhibited over a range of moisture contents referred to as plasticity limits—and the plasticity index (PI). The plastic limit is the lower water level at which soil begins to be malleable in a semisolid state, but molded pieces crumble easily when a little pressure is applied. At the point when the volume of the soil becomes nearly constant with further decreases in water content, the soil reaches the shrinkage state.

The upper plasticity limit (or liquid limit) is the water content at which the soil-water mixture changes from a liquid to a semifluid (or plastic) state and tends to flow when jolted. Obviously, an engineer charged with building a highway or building would not want to choose a soil for the foundation that tends to flow when wet.

The difference between the liquid limit and the plastic limit is the range of water content over which the soil is plastic and is called the plasticity index. Soils with the highest plasticity indices are unstable in bearing loads.

Several systems for classifying the stability of soil materials have been devised, but the best known (and probably the most useful) system is called the Unified System of Classification. This classification gives each soil type (fourteen classes) a two-letter designation, which is primarily defined on the basis of particle-size distribution, liquid limit, and plasticity index.

Cohesionless coarse-grained soils behave much differently than cohesive soils and are based on (from index properties) the size and distribution of particles in the soil. Other index properties (particle shape, in-place density, and relative density, for example) are important in describing cohesionless soils, because they relate to how closely particles can be packed together.

Soil Stress and Strain

If you are familiar with water pressure and its effect as you go deeper into the water (as when diving deep into a lake), then that the same concept applies to soil and pressure should not surprise you. Like water, the pressure within the soil increases as the depth increases. A soil, for example, that has a unit weight of 75 pounds per cubic feet exerts a pressure of 75 psi at one-foot depth and 225 psi at three feet, etc. As you might expect, as the pressure on a soil unit increases, the soil particles reorient themselves structurally to support the cumulative load. This consideration is important because the elasticity of the soil sample retrieved from beneath the load may not be truly representative, once it is delivered to the surface. In sampling, the importance of taking representative samples cannot be overstated.

The response of a soil to pressure (stress) is similar to what occurs when a load is applied to a solid object; the stress is transmitted throughout the material. The load subjects the material to pressure, which equals the amount of load divided by the surface area of the external face of the object over which it is applied. The response to this pressure or stress is called displacement or strain. Stress (like pressure), at any point within the object, can be defined as force per unit area.

Soil Compressibility

When a vertical load such as a building or material stockpile is placed above a soil layer, some settlement can be expected. Settlement is the vertical subsidence of the building (or load) as the soil is compressed.

Compressibility refers to the tendency of soil to decrease in volume under load. This compressibility is most significant in clay soils because of the inherent high porosity. Although the mechanics of compressibility and settlement are quite complex and beyond the scope of this text, you should know something about the actual evaluation process for these properties, which is accomplished in the consolidation test. This test subjects a soil sample to an increasing load. The change in thickness is measured after the application of each load increment.

Soil Compaction

The goal of compaction is to reduce void ratio and thus increase the soil density, which, in turn, increases the shear strength. This is accomplished by working the soil to reorient the soil grains into a more compact state. If water content is within a limited range (sufficient enough to lubricate particle movement), efficient compaction can be obtained.

The most effective compaction occurs when the soil placement layer (commonly called lift) is approximately eight inches. At this depth, the most energy is transmitted throughout the lift. Note that more energy must be dispersed, and the effort required to accomplish maximum density is greatly increased when the lift is greater than 10 inches in thickness.

For cohesive soils, compaction is best accomplished by blending or kneading the soil using sheepsfoot rollers and pneumatic tire rollers. These devices work to turn the soil into a denser state.

To check the effectiveness of the compactive effort, the in-place dry density of the soil (weight of solids per unit volume) is tested by comparing the dry density of field-compacted soil to a standard prepared in an environmental laboratory. Such a test allows a percent compaction comparison to be made.

Soil Failure

The construction and safety engineer must be concerned with soil structural implications involved with natural processes (such as frost heave, which could damage a septic system) and changes applied to soils during remediation efforts (e.g., when excavating to mitigate a hazardous materials spill in soil). Soil failure occurs whenever it cannot support a load. Failure of an overloaded foundation, collapse of the sides of an excavation, or slope failure on the sides of a dike, hill, or similar feature is termed structural failure.

The type of soil structural failure that probably occurs more frequently than any other is slope failure (commonly known in practice as cave-in). In a review of the Bureau of Labor Statistics annual report of on-the-job mishaps, including from eighty to one hundred fatalities every year—we see that cave-ins occurring in excavations accomplished in construction are more frequent occurrences than you might think, considering the obvious dangers inherent in excavation.

What is an excavation? How deep does an excavation have to be to be considered dangerous? Two good questions—and the answers could save your life or help you protect others when you become a safety engineer.

An excavation is any humanmade cut, cavity, trench, or depression in the earth's surface formed by earth removal. This can include excavations for anything from a remediation dig to sewer line installation.

No excavation activity should be accomplished without keeping personnel safety in mind. Any time soil is excavated, care and caution is advised. As a rule of thumb (and as law under 29 CFR 1926.650-.652), the Occupational Safety and Health Administration (OSHA) requires trench protection in any excavation five feet or more in depth.

Before digging begins, proper precautions must be taken. The responsible party in charge (the competent person, according to OSHA) must:

- Contact utility companies to ensure underground installations are identified and located.
- Ensure underground installations are protected, supported, or removed as necessary to safeguard workers.
- Remove or secure any surface obstacles (trees, rocks, and sidewalks, for example) that may create a hazard for workers.
- Classify the type of soil and rock deposits at the site as either stable rock, type A, type B, or type C soil. One visual and at least one manual analysis must make the soil classification.

Let's take a closer look at the requirement to classify the type of soil to be excavated. Before an excavation can be accomplished, the soil type must be determined. The soil must be classified as: stable rock, type A, type B, or type C soil. Remember, commonly you will find a combination of soil types at an excavation site. Soil classification (used in this manner) is used to determine the need for a protective system.

Exactly what do the soil classifications of stable rock, type A, type B, and type C soil mean?

- **stable rock:** A natural solid mineral material that can be excavated with vertical sides. Stable rock will remain intact while exposed, but keep in mind that even though solid rock is generally stable, it may become very unstable when excavated (in practice you never work in this kind of rock).
- **type A soil:** The most stable soil, and includes clay, silty clay, sandy clay, clay loam, and sometimes silty clay loam and sandy clay loam.
- **type B soil:** Moderately stable, and includes silt, silt loam, sandy loam, and sometimes silty clay loam and sand clay loam.
- **type C soil:** The least stable, and includes granular soils like gravel, sand, loamy sand, submerged soil, soil from which water is freely seeping, and submerged rock that is not stable.

How is soil tested? To test and classify soil for excavation, both visual and manual tests should be conducted. In visual soil testing, you should look at soil particle size and type. You'll see a mixture of soils, of course. You should check to see if the soil clumps when dug—if so it could be clay or silt. Type B or C soil can sometimes be identified by the presence of cracks in walls and spalling (breaks up in chips or fragments). If you notice layered systems with adjacent hazardous areas—buildings, roads, and vibrating machinery—you may require a professional engineer for classification. Standing water or water seeping through trench walls automatically classifies the soil as type C.

Manual soil testing is required before a protective system (e.g., shoring or shoring box) is selected. A sample taken from soil dug out into a spoil pile should be tested as soon as possible to preserve its natural moisture. Soil can be tested either on site or off site. Manual soil tests include a sedimentation test, wet shaking test, thread test, and ribbon test.

A sedimentation test determines how much silt and clay are in sandy soil. Saturated sandy soil is placed in a straight-side jar with about five inches of water. After the sample is thoroughly mixed (by shaking it) and allowed to settle, the percentage of sand is visible. A sample containing 80 percent sand, for example, will be classified as Type C.

The wet shaking test is another way to determine the amount of sand versus clay and silt in a soil sample. This test is accomplished by shaking a saturated sample by hand to gauge soil permeability based on the following facts: 1. shaken clay resists water movement through it; and 2. water flows freely through sand and less freely through silt.

The thread test is used to determine cohesion (remember, cohesion relates to stability—how well the grains hold together). After a representative soil sample is taken, it is rolled between the palms of the hands to about 1/8″ diameter and several inches in length (any child who has played in dirt has accomplished this at one time or another—nobody said soil science had to be boring). The rolled piece is placed on a flat surface, then picked up. If a sample holds together for two inches, it is considered cohesive.

The ribbon test is used as a backup for the thread test. It also determines cohesion. A representative soil sample is rolled out (using the palms of your hands) to 3/4″ in diameter and several inches in length. The sample is then squeezed between thumb and forefinger into a flat unbroken ribbon 1/8″ to 1/4″ thick, which is allowed to fall freely over the fingers. If the ribbon does not break off before several inches are squeezed out, the soil is considered cohesive.

Once soil has been properly classified, the correct protective system can be chosen (if necessary). This choice is based on both soil classification and site restrictions. There are two main types of protective systems: 1. sloping or benching and 2. shoring or shielding.

Sloping (or benching) is an excavation protective measure that cuts the walls of an excavation back at an angle to its floor. Examples of a sloped or angled cut and a bench system with one or more steps carved into the soil are shown in figure 11.17.A–B.

The angle used for sloping or benching is a ratio based on soil classification and site restrictions. In both systems, the flatter the angle, the greater the protection for workers. Reasonably safe side slopes for each of these soil types are presented in table 11.3.

Shoring and shielding are two protective measures that add support structure to an existing excavation (generally used in excavations with vertical sides but can be used with sloped or benched soil).

Shoring is a system designed to prevent cave-ins by supporting walls with vertical shores called uprights or sheeting. Wales are horizontal members along the sides of a shoring structure. Cross braces are supports placed horizontally between trench walls.

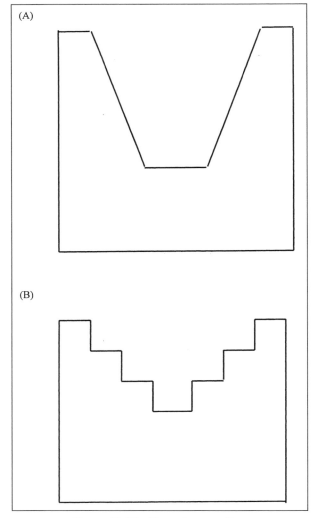

Figure 11.17 Sloped excavation.

Table 11.3 Maximum Safe Side Slopes in Excavations

Soil Type	Side Slope (Vertical to Horizontal)	Side Slope (Degrees from Horizontal)
A	75:1	53°
B	1:1	45°
C	1.5:1	34°

Source: OSHA Excavation Standard 29 CFR 1926.650-652.

Shielding is a system that uses a trench box or trench shield. They can be premanufactured or job-built under the specification of a licensed engineer. Shields are usually portable steel structures placed in the trench by heavy equipment. For deep excavations, trench boxes can be stacked and attached to each other with stacking lugs.

Soil Physics

Soil is a dynamic, heterogeneous body that is nonisotropic (does not have the same properties in all directions). As you might expect, because of these properties, various physical processes are active in soil at all times. This important point is made clear by Winegardner (1996) in the following: "all of the factors acting on a particular soil, in an established environment, at a specified time, are working from some state of imbalance to achieve a balance" (63).

Most soil specialists (and many other people) have little difficulty in understanding why soils are very important to the existence of life on earth. They know, for example, not only that soil is necessary (in a very direct sense) to sustain plant life (and thus other life forms that depend on plants), they also know that soil functions to store and conduct water, serves a critical purpose in soil engineering involved with construction, and acts as a sink and purifying medium for waste disposal systems.

The soil practitioner must be well-versed in the physical properties of soil. Specifically, he or she must have an understanding of those physical processes that are active in soil. These factors include physical interactions related to soil water, soil grains, organic matter, soil gases, and soil temperature. To gain this knowledge, the safety engineer must have training in basic geology, soil science, and/or engineering construction.

STRUCTURAL FAILURES

To this point, we have reviewed the basics involved with applied mechanics. This information is important to the safety engineer because without such knowledge, properly understanding the construction, function, and operation of workplace machines, equipment, and structures would be difficult. More importantly, a basic knowledge of applied mechanics also enables the safety engineer to understand that systems fail, what causes them to fail, and (more significantly) how to prevent such failures. Many types of failures are possible, and failures occur for many reasons. Structural failures are important to the safety engineer because when such failures occur, they typically cause damage and injuries (or worse) to workers and others. Structural failures can be caused by any of the following: design errors, faulty material, physical damage, overloading, poor workmanship, and poor maintenance and inspection practices.

- Design errors are not uncommon. They are usually the result of incorrect or poorly made assumptions. For example, the design engineer might assume some load or a maximum load for a design. However, the actual load may be much different in varying conditions. The 1981 Hyatt Regency Skywalk disaster (114 dead, over 200 injured) was caused by underestimating load, a badly designed and altered structural element design, and the failure to properly calculate and check the numbers (see Case Study following).
- Faulty materials cause structural failures for two main reasons: lack of uniformity of material and changes in properties (material strength, ductility, brittleness, and toughness) over time.
- Physical damage caused by usage, abuse, or unplanned events (natural or human-generated) are another cause of structural failure.
- Overloading and inadequate support are common causes of structural failure. A particular structure might have originally been designed to house an office complex, then later reconfigured for use as a machine shop. The added weight of machines and ancillaries is a change in the use environment, which may overload the original structure and eventually lead to failure because of inadequate structural support.
- Poor and faulty workmanship is a factor that must always be considered when studying structural failures. Improper assembly and maintenance of devices, machines, and structures have certainly caused their share of failures and/or collapses. In fact, one of the major reasons many companies have employed quality control procedures in their manufacturing and construction activities is to guard against poor and faulty workmanship.
- Poor maintenance, use, and inspection play an important role in structural failures. Obviously, exposures to various conditions during their use changes structures. Improper maintenance can affect a structure's useful life. Improper use can have the same effect. Inspection is important to ensure that maintenance and use are providing effective care and to guard against unexpected failure.

Case Study

On July 17, 1981, at 7:05 p.m., the Kansas City, Missouri, Hyatt Regency Hotel atrium held over 1,600 people. At that moment, two suspended structurally connected skywalks (bridges that connected two towers on the second- and fourth-floor levels) failed and fell, crushing the heavily occupied restaurant bar beneath them.

The death toll for this structural failure (the worst in U.S. history) was 114. Over two hundred more people were injured, many permanently disabled. Plaintiff's claims amounted to over $3 billion. (The hotel owner, Donald Hall [of Hallmark Cards], with an admirable sense of duty and decency, settled more than 90 percent of the claims).

Immediate theories on what caused this structural failure ranged from continued resonance to faulty materials to poor workmanship. A National Bureau of Standards investigation finally discovered the most probable cause—and their findings laid the blame squarely at the feet of the structural engineers.

The cause? A design change submitted by the contractor and approved by the design engineers and architect. In the original designs, both skywalks were supported by nuts at both second- and fourth-floor levels, threaded onto single continuous hanger rods, and spaced at regular intervals. Because of the single-rod design, while the roof trusses held the load of both, the welded box beams that supported the load of each skywalk were independent of the other.

The change? The contractor's design change shortened the hanger rods, added an extra hole to the fourth-floor beams, and hung the second-floor walkway by an independent second set of hanger rods from those box beam connections, thus putting the entire load of both walkways on the fourth-floor walkway.

This was not the only factor involved, however. In determining the dead load of the structure, the investigators discovered an 8 percent higher load than originally computed because of changes and additions to decking and flooring materials. The live load, however, was well within the limits. The obviously weak element in the design was the fourth-floor box beams, so the investigators tested both new duplicates and undamaged beams and hanger rods from the Hyatt Regency atrium.

The National Bureau of Standards report stated that the skywalks were underdesigned and that the design lacked redundancy. Six important points summarize the report:

- The collapse occurred under loads substantially less than those specified by the Kansas City Building Code.
- All the fourth-floor box-beam-hanger connections were candidates for the initiation of walkway collapse.
- The box-beam-hanger rod connections, the fourth-floor to ceiling hanger rods, and the third-floor-walkway hanger rods did not satisfy the design provisions of the Kansas City Building Code.
- The box-beam-hanger to rod connections under the original hanger rod detail (continuous rod) would not have satisfied the Kansas City Building Code.
- Neither the quality of the workmanship nor the materials used in the walkway system played a significant role in initiating the collapse. (Levy and Salvadori 1992, 229–30)

The bureau's findings also made clear that, although the original design would not have met the Kansas City Building Code either, the original design might not have failed under the minor load present that day. In the inevitable court case that followed, the licenses of the principal and the project manager of the firm responsible for the design were revoked. The attorney for the Missouri State licensing board said, "It wasn't a matter of doing something wrong, they just never did it at all. Nobody ever did any calculations to figure out whether or not the particular connection that held the skywalks up would work. It got built without anybody ever figuring out if it would be strong enough" (Levy and Salvadori 1992, 230).

REVIEW QUESTIONS

1. Why is "engineer a suitable designation for a safety engineer"? Where's the "engineering" in the profession?
2. What two learning style/interest-area qualities will help develop effective safety engineers?
3. Why is applied mechanics important to safety engineering?
4. How does force act on a body?
5. Name ten important physical or mechanical properties.
6. Define and explain the importance of factor of safety (safety factor).
7. How are forces classified?
8. What is the relationship between friction and velocity?
9. What reference sources can provide information on mechanical properties of materials? Why is this information important to safety engineers?

10. Explain Poisson's ratio.
11. What is friction, and how is molecular structure important to it?
12. Define gauge pressure.
13. What ratio do you use to calculate specific gravity?
14. What is Newton's Second Law? Define, and express mathematically.
15. Why is Newton's Second Law important knowledge for a safety engineer?
16. What is the relationship between centrifugal force and centripetal force?
17. What dangers does centrifugal force present?
18. Define stress. Define strain. How are they different?
19. How are stress forces categorized?
20. Name five examples of static applications.
21. What are the important considerations for welds?
22. What properties are important in dynamics, and why?
23. Define hydraulics and pneumatics, and describe why they are similar.
24. Why does the Leaning Tower of Pisa lean? Why is this important to safety engineering?
25. What are the mechanical characteristics of soil, and how do they affect construction?
26. What is important about soil consolidation for engineering?
27. What natural cycles affect soil's engineering properties? How?
28. What are seven important soil characteristics that engineers test for?
29. What are soil's three primary components?
30. How do you calculate the relationship between soil elements to determine soil volume? soil porosity?
31. What classification tests measure soil's index properties?
32. What's the difference between cohesionless and cohesive soils?
33. What changes occur in soil under pressure?
34. Why are engineers concerned with soil compressibility? How do you test for it?
35. How is soil compaction most effectively achieved?
36. Define soil failure and excavation, and describe the safety procedures that must be in place for safe excavation.
37. How are soils classified? What are the classifications? What tests are used to classify soil for evacuation?
38. Describe benching and sloping. What advantages does each bring to safe excavation?
39. What are shoring and shielding? Sketch examples.
40. What are soil's essential physical functions?
41. What two properties are important for determining beam load?
42. Why is a beam stronger when loaded along its thin dimension?
43. Define allowable load.
44. Why is housekeeping important to floor safety?
45. What information do you need to calculate floor load?
46. Define *column*. What happens to columns under load? under excessive load?

REFERENCES AND RECOMMENDED READING

AISC. 1980. *Manual of steel construction.* 8th ed. Chicago, IL: American Institute of Steel Construction, Inc..

ANSI. 1982. *Standard minimum design loads for buildings and other structures.* A58.1. New York: American National Standards Institute.

ASSE. 1991. *CSP refresher guide.* Des Plaines, IL: American Society of Safety Engineers, 1991.

ASSE. 1998. *The dictionary of terms used in the safety profession.* 3rd ed. Des Plaines, IL: American Society of Safety Engineers.

Bennett, C. S. 1967. *College physics.* New York: Harper and Row.

Brauer, R. L. 1994. *Safety and health for engineers.* New York: Van Nostrand Reinhold.

Buol, S. W., R. J. Southard, F. D. Hole, and R. J. McCracken. 1980. *Soil genesis and classification.* Ames, IA: Iowa State University, 1980.

Giachino, J. W., and W. Weeks. *Welding skills.* Homewood, IL: American Technical Publishers, 1985.

Heisler, S. I. *The Wiley engineer's desk reference.* New York: John Wiley and Sons, 1984.

Hitchcock, R. T., and R. M. Patterson. 1995. *Radio-frequency and ELF electromagnetic energies.* New York: Van Nostrand Reinhold.

Kehew, A. E. *Geology for engineers and environmental scientists.* Englewood Cliffs, NJ: Prentice-Hall, 1995.

Levy, M., and M. Salvadori. 1992. *Why buildings fall down: How structures fail.* New York: W. W. Norton and Co.

Mansdorf, S. Z. 1993. *Complete manual of industrial safety.* Englewood Cliffs, NJ: Prentice Hall.

Merritt, F. S., ed. *Standard handbook for civil engineers.* 3rd ed. New York: McGraw-Hill, 1983.

NIOSH. 2005. *Common hearing loss prevention terms.* National Institute for Occupational Safety and Health, Atlanta, GA.

OSHA. *Excavation standard.* 29 CFR 1926.650-652.

Shleien, B., L. Slaback, and B. Birky. 1998. *Handbook of health physics and radiological health.* Baltimore: Lippincott Williams and Wilkins.

Tapley, B., ed. 1990. *Eshbach's handbook of engineering fundamentals.* 4th ed. New York: John Wiley and Sons.

United States Department of Energy. 2009. PNNL hoisting and rigging manual—Slings. http://www.pnl.gov/contracts/hoist_rigging/slings.asp (accessed January 5, 2009).

Urquhart, L. *Civil engineering handbook.* 4th ed. New York: McGraw-Hill, 1959.

Winegardner, D. C. 1996. *An introduction to soils for environmental professionals.* Boca Raton, FL: CRC/Lewis Publishers.

Fire Science Concepts

Fuel + Oxidative Agent → Oxidation Products + Heat

Although technical knowledge about flame, heat, and smoke continues to grow, and although additional information continues to be acquired concerning the ignition, combustibility, and flame propagation of various solids, liquids, and gases, it still is not possible to predict with any degree of accuracy the probability of fire initiation or consequences of such initiation. Thus, while the study of controlled fires in laboratory situations provides much useful information, most unwanted fires happened and develop under widely varying conditions, making it virtually impossible to compile complete bodies of information from actual unwanted fire situations. This fact is further complicated because the progress of any unwanted fire varies from the time of discovery to the time when control measures are applied.

—Cote and Bugbee, *Principles of Fire Protection*, 1991

They will take your hand and lead you to the pearls of the desert, those secret wells swallowed by oyster crags of wadi, underground caverns that bubble rusty salt water you would sell your own mothers to drink.

—Holman 1998

INTRODUCTION

Industrial facilities are not immune to fire and its terrible consequences. Each year fire-related losses in the United States are considerable. According to conservative figures reported by Brauer (1994), about 1 million fires involving structures and about 8,000 deaths occur each year. The total annual property loss is more than $7 billion. Complicating the fire problem is the point that Cote and Bugbee (1991) made earlier—the unpredictability of fire. Fortunately, facility safety engineers are aided in their efforts in fire prevention and control by the authoritative and professional guidance readily available from the National Fire Protection Association (NFPA), the National Safety Council (NSC), Fire Code agencies, local fire authorities, and OSHA regulations.

Along with providing fire prevention guidance, OSHA regulates several aspects of fire prevention and emergency response in the workplace. Emergency response, evacuation, and fire prevention plans are required under OSHA's 29 CFR 1910.38. The requirements for fire extinguishers and worker training are addressed in 29 CFR 1910.157. In conjunction with state and municipal authorities, OSHA has listed several fire safety requirements for general industry.

All of the advisory and regulatory authorities approach fire safety in much the same manner. For example, they all agree that electrical short circuits or malfunctions usually start fires in the workplace. Other leading causes of workplace fires are friction heat, welding and cutting of metals, improperly stored flammable/combustible materials, open flames, and cigarette smoking.

In this chapter, we not only discuss the fundamental science and engineering related to fires to develop an appreciation of the principles on which fire regulations and fire protection systems are based but we also discuss basic water hydraulics. This makes sense when you consider that the primary method of fighting fires is by discharging an appropriate amount of water until the fire is extinguished. The water may have an additive, such as concentrated foam or a wetting agent, added to it, but we are still supplying and pumping water for fire fighting. Thus, the supply and movement of water for fire protection is an applied branch of mathematics known as *hydraulics* that every safety engineer needs to be with familiar with.[1]

Key Terms

aerosol: The liquid and solid particulates in smoke.

autoignition temperature: The lowest temperature at which a vapor/air mixture within its flammability limits can undergo spontaneous ignition.

automatic suppression system: A fire suppression system that activates upon detection of a fire without the need for human intervention.

building code: A set of requirements intended to ensure that an acceptable level of safety (including fire safety) is incorporated into a building at the time of construction.

chemical asphyxiants: Toxicants that depress the central nervous system, causing loss of consciousness and ultimately death.

class A fire: Fire involving common solids such as wood, paper, and some plastics and textiles.

class B fire: Fire involving liquids or solids (such as thermoplastics) that produce combustible liquids when they burn.

class C fire: Fire involving class A or class B materials as well as electrical equipment.

class D fire: Fire involving combustible metals.

code: Comprehensive set of requirements intended to address fire safety in a facility. Codes may reference numerous standards.

combustible liquid: A liquid with flash point equal to or greater than 100°F (38°C).

conductive heat transfer: The transfer of heat within solid materials or between solid materials in contact.

convected heat: The portion of heat released in a fire that is carried away by the hot products of combustion.

convective heat transfer: The transfer of heat between a gas and a solid surface at different temperatures.

deflagration: The rapid propagation of a flame through a vapor/air mixture following ignition.

extinguish: Completely put out a fire.

fire: A process entailing rapid oxidative, exothermic reactions in which part of the released energy sustains the process.

fire dynamics: The interaction among the complex phenomena involved in a building fire.

fire-fighting foam: A fire-fighting medium that is created by adding a foaming agent to a liquid (usually water).

fire point: The lowest temperature at which flaming can be sustained at the liquid's surface.

fire resistance: Ability of a building assembly to withstand the passage of flame and the transmission of excessive heat and to stay in place when exposed to fire.

fire signature: A property of fire (temperature, smoke concentration, etc.) that issued to detect the presence of fire.

fire tetrahedron: A schematic representation of fire in which the four elements required to initiate and maintain fire (fuel, oxidant, heat, and chain reactions) are depicted as the four corners of a tetrahedron.

flame detector: A detector that is activated by electromagnetic radiation emitted by flames.

flammable liquid: A liquid whose flash point is less than 100°F (38°C).

flashover: The very sudden transition from the burning of a few objects in a room to full involvement of all combustibles in the room.

flash point: The lowest temperature at which flames will flash across a liquid's surface. The measurement of a flash point is usually done in a loosely sealed vessel referred to as a closed cup.

halon: A hydrocarbon in which some or all of the hydrogen atoms are replaced by fluorine, chlorine, or bromine atom.

heat detector: A detector that activates when its sensing element reaches a critical temperature or rate-of-rise temperature.

heat of combustion: The heat released when a combustible material undergoes complete combustion.

heat transfer: The flow of heat from regions of high temperatures to regions of low temperature.

lower flammability limit (LFL): The lowest concentration of a vapor/air mixture that can be ignited by a pilot.

piloted ignition: Ignition of a vapor/air mixture by a small ignition source.

smoke: The airborne particulates (solid and liquid) and gases produced when a material undergoes pyrolysis or combustion.

thermal decomposition (pyrolysis): A process whereby chemical bonds within macromolecules forming a solid are broken by heat and flammable vapors are released.

thermoplastic: A polymeric solid that melts at a temperature lower than its ignition temperature.

upper flammability limit (UFL): The highest concentration of a vapor/air mixture that can be ignited by a pilot.

FIRE TETRAHEDRON

Figure 12.1 depicts the components required for the existence of a fire and thereby suggests how to prevent and suppress fires. Simply, fire is a process entailing a rapid oxidative and exothermic reactions in which part of the released energy sustains the process. For fire to occur, a fuel is necessary to provide a source of material for the exothermic reaction. Oxygen in the air (or other oxidant) must be present. Heat must be present for the fire to be initiated (or sustained). This heat can be in the form of a spark, a flame, or a heated environment. Finally, the relative concentration of fuel vapors and oxygen entering the flame must be appropriate to initiate or sustain the complex chain reactions (free radicals) that typify flame chemistry.

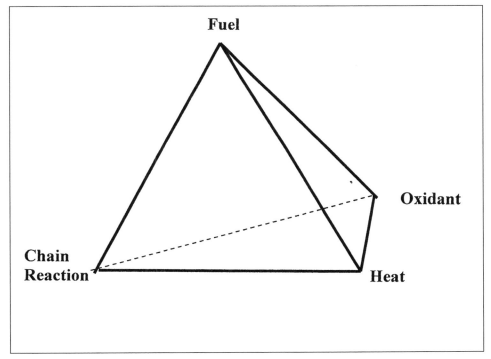

Figure 12.1   The fire tetrahedron. The four elements required to initiate and maintain fire are shown as the four corners of the fire tetrahedron.

Did You Know?

If just one of the four components of the fire tetrahedron is absent, fire cannot be initiated or sustained.

Note: Because oxygen is naturally present in most earth environments, fire hazards usually involve the mishandling of fuel or heat.

FIRE: A CHEMICAL REACTION

To gain a better perspective of the chemical reaction known as *fire*, remember that the combustion reaction normally occurs in the gas phase; generally, the oxidizer is air. This is represented (along with exothermic reactions and released energy) according to the following word equation:

$$\text{Fuel} + \text{Oxidizing Agent} \rightarrow \text{Oxidation Products} + \text{Heat}$$

If the fuel is C_xH_y (i.e., a hydrocarbon), and the oxidizing agent is oxygen, O_2, combustion of the fuel is described by the chemical equation:

$$C_xH_y + (x + y/4)\, O_2 \rightarrow x\, CO_2 + y/2\, H_2O + \Delta H_c$$

This equation assumes complete combustion; that is, the fuel is entirely consumed and the products of combustion include only carbon dioxide, CO_2, water vapor, H_2O, plus heat. The heat released is an oxidative reaction, ΔH_c, and is referred to as the heat of combustion.

If a flammable gas is mixed with air, there is a minimum gas concentration below which ignition will not occur. That concentration is known as the *lower flammable limit (LFL)*. When trying to visualize LFL and its counterpart, *UFL (upper flammable limit)*, it helps to use an example that most people are familiar with—the combustion process that occurs in the automobile engine. When an automobile engine has a gas/air mixture that is below the LFL, the engine will not start because the mixture is too lean. When the same engine has a gas/air mixture that is above the UFL, it will not start because the mixture is too rich (the engine is flooded). However, when the gas/air mixture is between the LFL and UFL levels, the engine should start (Spellman 1996).

TYPE OF FUEL

Vapors, liquids, or solids can provide the fuel for a flame. However, since flaming combustion is a gas phase phenomenon, liquids and solids must first be vaporized to generate vapors for the flames. As the combustion of vapors is common to flames associated with the burning of vapors, liquids, and solids, the flaming combustion of vapors is discussed first.

- **vapor:** Two scenarios exist in which the fuel for flames is in the form of vapor. If the vapor and the oxidant are intimately mixed before combustion, the flame is referred to as a premixed flame. Not all concentrations of vapor in air will ignite. As mentioned, a vapor of a specified material will ignite with air within upper and lower flammability limits (UFL and LFL). Again, above the UFL, the mixture is too rich (the carburetor is flooded) in fuel to sustain combustion, and below the LFL too little fuel is present to maintain heat generation at a level high enough to sustain the reaction. Because a vapor mixture cannot be ignited if it remains below its LFL (mixture is too lean), a common fire prevention measure is to provide adequate mechanical ventilation in areas where combustible vapors may be present. This measure ensures that the vapor concentration never reaches the LFL.
- **liquid:** Fires involving liquids are different from those involving gases because vaporization of the liquid must occur before fire can begin. The thermodynamic properties of the liquid (e.g., equilibrium vapor pressure as a function of temperature) as well as the heat and mass transfer characteristics of the situation determine the potential for combustion. A liquid can only be ignited by a pilot if its vaporization rate is sufficient to ensure that the vapor/air mixture at the liquid's surface is within the flammability limits of the fuel.

  Liquids are classified according to the lowest temperature at which air and vapor mixtures will combust at its surface. This temperature is called the flash point of the liquid. At the flash point, flames flash across the liquid surface, but sustained flaming may not result. The fire point is a somewhat higher temperature at which burning will be sustained once the vapors have been ignited. The difference between the flash point and the fire point relates to the complex manner in which energy is generated in the reaction in the gaseous phase and transferred back to the liquid surface, thereby vaporizing more potentially combustible material.

Flammable liquids have a flash point below 100°F. Both flammable and combustible liquids are divided into the three classifications shown below and in figure 12.2 (NFPA 1981).

Flammable Liquids

Class I-A	Flash point below 73°F, boiling point below 100°F
Class I-B	Flash point below 73°F, boiling point at or above 100°F
Class I-C	Flash point at or above 73°F, but below 100°F

Combustible Liquids

Class II	Flash point at or above 100°F, but below 140°F
Class III-A	Flash point at or above 140°F, but below 200°F
Class III-B	Flash point at or above 200°F

- **solid:** The mechanisms involved with burning of solids are more varied and more complex than those for gas phase or liquid fuels. As with liquids, flaming combustion involves conversion of the solid to a vapor, which then becomes involved in the exothermic reaction that is characteristic of the flame. In addition, some solids can undergo smoldering in which the solid material is oxidized directly.

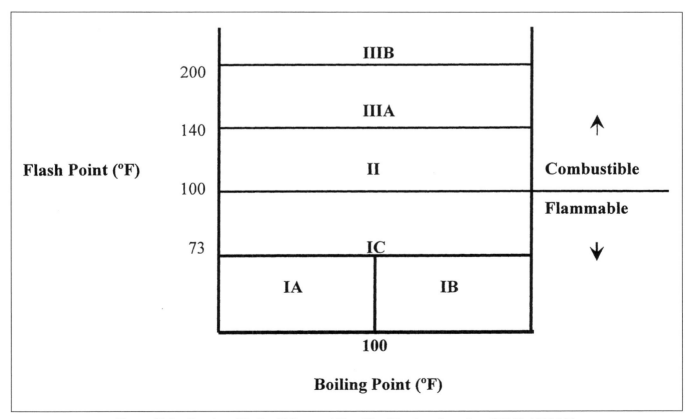

Figure 12.2 Combustible liquid flammability. Classifications based on OSHA 1910.106.

WATER HYDRAULICS

At the present time, water is by far the most commonly used fire suppression agent. Although limitations are associated with the use of water for this purpose, water has many strong advantages. In addition to its widespread availability and low cost, water has a number of physical properties that make it a good choice to suppress or extinguish fires. The objective in suppressing fires with water is to deliver water to the fire in such a state that the fire can be quenched by cooling; quenching by cooling is one of water's most significant physical properties.

Because of the importance of water in fire science, it is important for safety engineers to have a solid foundational knowledge of water hydraulics. This is the case, of course, because the primary concern of safety engineers and fire protection/suppression professionals is moving water from one location to another.

What Is Water Hydraulics?

The word *hydraulic* is derived from the Greek words *hydro* (meaning "water") and *aulis* (meaning "pipe"). Originally, the term *hydraulics* referred only to the study of water at rest and in motion (flow of water in pipes or channels). Today it is taken to mean the flow of any "liquid" in a system.

What is a liquid? In terms of hydraulics, a liquid can be either oil or water. In fluid power systems used in modern industrial equipment, the hydraulic liquid of choice is oil. Some common examples of hydraulic fluid power systems include automobile braking and power steering systems, hydraulic elevators, and hydraulic jacks or lifts. Probably the most familiar hydraulic fluid power systems in many industrial operations are used on dump trucks, front-end loaders, graders, and earth-moving and excavations equipment. In this text, we are concerned with liquid water.

Experience has shown that many safety practitioners (and others) find the study of water hydraulics difficult and puzzling (especially on certification examination questions), but we know from our experience that it is not mysterious or difficult. It is the function or output of practical applications of the basic principles of water physics.

Basic Concepts

Air Pressure (at sea level) = 14.7 pounds per square inch (psi)

The relationship shown above is important because our study of hydraulics begins with air. A blanket of air, many miles thick, surrounds the earth. The weight of this blanket on a given square inch of the earth's surface will vary according to the thickness of the atmospheric blanket above that point. As shown above, at sea level, the pressure exerted is 14.7 pounds per square inch (psi). On a mountaintop, air pressure decreases because the blanket is not as thick.

$$1 \text{ ft}^3 \text{ H}_2\text{O} = 62.4 \text{ lb}$$

The relationship shown above is also important: both cubic feet and pounds are used to describe a volume of water. There is a defined relationship between these two methods of measurement. The specific weight of water is defined relative to a cubic foot. One cubic foot of water weighs 62.4 pounds (see figure 12.3). This relationship is true only at a temperature of 4°C and at a pressure of one atmosphere [known as standard temperature and pressure (STP)—14.7 lbs per square inch at sea level containing 7.48 gallons]. The weight varies so little that, for practical purposes, this weight is used from a temperature 0°C to 100°C. One cubic inch of water weighs 0.0362 pounds. Water one foot deep will exert a pressure of 0.43 pounds per square inch on the bottom area (12 in x 0.0362 lb/in³). A column of water two feet high exerts 0.86 psi, one ten-feet high exerts 4.3 psi, and one fifty-five-feet high exerts

$$55 \text{ ft x } 0.43 \text{ psi/ft} = 23.65 \text{ psi}$$

A column of water 2.31 feet high will exert 1.0 psi. To produce a pressure of 50 psi requires a water column:

$$50 \text{ psi x } 2.31 \text{ ft/psi} = 115.5 \text{ ft}$$

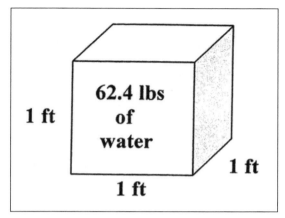

Figure 12.3   One cubic foot of water weighs 62.4 lbs.

Note: The important points being made here are

1. 1 ft³ H_2O = 62.4 lb (see figure 12.3)
2. A column of water 2.31 ft high will exert 1.0 psi

Another relationship is also important:

$$1 \text{ gallon } H_2O = 8.34 \text{ pounds}$$

As mentioned, at standard temperature and pressure, one cubic foot of water contains 7.48 gallons. With these two relationships, we can determine the weight of one gallon of water. This is accomplished by

$$\text{wt. of gallon of water} = 62.4 \text{ lb} \div 7.48 \text{ gal} = 8.34 \text{ lb/gal}$$

Thus,

$$1 \text{ gallon } H_2O = 8.34 \text{ pounds}$$

Note: This information allows cubic feet to be converted to gallons by simply multiplying the number of cubic feet by 7.48 gal/ft³.

Example 12.1
Problem:
Find the number of gallons in a reservoir that has a volume of 855.5 ft³.

Solution:

$$855.5 \text{ ft}^3 \times 7.48 \text{ gal/ft}^3 = 6,399 \text{ gallons (rounded)}$$

Note: *Head* is used to designate water pressure in terms of the height of a column of water in feet. For example, a 10 ft column of water exerts 4.3 psi. This can be called 4.3 psi pressure or 10 feet of head.

Stevin's Law

Stevin's Law deals with water at rest. Specifically, the law states: "The pressure at any point in a fluid at rest depends on the distance measured vertically to the free surface and the density of the fluid." Stated as a formula, this becomes

$$p = w \times h \tag{12.1}$$

Where

p = pressure in pounds per square foot (psf)
w = density in pounds per cubic foot (lb/ft^3)
h = vertical distance in feet

Example 12.2
Problem:
What is the pressure at a point 18 feet below the surface of a reservoir?

Solution:

Note: To calculate this, we must know that the density of the water, w, is 62.4 pounds per cubic foot.

p = w x h
  = 62.4 lb/ft^3 x 18 ft
  = 1123 lb/ft^2 or 1123 psf

Practitioners in the field generally measure pressure in pounds per square *inch* rather than pounds per square *foot*; to convert, divide by 144 in^2/ft^2 (12 in x 12 in = 144 in^2):

$$P = \frac{1123 \text{ psf}}{144 \text{ in}^2/\text{ft}^2} = 7.8 \text{ lb/in}^2 \text{ or psi (rounded)}$$

PROPERTIES OF WATER

Table 12.1 shows the relationship between temperature, specific weight, and density of water.

Density and Specific Gravity

When we say that iron is heavier than aluminum, we say that iron has greater density than aluminum. In practice, what we are really saying is that a given volume of iron is heavier than the same volume of aluminum.

> **Note:** What is density? *Density* is the *mass per unit volume* of a substance.

Suppose you had a tub of lard and a large box of cold cereal, each having a mass of 600 grams. The density of the cereal would be much less than the density of the lard because the cereal occupies a much larger volume than the lard occupies.

Table 12.1 Water Properties (Temperature, Specific Weight, and Density)

Temperature (°F)	Specific Weight (lb/ft^3)	Density (slugs/ft^3)	Temperature (°F)	Specific Weight (lb/ft^3)	Density (slugs/ft^3)
32	62.4	1.94	130	61.5	1.91
40	62.4	1.94	140	61.4	1.91
50	62.4	1.94	150	61.2	1.90
60	62.4	1.94	160	61.0	1.90
70	62.3	1.94	170	60.8	1.89
80	62.2	1.93	180	60.6	1.88
90	62.1	1.93	190	60.4	1.88
100	62.0	1.93	200	60.1	1.87
110	61.9	1.92	210	59.8	1.86
120	61.7	1.92			

The density of an object can be calculated by using the formula:

$$\text{Density} = \frac{\text{Mass}}{\text{Volume}} \qquad (12.2)$$

The most common measures of density are pounds per cubic foot (lb/ft³) and pounds per gallon (lb/gal).

- 1 cu ft of water weights 62.4 lbs—Density = 62.4 lb/cu/ft
- One gallon of water weighs 8.34 lbs—Density = 8.34 lb/gal

The density of a dry material, such as cereal, lime, soda, and sand, is usually expressed in pounds per cubic foot. The density of a liquid, such as liquid alum, liquid chlorine, or water, can be expressed either as pounds per cubic foot or as pounds per gallon. The density of a gas, such as chlorine gas, methane, carbon dioxide, or air, is usually expressed in pounds per cubic foot.

As shown in table 12.1, the density of a substance like water changes slightly as the temperature of the substance changes. This occurs because substances usually increase in volume (size—they expand) as they become warmer. Because of this expansion with warming, the same weight is spread over a larger volume, so the density is lower when a substance is warm than when it is cold.

While density is the mass per unit of a volume, specific gravity is the weight (or density) of a substance compared to the weight (or density) of an equal volume of water. [Note: The specific gravity of water is 1.] This relationship is easily seen when a cubic foot of water, which weighs 62.4 pounds as shown earlier, is compared to a cubic foot of aluminum, which weights 178 pounds. Aluminum is 2.7 times as heavy as water.

It is not that difficult to find the specific gravity of a piece of metal. All you have to do is to weigh the metal in air, then weigh it under water. Its loss of weight is the weight of an equal volume of water. To find the specific gravity, divide the weight of the metal by its loss of weight in water.

$$\text{Specific Gravity} = \frac{\text{Weight of Substance}}{\text{Weight of Equal Volume of Water}} \qquad (12.3)$$

Example 12.3
Problem:
Suppose a piece of metal weighs 150 pounds in air and 85 pounds under water. What is the specific gravity?

Solution:
Step 1: 150 lb - 85 lb = 65 lb loss of weight in water
Step 2:

$$\text{specific gravity} = \frac{150}{65} = 2.3$$

Note: In a calculation of specific gravity, it is *essential* that the densities be expressed in the same units.

As stated earlier, the specific gravity of water is one (1), which is the standard, the reference that all other liquid or solid substances are compared. Specifically, any object that has a specific gravity greater than 1 will sink in water (rocks, steel, iron, grit, floc, sludge). Substances with a specific gravity of less than 1 will float (wood, scum, and gasoline). Considering the total weight and volume of a ship, its specific gravity is less than 1; therefore, it can float.

The most common use of specific gravity is in gallons-to-pounds conversions. In many cases, the liquids being handled have a specific gravity of 1.00 or very nearly 1.00 (between 0.98 and 1.02), so 1.00 may be used in the calculations without introducing significant error. However, in calculations involving a liquid with a specific gravity of less than 0.98 or greater than 1.02, the conversions from gallons to pounds must consider specific gravity. The technique is illustrated in the following example.

Example 12.4
Problem:
There are 1,455 gallons of a certain liquid in a basin. If the specific gravity of the liquid is 0.94, how many pounds of liquid are in the basin?

Solution:
Normally, for a conversion from gallons to pounds, we would use the factor 8.34 lb/gal (the density of water) if the substance's specific gravity is between 0.98 and 1.02. However, in this instance the substance has a specific gravity outside this range, so the 8.34 factor must be adjusted.
    Multiply 8.34 lb/gal by the specific gravity to obtain the adjusted factor:

Step 1:   (8.34 lb/gal) (0.94) = 7.84 lb/gal (rounded)
Step 2:   Then convert 1,455 gal to pounds using the corrected factor:
          (1,455 gal) (7.84 lb/gal) = 11,407 lb (rounded)

Principles of Water Flow in Pipes and Hoses

Principle 1: If the pipe or hose size remains constant, water velocity within a system will be constant.

Principle 2: Within the same system, an increase in pipe or hose diameter will result in a reduction in water velocity.

Principle 3: Within the same system, a reduction in pipe or hose size will result in an increase in water velocity.

Principle 4: If pipe or hose size within a system remains constant, water flowing uphill will travel with the same velocity as water flowing downhill.

Force and Pressure

Water exerts force and pressure against the walls of its container, whether it is stored in a tank or flowing in a pipeline. There is a difference between force and pressure, though they are closely related. Force and pressure are defined below.
    Force is the push or pull influence that causes motion. In the English system, force and weight are often used in the same way. The weight of a cubic foot of water is 62.4 pounds. The force exerted on the bottom of a one-foot cube is 62.4 pounds (see figure 12.3). If we stack two cubes on top of one another, the force on the bottom will be 124.8 pounds.
    Pressure is a force per unit of area. In equation form, this can be expressed as:

$$P = \frac{F}{A} \tag{12.4}$$

Where
        P = pressure
        F = force
        A = area over which the force is distributed

    Earlier we pointed out that pounds per square inch or pounds per square foot are common expressions of pressure. Again, the pressure on the bottom of the cube is 62.4 pounds per square foot (see figure 12.3). It is normal to express pressure in pounds per square inch (psi). We pointed out that this is easily accomplished by determining the weight of one square inch of a cube one foot high. If we have a cube that is 12 inches on each side, the number of square inches on the bottom surface of the cube is 12 x 12 = 144 in². Dividing the weight by the number of square inches determines the weight on each square inch.

$$psi = \frac{62.4 \text{ lb/ft}}{144 \text{ in}^2} = 0.433 \text{ psi/ft}$$

This is the weight of a column of water one inch square and one foot tall. If the column of water were two feet tall, the pressure would be 2 ft x 0.433 psi/ft = 0.866 psi.

Note: 1 foot of water = 0.433 psi

With the above information, feet of head can be converted to psi by multiplying the feet of head times 0.433 psi/ft.

Example 12.5
Problem:
A tank is mounted at a height of 90 feet. Find the pressure at the bottom of the tank.

Solution:

$$90 \text{ ft x } 0.433 \text{ psi/ft} = 39 \text{ psi (rounded)}$$

Note: To convert psi to feet, you would divide the psi by 0.433 psi/ft.

Example 12.6
Problem:
Find the height of water in a tank if the pressure at the bottom of the tank is 22 psi.

Solution:

$$\text{height in feet} = \frac{22 \text{ psi}}{0.433 \text{ psi/ft}} = 51 \text{ ft (rounded)}$$

Note: One of the problems encountered in a hydraulic system is storing the liquid. Unlike air, which is readily compressible and is capable of being stored in large quantities in relatively small containers, a liquid such as water cannot be compressed. Therefore, it is not possible to store a large amount of water in a small tank—62.4 lbs of water occupies a volume of one cubic foot, regardless of the pressure applied to it.

HYDROSTATIC PRESSURE

Before safety engineers can begin to understand the principles of water movement for the purpose of fire extinguishments, they must first master the principles associated with water at rest, also known as *hydrostatics*.

Figure 12.4 shows a number of differently shaped, connected, open containers of water. Note that the water level is the same in each container, regardless of the shape or size of the container. This occurs because pressure is developed, within water (or any other liquid), by the weight of the water above. If the water level in any one container were to be momentarily higher than that in any of the other containers, the higher pressure at the bottom of this container would cause some water to flow into the container having the lower liquid level. In addition, the pressure of the water at any level (such as Line T) is the same in each of the containers. Pressure increases because of the weight of the water. The farther down from the surface, the more pressure is created. This illustrates that the *weight*, not the volume, of water contained in a vessel determines the pressure at the bottom of the vessel.

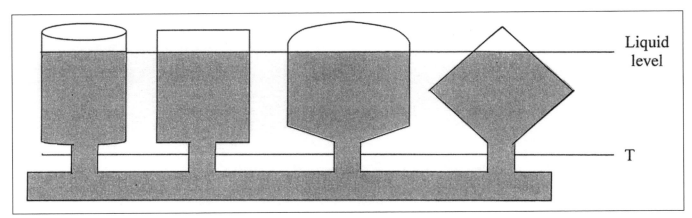

Figure 12.4 Hydrostatic pressure.

Nathanson (1997) points out some very important principles that always apply for hydrostatic pressure.

1. The pressure depends only on the depth of water above the point in question (not on the water surface area).
2. The pressure increases in direct proportion to the depth.
3. The pressure in a continuous volume of water is the same at all points that are at the same depth.
4. The pressure at any point in the water acts in all directions at the same depth.

Effects of Water under Pressure

Hauser (1996) points out that water under pressure and in motion can exert tremendous forces inside a pipeline. One of these forces, called hydraulic shock or *water hammer,* is the momentary increase in pressure that occurs when there is a sudden change of direction or velocity of the water.

When a rapidly closing valve suddenly stops water flowing in a pipeline, pressure energy is transferred to the valve and pipe wall. Shock waves are set up within the system. Waves of pressure move in horizontal yo-yo fashion—back and forth—against any solid obstacles in the system. Neither the water nor the pipe will compress to absorb the shock, which may result in damage to pipes, valves, and shaking of loose fittings.

Another effect of water under pressure is called thrust. *Thrust* is the force that water exerts on a pipeline as it rounds a bend. As shown in figure 12.5, thrust usually acts perpendicular (at 90°) to the inside surface it pushes against. As stated, it affects bends, but also reducers, dead ends, and tees. Uncontrolled, the thrust can cause movement in the fitting or pipeline, which will lead to separation of the pipe coupling away from both sections of pipeline or at some other nearby coupling upstream or downstream of the fitting.

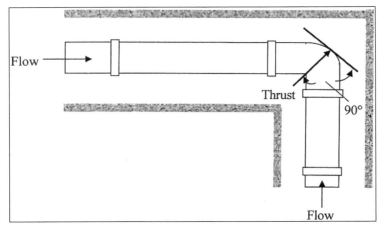

Figure 12.5 Shows direction of thrust in a pipe in a trench (viewed from above).

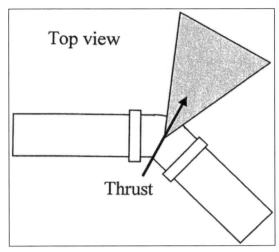

Figure 12.6   Thrust block.

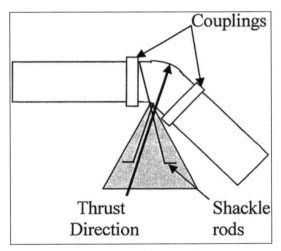

Figure 12.7   Thrust anchor.

There are two types of devices commonly used to control thrust in larger pipelines: thrust blocks and/or thrust anchors. A *thrust block* is a mass of concrete cast in place onto the pipe and around the outside bend of the turn. An example is shown in figure 12.6. These are used for pipes with tees or elbows that turn left or right or slant upward. The thrust is transferred to the soil through the larger bearing surface of the block.

A *thrust anchor* is a massive block of concrete, often a cube, cast in place below the fitting to be anchored (see figure 12.7). As shown in figure 12.7, imbedded steel shackle rods anchor the fitting to the concrete block, effectively resisting upward thrusts.

The size and shape of a thrust control device depends on pipe size, type of fitting, water pressure, water hammer, and soil type.

HEAD

Head is defined as the vertical distance the water must be lifted from the supply tank to the discharge, or as the height a column of water would rise due to the pressure at its base. Specifically, head is pressure expressed in units of feet of water instead of pounds per square inch (psi). A perfect vacuum plus atmospheric pressure of 14.7 psi would lift the water 34 feet. If the top of the sealed tube is opened to the atmosphere and the reservoir is enclosed, the pressure in the reservoir is increased; the water will rise in the tube. Because atmospheric pressure is essentially universal, we usually ignore the first 14.7 psi of actual pressure measurements and measure only the difference between the water pressure and the atmospheric pressure; we call this *gauge pressure*. For example, water in an open reservoir is subjected to the 14.7 psi of atmospheric pressure, but subtracting this 14.7 psi leaves a gauge pressure of 0 psi. This shows that the water would rise 0 feet above the reservoir surface. If the gauge pressure in a water main were 120 psi, the water would rise in a tube connected to the main:

$$120 \text{ psi} \times 2.31 \text{ ft/psi} = 277 \text{ ft (rounded)}$$

The *total head* includes the vertical distance the liquid must be lifted (static head), the loss to friction (friction head), and the energy required to maintain the desired velocity (velocity head).

$$\text{Total Head} = \text{Static Head} + \text{Friction Head} + \text{Velocity Head} \qquad (12.5)$$

Static head is the actual *vertical* distance the liquid must be lifted.

$$\text{Static Head} = \text{Discharge Elevation - Supply Elevation} \qquad (12.6)$$

Example 12.7
Problem:
The supply tank is located at elevation 118 feet. The discharge point is at elevation 215 feet. What is the static head in feet?

Solution:
$$\text{Static Head, ft} = 215 \text{ ft - } 118 \text{ ft} = 97 \text{ feet}$$

Friction head is the equivalent distance of the energy that must be supplied to overcome friction. Engineering references include tables showing the equivalent vertical distance for various sizes and types of pipes, fittings, and valves. The total friction head is the sum of the equivalent vertical distances for each component.

$$\text{Friction Head, ft} = \text{Energy Losses Due to Friction} \qquad (12.7)$$

Velocity head is the equivalent distance of the energy consumed in achieving and maintaining the desired velocity in the system.

$$\text{Velocity Head, ft} = \text{Energy Losses to Maintain Velocity} \qquad (12.8)$$

Total Dynamic Head (Total System Head)

$$\text{Total Head} = \text{Static Head} + \text{Friction Head} + \text{Velocity Head} \qquad (12.9)$$

Pressure/Head

The pressure exerted by water is directly proportional to its depth or head in the pipe, tank, or channel. If the pressure is known, the equivalent head can be calculated.

$$\text{Head, ft} = \text{Pressure, psi} \times 2.31 \text{ ft/psi} \qquad (12.10)$$

Example 12.8
Problem:
The pressure gauge on the discharge line from the influent pump reads 72.3 psi. What is the equivalent head in feet?

Solution:

$$\text{Head, ft} = 72.3 \times 2.31 \text{ ft/psi} = 167 \text{ ft}$$

Head/Pressure

If the head is known, the equivalent pressure can be calculated by

$$\text{Pressure, psi} = \frac{\text{Head, ft}}{2.31 \text{ ft/psi}} \qquad\qquad (12.11)$$

Example 12.9
Problem:
The tank is 22 feet deep. What is the pressure in psi at the bottom of the tank when it is filled with water?

Solution:

$$\text{Pressure, psi} = \frac{22 \text{ ft}}{2.31 \text{ ft/psi}} = 9.52 \text{ psi (rounded)}$$

POTENTIAL ENERGY

Hydrostatic pressure is potential energy. Potential energy is stored energy that can perform work once it is released. The static body of water within a water distribution system is subject to potential energy due to elevation and potential energy due to external pressure sources, such as pumps. The potential energy possessed by water due to its elevation above a reference elevation is given by the following equation.

$$\text{P.E.} = mgh \qquad\qquad (12.12)$$

Where

P.E. = potential energy (ft-lb)
m   = mass (slugs = W [lb]/g [ft/s^2])
g   = the gravitational constant (32.2 ft/s^2)
h   = the height or elevation (ft)

Example 12.10
Problem:
How much potential energy is stored in a 200,000 gal. water gravity tank whose average elevation above ground level is 125 ft? Recall: Density of water = 62.4 lb/ft^3; 7.48 gals = 1 ft^3.

Solution:

W   = (62.4 lb/ft^3)(200,000 gal) / 7.48 gal/ft^3)
    = 1,668,449 lbs
h   = 125 ft
P.E. = mgh = Wh
P.E. = 2.08 x 10^8

Principles of Hydrostatics

Principle 1: The pressure at a point in a liquid is applied equally in every direction.

Principle 2: Pressure applied on a confined liquid from an external source will be transmitted equally in all directions throughout the liquid without a reduction in magnitude.

Principle 3: The pressure created by a liquid in an open container is directly proportional to the depth of the liquid.

Principle 4: The pressure created by a liquid in an open container is proportional to the density of the liquid.

Principle 5: The pressure at the bottom of a container is not affected by the shape or volume of a container.

FLOW/DISCHARGE RATE: WATER IN MOTION

The study of fluid flow (hydrokinetics) is much more complicated than that of fluids at rest, but it is important to have an understanding of these principles because the water in an "activated" fire suppression system is in motion.

Discharge (or flow) is the quantity of water passing a given point in a pipe or channel during a given period. Stated another way for open channels: The flow rate through an open channel is directly related to the velocity of the liquid and the cross-sectional area of the liquid in the channel.

$$Q = A \times V \tag{12.13}$$

Where

Q = flow—discharge in cubic feet per second (cfs)
A = cross-sectional area of the pipe or channel (ft^2)
V = water velocity in feet per second (fps or ft/sec)

Example 12.11
Problem:
The channel is 6 feet wide and the water depth is 3 feet. The velocity in the channel is 4 feet per second. What is the discharge or flow rate in cubic feet per second?

Solution:

$$\text{Flow, cfs} = 6 \text{ ft} \times 3 \text{ ft} \times 4 \text{ ft/second} = 72 \text{ cfs}$$

Discharge or flow can be recorded as gallons/day (gpd), gallons/minute (gpm), or cubic feet (cfs). Flows directly from many waterworks plants are large and are often referred to in million gallons per day (MGD). The discharge or flow rate can be converted from cfs to other units such as gallons per minute (gpm) or million gallons per day (MGD) by using appropriate conversion factors.

Example 12.12
Problem:
A pipe 12 inches in diameter has water flowing through it at 10 feet per second. What is the discharge in a. cfs, b. gpm, and c. MGD? Before we can use the basic formula (12.14), we must determine the area A of the pipe. The formula for the area of a circle is

$$A = \pi \times \frac{D^2}{4} = \pi \times r^2 \tag{12.14}$$

(π is the constant value 3.14159 or simply 3.14)

Where

D = diameter of the circle in feet
r = radius of the circle in feet

Therefore, the area of the pipe is:

$$A = \pi \frac{D^2}{4} = 3.14 \times \frac{(1 \text{ ft})^2}{4} = 0.785 \text{ ft}^2$$

Now we can determine the discharge in cfs (part a):

$$Q = V \times A = 10 \text{ ft/sec} \times 0.785 \text{ ft}^2 = \textbf{7.85} \text{ ft}^3/\text{sec or cfs}$$

For part b, we need to know that 1 cubic foot per second is 449 gallons per minute, so 7.85 cfs x 449 gpm/cfs = **3525 gpm** (rounded).

Finally, for part c, one million gallons per day is 1.55 cfs, so

$$\frac{7.85 \text{ cfs}}{1.55 \frac{\text{cfs}}{\text{MGD}}} = 5.06 \text{ MGD}$$

Note: Flow may be *laminar* (streamlined—see figure 12.8) or *turbulent* (see figure 12.9). Laminar flow occurs at extremely low velocities. The water moves in straight parallel lines, called *streamlines*, or *laminae*, which slide upon each other as they travel, rather than mixing up. Normal pipe flow is turbulent flow, which occurs because of friction encountered on the inside of the pipe. The outside layers of flow are thrown into the inner layers; the result is that all the layers mix and are moving in different directions and at different velocities. However, the direction of flow is forward.

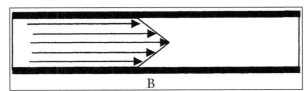

Figure 12.8   Laminar (streamline) flow.

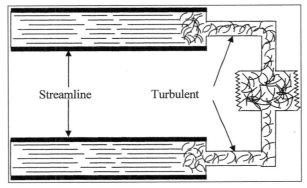

Figure 12.9   Turbulent flow.

Note: Flow may be steady of unsteady. For our purposes, we consider steady state flow only; that is, most of the hydraulic calculations in this manual assume steady state flow.

Kinetic Energy

As mentioned, we typically associate potential energy with a body of water at rest. Once the water is set into motion, its velocity creates *kinetic energy*. It is important to point out, however, that even when water is in motion, it is possible that some potential energy will still be present.

Area/Velocity

The *law of continuity* states that the discharge at each point in a pipe or channel is the same as the discharge at any other point (if water does not leave or enter the pipe or channel). That is, under the assumption of steady state flow, the flow that enters the pipe or channel is the same flow that exits the pipe or channel. In equation form, this becomes

$$Q_1 = Q_2 \text{ or } A_1V_1 = A_2V_2 \qquad (12.15)$$

Note: In regard to the area/velocity relationship, the equation (12.15) also makes clear that for a given flow rate the velocity of the liquid varies indirectly with changes in the cross-sectional area of the channel or pipe.

Example 12.13
Problem:
A pipe 12 inches in diameter is connected to a 6 inch diameter pipe. The velocity of the water in the 12 inch pipe is 3 fps. What is the velocity in the 6 inch pipe?

Solution:
Using the equation $A_1V_1 = A_2V_2$, we need to determine the area of each pipe:

$$12 \text{ in.: } A = \pi \times \frac{D^2}{4}$$

$$= 3.14 \times \frac{(1 \text{ ft})^2}{4}$$

$$= 0.785 \text{ ft}^2$$

$$6 \text{ in.: } A = 3.14 \times \frac{(0.5)^2}{4}$$

$$= 0.196 \text{ ft}^2$$

The continuity equation now becomes

$$(0.785 \text{ ft}^2) \times (3 \frac{\text{ft}}{\text{sec}}) = (0.196 \text{ ft}^2) \times V_2$$

Solving for V_2:

$$V_2 = \frac{(0.785 \text{ ft}^2) \times (3 \text{ ft/sec})}{(0.196 \text{ ft}^2)}$$

$$= 12 \text{ ft/sec or fps}$$

Pressure/Velocity

In a closed pipe flowing full (under pressure), the pressure is indirectly related to the velocity of the liquid. This principle, when combined with the principle discussed in the previous section, forms the basis for several flow measurement devices (Venturi meters and rotameters).

$$\text{Velocity}_1 \times \text{Pressure}_1 = \text{Velocity}_2 \times \text{Pressure}_2 \qquad (12.16)$$

or

$$V_1 P_1 = V_2 P_2$$

PIEZOMETRIC SURFACE AND BERNOULLI'S THEOREM

To keep water distribution systems operating properly and efficiently, you must understand the basics of hydraulics—the laws of force, motion, and others. As stated previously, many applications of hydraulics in water distribution systems involve water in motion—in pipes under pressure or in open channels under the force of gravity. The volume of water flowing past any given point in the pipe or channel per unit time is called the flow rate or discharge—or just *flow*.

In regard to flow, *continuity of flow* and the *continuity equation* have been discussed (i.e., equation 12.16). Along with the continuity of flow principle and continuity equation, the law of conservation of energy, piezometric surface, and Bernoulli's theorem (or principle) are also important to our study of water hydraulics.

Conservation of Energy

Many of the principles of physics are important to the study of hydraulics. When applied to problems involving the flow of water, few of the principles of physical science are more important and useful to us than the Law of Conservation of Energy. Simply, the Law of Conservation of Energy states that energy can neither be created nor destroyed, but it can be converted from one form to another. In a given closed system, the total energy is constant. What this means is that any change in the potential energy of a system must be matched by a corresponding change in kinetic energy.

Energy Head

As mentioned, two types of energy, kinetic and potential, and three forms of mechanical energy exist in hydraulic systems: potential energy due to elevation, potential energy due to pressure, and kinetic energy due to velocity. Energy has the units of foot-pounds (ft-lbs). It is convenient to express hydraulic energy in terms of *energy head*, in feet of water. This is equivalent to foot-pounds per pound of water (ft lb/lb = ft).

Piezometric Surface

We have seen that when a vertical tube, open at the top, is installed onto a vessel of water, the water will rise in the tube to the water level in the tank. The water level to which the water rises in a tube is the *piezometric surface*. That is, the piezometric surface is an imaginary surface that coincides with the level of the water to which water in a system would rise in a *piezometer* (an instrument used to measure pressure).

The surface of water that is in contact with the atmosphere is known as *free water surface*. Many important hydraulic measurements are based on the difference in height between the free water surface and some point in the water system. The *piezometric surface* is used to locate this free water surface in a vessel, where it cannot be observed directly.

To understand how a piezometer actually measures pressure, consider the following example.

If a clear, see-through pipe is connected to the side of a clear glass or plastic vessel, the water will rise in the pipe to indicate the level of the water in the vessel. Such a see-through pipe, the piezometer, allows you to see the level of the top of the water in the pipe; this is the piezometric surface.

In practice, a piezometer is connected to the side of a tank or pipeline. If the water-containing vessel is not under pressure (as is the case in figure 12.10), the piezometric surface will be the same as the free water surface in the vessel, just as it would if a drinking straw (the piezometer) were left standing a glass of water.

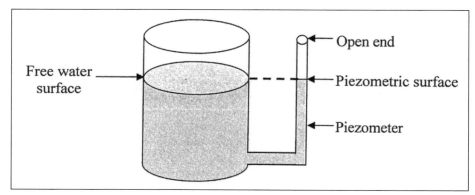

Figure 12.10   A container not under pressure in which the piezometric surface is the same as the free water surface in the vessel.

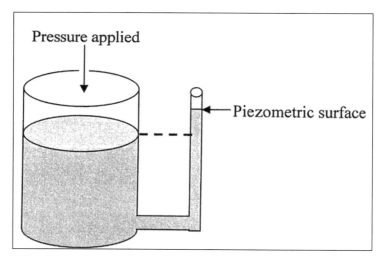

Figure 12.11   A container under pressure in which the piezometric surface is above the level of the water in the tank.

When pressurized in a tank and pipeline system, as they often are, the pressure will cause the piezometric surface to rise above the level of the water in the tank. The greater the pressure, the higher the piezometric surface (see figure 12.11). An increased pressure in a water pipeline system is usually obtained by elevating the water tank.

Note: In practice, piezometers are not installed on water towers because water towers are hundreds of feet high, or on pipelines. Instead, pressure gages are used to record pressure in feet of water or in psi.

Water only rises to the water level of the main body of water when it is at rest (static or standing water). The situation is quite different when water is flowing. Consider, for example, an elevated storage tank supplying a distribution system pipeline. When the system is at rest, all valves closed, all the piezometric surfaces are the same height as the free water surface in storage. On the other hand, when the valves are opened and the water begins to flow, the piezometric surface changes. This is an important point because as water continues to flow down a pipeline, less and less pressure is exerted. This happens because some pressure is lost (used up) keeping the water moving over the interior surface of the pipe (friction). The pressure that is lost is called *head loss*.

Head Loss

Head loss is best explained by example. Figure 12.12 shows an elevated storage tank feeding a distribution system pipeline. When the valve is closed (figure 12.12 [A]), all the piezometric surfaces are the same height as the free water surface in storage. When the valve opens and water begins to flow (figure 12.12 [B]), the piezometric surfaces *drop*. The farther along the pipeline, the lower the piezometric surface, because some of the pressure is used up keeping the water moving over the rough interior surface of the pipe. Thus, pressure is lost and is no longer available to push water up in a piezometer; this is the head loss.

Hydraulic Grade Line (HGL)

When the valve is opened in figure 12.12, flow begins with a corresponding energy loss due to friction. The pressures along the pipeline can measure this loss. In figure 12.12 (B), the difference in pressure heads between sections 1, 2, and 3 can be seen in the piezometer tubes attached to the pipe. A line connecting the water surface in the tank with the water levels at section 1, 2, and 3 shows the pattern of continuous pressure loss along the pipeline. This is called the *hydraulic grade line (HGL)* or *hydraulic gradient* of the system. [Note: It is important to point out that in a static water system, the HGL is always horizontal. The HGL is a very useful graphical aid when analyzing pipe flow problems.]

Note: Changes in the piezometric surface occur when water is flowing.

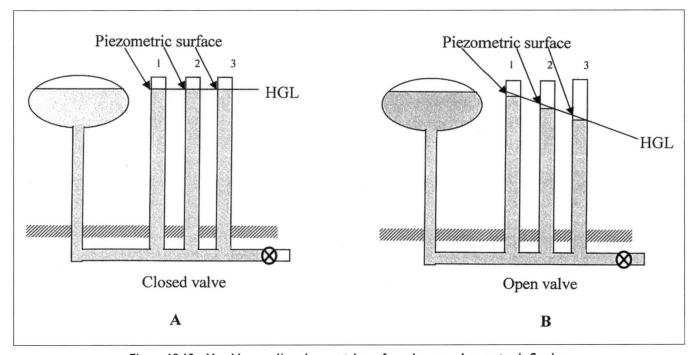

Figure 12.12 Head loss and/or piezometric surface changes when water is flowing.

Bernoulli's Theorem

When applied to the flow of water through a hydraulic system, the Law of Conservation of Energy takes on a derivative form known as *Bernoulli's theorem*. Nathanson (1997) points out that Swiss physicist and mathematician Samuel Bernoulli developed the calculation for the total energy relationship from point to point in a steady state fluid system in the 1700s. Before discussing Bernoulli's energy equation, it is important to understand the basic principle behind Bernoulli's equation.

As mentioned, water (and any other hydraulic fluid) in a hydraulic system possesses two types of energy—kinetic and potential. *Kinetic energy* is present when the water is in motion. The faster the water moves, the more kinetic energy is used. *Potential energy* is a result of the water pressure. The *total energy* of the water is the sum of the kinetic and potential energy. Bernoulli's principle states that the total energy of the water (fluid) always remains constant. Therefore, when the water flow in a system increases, the pressure must decrease. When water starts to flow in a hydraulic system, the pressure drops. When the flow stops, the pressure rises again. The pressure gages shown in figure 12.13 indicate this balance more clearly.

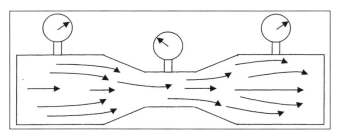

Figure 12.13   Bernoulli's principle.

Note: The basic principle explained above ignores friction losses from point to point in a fluid system employing steady state flow.

Bernoulli's Equation

In a hydraulic system, total energy head is equal to the sum of three individual energy heads. This can be expressed as

Total Head = Elevation Head + Pressure Head + Velocity Head

Where

elevation head—pressure due to the elevation of the water
pressure head—the height of a column of water that a given hydrostatic pressure in a system could support
velocity head—energy present due to the velocity of the water

This can be expressed mathematically as

$$E = z + \frac{p}{w} + \frac{v^2}{2g} \tag{12.17}$$

Where

E = total energy head
z = height of the water above a reference plane, ft
p = pressure, psi
w = unit weight of water, 62.4 lb/ft^3
v = flow velocity, ft/s
g = acceleration due to gravity, 32.2 ft/s^2

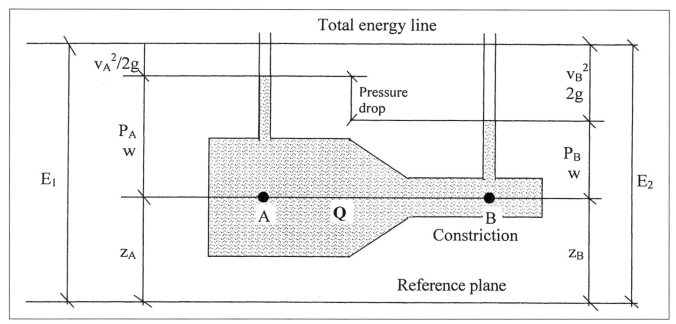

Figure 12.14 **The result of the law of conservation. Since the velocity and the kinetic energy of the water flowing in the constricted section must increase, the potential energy may decrease. This is observed as a pressure drop in the constriction.**

Consider the constriction in the section of pipe shown in figure 12.14. We know, based on the law of energy conservation, that the total energy head at section A, E_1, must equal the total energy head at section B, E_2, and using equation 12.17, we get Bernoulli's equation.

$$z_A = \frac{P_A}{w} + \frac{v_A^2}{2g} = z_B + \frac{P_B}{w} + \frac{V_B^2}{2g} \qquad (12.18)$$

The pipeline system shown in figure 12.14 is horizontal; therefore, we can simplify Bernoulli's equation because $z_A = z_B$.

Because they are equal, the elevation heads cancel out from both sides, leaving:

$$\frac{P_A}{w} + \frac{v_A^2}{2g} + \frac{P_B}{w} + \frac{v_B^2}{2g} \qquad (12.19)$$

As water passes through the constricted section of the pipe (section B), we know from continuity of flow that the velocity at section B must be greater than the velocity at section A because of the smaller flow area at section B. This means that the velocity head in the system increases as the water flows into the constricted section. However, the total energy must remain constant. For this to occur, the pressure head, and therefore the pressure, must drop. In effect, pressure energy is converted into kinetic energy in the constriction.

The fact that the pressure in the narrower pipe section (constriction) is less than the pressure in the bigger section seems to defy common sense. However, it does follow logically from continuity of flow and conservation of energy. The fact that there is a pressure difference allows measurement of flow rate in the closed pipe.

Example 12.14
Problem:
In figure 12.14, the diameter at section A is 8 in. and at section B, it is 4 in. The flow rate through the pipe is 3.0 cfs and the pressure at section A is 100 psi. What is the pressure in the constriction at section B?

Solution:
    Step 1: Compute the flow area at each section, as follows:

$$A_A = \frac{\pi(0.666 \text{ ft})^2}{4} = 0.349 \text{ ft}^2 \text{ (rounded)}$$

$$A_B = \frac{\pi(0.333 \text{ ft})^2}{4} = 0.087 \text{ ft}^2$$

Step 2: From Q = A x V or V = Q/A, we get

$$V_A = \frac{3.0 \text{ ft}^3/\text{s}}{0.349 \text{ ft}^2} = 8.6 \text{ ft/s (rounded)}$$

$$V_B = \frac{3.0 \text{ ft}^3/\text{s}}{0.087 \text{ ft}^2} = 34.5 \text{ ft/s (rounded)}$$

Step 3: We get

$$\frac{100 \times 144}{62.4} + \frac{8.6^2}{2 \times 32.2} = \frac{p_B \times 144}{62.4} + \frac{34.5^2}{2 \times 32.2}$$

Note: The pressures are multiplied by 144 in²/ft² to convert from psi to lb/ft² to be consistent with the units for w; the energy head terms are in feet of head.

Continuing, we get

$$231 + 1.15 = 2.3p_B + 18.5$$

and

$$p_B = \frac{232.2 - 18.5}{2.3} = \frac{213.7}{2.3} = 93 \text{ psi (rounded)}$$

Example 12.15
Problem:
In figure 12.15, if the pressure at P_1 reads 80 psi and the gauge at P_2 reads 40 psi, what is the water velocity at P_2 if v_1 is 15 feet per second (fps)?

Solution:

$$(2.31)(P_1) + \frac{(v_1)^2}{(2)(g)} + h_1 = (2.31)(P_2) + \frac{(v_2)^2}{(2)(g)} + h_2$$

In this case, h_1 and h_2 are the same because there is no change in elevation; thus, through algebra we can synthesize Bernoulli's equation as follows:

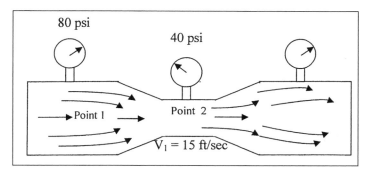

Figure 12.15   [example 12.15].

$$(2.31)(P_1 - P_2) + \frac{(v_1)^2}{(2)(g)} = \frac{(v_2)^2}{(2)(g)}$$

$(2.31)(2)(g)(P_1 - P_2) + (v_1)^2 = (v_2)^2$

$v_2 = \sqrt{(2.31)(2)(g)(P_1 - P_2)(v_1)^2}$

$v_2 = \sqrt{(2.31 \text{ ft/psi})(64.4 \text{ ft/sec}^2)(80 - 40) + 15^2 \text{ ft}^2/\text{sec}^2}$

$v_2 = \sqrt{(2.31 \text{ ft/psi})(64.4 \text{ ft/sec}^2)(40 \text{ psi}) + 225 \text{ ft}^2/\text{sec}^2}$

$v^2 = 78.6 \text{ ft/sec}$

FRICTION HEAD LOSS

Materials or substances capable of flowing cannot flow freely. Nothing flows without encountering some type of resistance. Consider electricity, the flow of free electrons in a conductor. Whatever type of conductor used (i.e., copper, aluminum, silver, etc.) offers some resistance. In hydraulics, the flow of water is analogous to the flow of electricity. Within a pipe, flowing water, like electron flow in a conductor, encounters resistance. However, resistance to the flow of water is generally termed friction loss (or more appropriately, head loss).

Flow in Pipelines

The problem of water flow in pipelines—the prediction of flow rate through pipes of given characteristics, the calculation of energy conversions therein, and so forth—is encountered in many applications of fire service water distribution operations and practice. The subject of pipe flow presented herein embraces those problems in which pipes flow completely full and also partially full.

The solution of practical pipe flow problems resulting from application of the energy principle, the equation of continuity, and the principle and equation of water resistance are also discussed. Resistance to flow in pipes is not only the result of long reaches of pipe but is also offered by pipe fittings, such as bends and valves, which dissipate energy by producing relatively large-scale turbulence.

In order to gain understanding of what friction head loss is all about, it is necessary to review a few terms presented earlier and to introduce some new terms pertinent to the subject (Lindeburg 1986).

- **laminar and turbulent flow:** *Laminar flow* is ideal flow; that is, water particles moving along straight, parallel paths, in layers or streamlines. Moreover, in laminar flow there is no turbulence in the water and no friction loss. This is not typical of normal pipe flow because the water velocity is too great. *Turbulent flow* (characterized as "normal" for a typical water system) occurs when water particles move in a haphazard fashion and continually cross each other in all directions, resulting in pressure losses along a length of pipe.

- **hydraulic grade line (HGL):** Recall that the hydraulic grade line (HGL) (shown in figure 12.16) is a line connecting two points to which the liquid would rise at various places along any pipe or open channel if piezometers were inserted in the liquid. It is a measure of the pressure head available at these various points.

Note: When water flows in an open channel, the HGL coincides with the profile of the water surface.

- **energy grade line:** The total energy of flow in any section with reference to some datum (i.e., a reference line, surface, or point) is the sum of the elevation head z, the pressure head y, and the velocity head $V^2/2g$. Figure 12.16 shows the *energy grade line* or *energy gradient*, which represents the energy from section to section. In the absence of frictional losses, the energy grade line remains horizontal, although the relative distribution of energy may vary among the elevation, pressure, and velocity heads. In all real systems, however, losses of energy occur because of resistance to flow, and the resulting energy grade line is *sloped* (i.e., the energy grade line is the slope of the specific energy line).
- **specific energy (E):** Sometimes called *specific head*, specific energy is the sum of the pressure head y and the velocity head $V^2/2g$. The specific energy concept is especially useful in analyzing flow in open channels.
- **steady flow:** This occurs when the discharge or rate of flow at any cross section is constant.
- **uniform and nonuniform flow:** *Uniform flow* occurs when the depth, cross-sectional area, and other elements of flow are substantially constant from section to section. *Nonuniform flow* occurs when the slope, cross-sectional area, and velocity change from section to section. The flow through a Venturi section used for measuring flow is a good example.
- **varied flow:** Flow in a channel is considered varied if the depth of flow changes along the length of the channel. The flow may be gradually varied or rapidly varied (i.e., when the depth of flow changes abruptly) as shown in figure 12.17.
- **slope (gradient):** The head loss per foot of channel.

MAJOR HEAD LOSS

Major head loss consists of pressure decreases along the length of pipe caused by friction created as water encounters the surfaces of the pipe. It typically accounts for most of the pressure drop in a pressurized or dynamic water system.

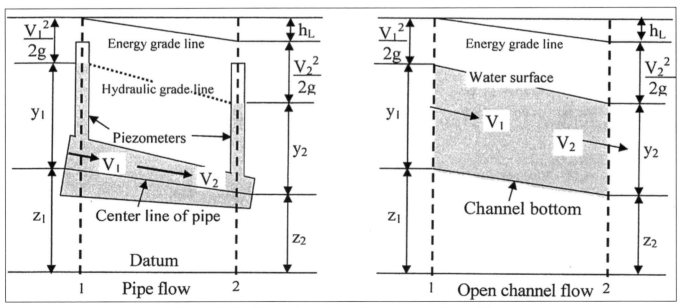

Figure 12.16   Comparison of pipe flow and open-channel flow.

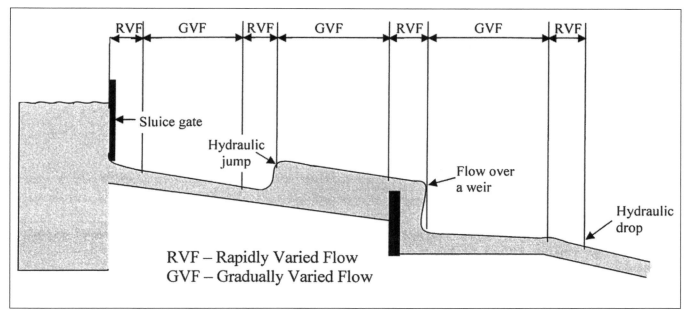

Figure 12.17 Varied flow.

Four components contribute to major head loss: roughness, length, diameter, and velocity.

1. **Roughness**
   Even when new, the interior surfaces of pipes are rough. The roughness varies, of course, depending on pipe material, corrosion (tuberculation and pitting), and age. Because normal flow in a water pipe is turbulent, the turbulence increases with pipe roughness, which, in turn, causes pressure to drop over the length of the pipe.
2. **Pipe Length**
   With every foot of pipe length, friction losses occur. The longer the pipe, the more head loss. Friction loss because of pipe length must be factored into head loss calculations.
3. **Pipe Diameter**
   Generally, small diameter pipes have more head loss than large diameter pipes. This is the case because in large diameter pipes less of the water actually touches the interior surfaces of the pipe (encountering less friction) than in a small diameter pipe.
4. **Water Velocity**
   Turbulence in a water pipe is directly proportional to the speed (or velocity) of the flow. Thus, the velocity head also contributes to head loss.

Note: For the same diameter pipe, when flow increases, head loss increases.

Calculating Major Head Loss

Darcy, Weisbach, and others developed the first practical equation used to determine pipe friction in about 1850. This equation is rarely used in fire science engineering applications; however, it is recommended for testing foam-proportioning systems according to NFPA 11, *Standard for Low-Expansion Foam.* The equation or formula now known as the Darcy-Weisbach equation for circular pipes is

$$h_f = f \frac{LV^2}{D2g} \qquad (12.20)$$

In terms of the flow rate Q, the equation becomes

$$h_f = \frac{8fLQ^2}{\pi^2 gD^5}$$

(12.21)

Where

h_f = head loss, (ft)
f = coefficient of friction
L = length of pipe, (ft)
V = mean velocity, (ft/s)
D = diameter of pipe, (ft)
g = acceleration due to gravity, (32.2 ft/s^2)
Q = flow rate, (ft^3/s)

The Darcy-Weisbach formula as such was meant to apply to the flow of any fluid, and into this friction factor was incorporated the degree of roughness and an element called the Reynolds Number, which was based on the viscosity of the fluid and the degree of turbulence of flow.

The Darcy-Weisbach formula is used primarily for determining head loss calculations in pipes. For making this determination in open channels, the Manning equation was developed during the later part of the nineteenth century. Later, this equation was used for both open channels and closed conduits.

In the early 1900s, a more practical equation, the Hazen-Williams equation, was developed for use in making calculations related to water pipes. Fire science engineering uses this formula more commonly than the Darcy-Weisbach formula because it is easier to understand and use.

$$Q = 0.435 \times CD^{2.63} \times S^{0.54}$$

(12.22)

Where

Q = flow rate, (ft^3/s)
C = coefficient of roughness (C decreases with roughness)
D = hydraulic radius R, (ft)
S = slope of energy grade line, (ft/ft)

Example 12.16
Problem:
How much pressure is lost to friction as 1,450 gpm travel through 1,000 feet of six-inch Schedule 40 steel pipe having a C of 100?

Solution:

$$P_f = \frac{(4.52)(Q)^{1.85}}{(C)^{1.85}(D)^{4.87}}$$

$$P_f = \frac{(4.52)(1450)^{1.85}}{(100)^{1.85}(6.065)^{4.87}}$$
$$P_f = 0.098 \text{ psi per foot}$$

Then the total loss over the 1,000 feet length is

$$P_f = (0.098 \text{ psi/ft})(1000 \text{ ft})$$
$$P_f = 98 \text{ psi}$$

C FACTOR

C factor, as used in the Hazen-Williams equation, designates the coefficient of roughness. *C* does not vary appreciably with velocity, and by comparing pipe types and ages, it includes only the concept of roughness, ignoring fluid viscosity and Reynolds Number.

Based on experience (experimentation), accepted tables of *C* factors have been established for pipe (see table 12.2; C factors are also listed in NFPA-24). Generally, *C* factor decreases by one with each year of pipe age. Flow for a newly designed system is often calculated with a *C* factor of 100, based on averaging it over the life of the pipe system.

Note: A high *C* factor means a smooth pipe. A low *C* factor means a rough pipe.

Note: An alternate to calculating the Hazen-Williams equation, called an alignment chart, has become quite popular for fieldwork. The alignment chart can be used with reasonable accuracy.

Note: In practice, if minor head loss is less than 5 percent of the total head loss, it is usually ignored.

Slope

Slope is defined as the head loss per foot. Slope is the amount of incline of the pipe and is calculated as feet of drop per foot of pipe length (ft/ft). Slope is designed to be just enough to overcome frictional losses, so that the velocity remains constant, the water keeps flowing, and solids will not settle in the conduit. In piped systems, where pressure loss for every foot of pipe is experienced, slope is not provided by slanting the pipe but instead by pressure added to overcome friction.

Minor Head Loss

In addition to the head loss caused by friction between the fluid and the pipe wall, losses also are caused by turbulence created by obstructions (i.e., valves and fittings of all types) in the line, changes in direction, and changes in flow area.

Table 12.2 C Factors (Adaptation from Lindeburg 1986)

Type of Pipe	C Factor
Asbestos Cement	140
Brass	140
Brick Sewer	100
Cast Iron	
10 years old	110
20 years old	90
Ductile Iron (cement-lined)	140
Concrete or Concrete-lined	
Smooth, steel forms	140
Wooden forms	120
Rough	110
Copper	140
Fire Hose (rubber-lined)	135
Galvanized Iron	120
Glass	140
Lead	130
Masonry Conduit	130
Plastic	150
Steel	
Coal-tar enamel lined	150
New unlined	140
Riveted	110
Tin	130
Vitrified	120
Wood Stave	120

BASIC PIPING HYDRAULICS

Water, regardless of the source, is typically conveyed to the waterworks for treatment and distributed to the users. Conveyance from the source to the point of treatment occurs by aqueducts, pipelines, or open channels, but the treated water is normally distributed in pressurized closed conduits. After use, whatever the purpose, the water becomes wastewater, which must be disposed of somehow but almost always ends up being conveyed back to a treatment facility before being outfalled to some water body to begin the cycle again.

We call this an urban water cycle, because it provides a human-generated imitation of the natural water cycle. Unlike the natural water cycle, however, without pipes, the cycle would be nonexistent or, at the very least, short-circuited.

For use as water mains in a distribution system for fire service flow, pipes must be strong and durable in order to resist applied forces and corrosion. The pipe is subjected to internal pressure from the water and to external pressure from the weight of the backfill (soil) and vehicles above it. The pipe may also have to withstand water hammer. Damage due to corrosion or rusting may also occur internally because of the water quality or externally because of the nature of the soil conditions.

Pipes used in a fire flow system must be strong and durable to resist the abrasive and corrosive properties of the water. Like other buried pipes, fire service flow pipes must also be able to withstand stresses caused by the soil backfill material and the effect of vehicles passing above the pipeline. Joints between pipe sections should be flexible, but tight enough to prevent excessive leakage.

Of course, pipes must be constructed to withstand the expected conditions of exposure, and pipe configuration systems for water distribution and fire service flow systems must be properly designed and installed in terms of water hydraulics. Because safety engineers should have a basic knowledge of water hydraulics related to commonly used standard piping configurations, piping basics are briefly discussed in this section.

PIPING NETWORKS

It would be far less costly and make for more efficient operation if municipal water distribution systems were built with separate single-pipe networks extending from waterworks to water storage facilities. Unfortunately, this ideal single-pipe scenario is not practical for real-world applications. Instead of a single piping system, a network of pipes is laid under the streets. Each of these piping networks is composed of different materials that vary (sometimes considerably) in diameter, length, and age. These networks range in complexity to varying degrees, and each of these joined-together pipes contribute energy losses to the system.

Water flow networks may consist of pipes arranged in series, parallel, or some complicated combination. In any case, an evaluation of friction losses for the flows is based on energy conservation principles applied to the flow junction points. Methods of computation depend on the particular piping configuration. In general, however, they involve establishing a sufficient number of simultaneous equations or employing a friction-loss formula where the friction coefficient depends only on the roughness of the pipe (e.g., Hazen-Williams equation). [Note: Demonstrating the procedure for making these complex computations is beyond the scope of this text. We only present engineers with "need to know" aspects of complex or compound piping systems in this text.]

When two pipes of different sizes or roughness are connected in series (see figure 12.18), head loss for a given discharge, or discharge for a given head loss, may be calculated by applying the appropriate equation between the bonding points, taking into account all losses in the interval. Thus, head losses are cumulative.

Series pipes may be treated as a single pipe of constant diameter to simplify the calculation of friction losses. The approach involves determining an "equivalent length" of a constant diameter pipe that has the same friction loss and discharge characteristics as the actual series pipe system. In addition, application of the continuity equation to the solution allows the head loss to be expressed in terms of only one pipe size.

Note: In addition to the head loss caused by friction between the water and the pipe wall, losses also are caused by minor losses: obstructions in the line, changes in directions, and changes in flow area. In practice, the method of equivalent length is often used to determine these losses. The method of equivalent length uses a table to convert each valve or fitting into an equivalent length of straight pipe.

In making calculations involving pipes in series, remember these two important basic operational tenets:

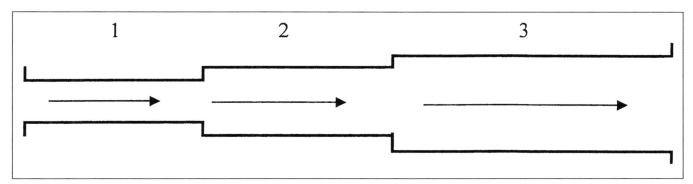

Figure 12.18 Pipes in series.

1. The same flow passes through all pipes connected in series.
2. The total head loss is the sum of the head losses of all of the component pipes.

In some operations involving series networks where the flow is given and the total head loss is unknown, we can use the Hazen-Williams equation to solve for the slope and the head loss of each pipe as if they were separate pipes. Adding up the head losses to get the total head loss is then a simple matter.

Other series network calculations may not be as simple to solve using the Hazen-Williams equation. For example, one problem we may be faced with is what diameter to use with varying sized pipes connected together in a series combination. Moreover, head loss is applied to both pipes (or other multiples), and it is not known how much loss originates from each one; thus, determining slope would be difficult—but not impossible.

In such cases the equivalent pipe theory, as mentioned earlier, can be used. Again, one single "equivalent pipe" is created that will carry the correct flow. This is practical because the head loss through it is the same as that in the actual system. The equivalent pipe can have any *C* factor and diameter, just as long as those same dimensions are maintained all the way through to the end. Keep in mind that the equivalent pipe must have the correct length, so that it will allow the correct flow-through, which yields the correct head loss (the given head loss) (Lindeburg 1986).

Two or more pipes connected (as in figure 12.19) so that flow is first divided among the pipes and is then rejoined comprise a parallel pipe system. A parallel pipe system is a common method for increasing the capacity of an existing line. Determining flows in pipes arranged in parallel are also made by application of energy conservation principles—specifically, energy losses through all pipes connecting common junction points must be equal. Each leg of the parallel network is treated as a series piping system and converted to a single equivalent length pipe. The friction losses through the equivalent length parallel pipes are then considered equal and the respective flows determined by proportional distribution.

Note: Computations used to determine friction losses in parallel combinations may be accomplished using a simultaneous solution approach for a parallel system that has only two branches. However, if the parallel system has three or more branches, a modified procedure using the Hazen-Williams loss equation is easier.

FLOW MEASUREMENT

For many years, *differential pressure flowmeters* have been the most widely applied flow-measuring device for water flow in pipes that required accurate measurement at reasonable cost. The differential pressure type of flowmeter makes up the largest segment of the total flow measurement devices currently being used.

This type of device has a flow restriction in the line that causes a differential pressure of "head" to be developed between the two measurement locations. Differential pressure flowmeters are also known as head meters, and, of all the head meters, the *orifice flowmeter* is the most widely applied device. For example, specific orifices that have standard diameters are used as sprinklers.

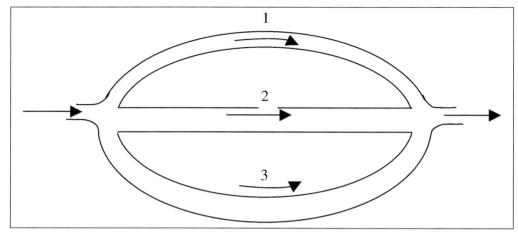

Figure 12.19   Pipes in parallel.

WATER HYDRAULICS CALCULATIONS

When safety engineers sit for various professional practice certification or licensure examinations, they should not be surprised to encounter several mathematical operations, including water hydraulics calculations, on such tests. In this section, we have included a representative sample of the most common types of math questions.

Pressure Calculation

$$P = \frac{F}{a}$$

Where

P = pressure (pounds per square inch—psi)
F = hydrostatic force (lbs)
a = area (in²)

Example 12.17
Problem:
A 200,000 gallon suction tank for a fire pump has a base diameter of 30 ft. What does a pressure gage on the suction side of the fire pump read under no-flow conditions? (Note: Assume the level of the gage is at the same elevation as the base of the tank. Also, there are 7.48 gal/ft³.)

Solution:

$$P = F/a$$
$$P = ?$$
$$F = 200,000 \text{ gals x } \text{ft}^3/7.48 \text{ gal x } 62.4 \text{ lb/ft}^3$$
$$= 1,668,449 \text{ lbs}$$
$$a = \pi r^2 = (3.14)(15\text{ft x } 12\text{in/ft})^2 = 101,736 \text{ in}^2$$
$$P = 1,668,449 \text{ lbs}/101,736 \text{ in}^2$$
$$= 16.4 \text{ psi}$$

How high is the tank?

Answer: 16.4 psi/0.433 psi/ft of head = 37.9 ft

Velocity Head (Torricelli's Equation) Calculation

$$h_v = \frac{v^2}{2\,g}$$

Where

h_v = velocity head (ft)
v = velocity (ft/s)
g = gravitational constant (32.2 ft/s²)

Example 12.18
Problem:
A water test is conducted at the base of a gravity tank whose water level is 140 ft. above grade. Neglecting friction losses, what will the velocity of a stream of water be from an open four-inch hydrant butt at ground level?

Solution:

$h_v = v^2/2g$
$h_v = 140$ ft
$v = ?$
$g = 32.2$ ft/s²
$v^2 = 2gh$
   $= [(2)(32.2 \text{ ft/s}^2)(140\text{ft})]^{1/2}$
$V = 95$ ft/s

(Note that we would get the same result if we applied the $v_2 = v_o^2 + 2as$ equation.)

Pipe Head Pressure

$$h_p = p/w$$

Where

h_p = pressure heat (ft)
p = pressure (lbs/ft² or psf)
w = density (lbs/ft³)

Example 12.19
Problem:
A pressure gage at the base of a gravity tank reads 70 psi. How high is the level of the water in the tank?

Solution:

$h_p = p/w$
$h_p = ?$
$p = 70$ psi x 144 si/sf = 10,080 psf
$w = 62.4$ lbs/ft³ (from applicable table)
$h_p = 10,080$ lbs/ft² /62.4 lbs/ft³
   $= 161.6$ ft

Note that to determine the water level in the tank we could also have used the operation 70 psi/.433 psi/ft = 161.6 ft.

Pipe Velocity Pressure

$$P_v = \frac{Q^2}{891d^4}$$

Where

P_v = velocity pressure (psi)
Q = flow rate (gpm)
d = diameter in inches (in)

Example 12.20
Problem:
What is the velocity pressure when 1400 gpm is flowing through an eight-inch diameter pipe?

Solution:

$P_v = Q^2/891d^4$
$P_v = ?$
$Q = 1400$ gpm
$d = 8$ inches
$P_v = 1400^2 / 891(8)^4$
     $= 0.54$ ft/s

Water Supply Flow Calculation

$$Q_2 = Q_1 \left[\frac{(S - R_2)^{0.54}}{(S - R_1)^{0.54}}\right]$$

Where

R_2 = the residual pressure for the predicted flow rate (psi)
Q_1 = known flow rate from a group of open hydrants with a measured residual pressure (gpm)
Q_2 = predicted flow rate at a different residual pressure, R_2, usually 20 psi (gpm)
S = the static pressure in a hydrant system with no flow from any hydrants (psi)
R_1 = the residual pressure measured at a nonflowing hydrant during a flow test (psi)

Example 12.21
Problem:
The static pressure of a hydrant system is 60 psi. When the flow from several hydrants sums up to 2200 gpm, the pressure at the nonflowing hydrant reduces to 40 psi. 1. What is the expected flow for a residual pressure of 20 psi (the minimum for fire engine use)? 2. What would the static pressure need to be to get a flow of 4000 gpm at 20 psi?

Part 1

S = 60 psi
Q_1 = 2200 gpm
R_1 = 40 psi
R_2 = 20 psi

Hence $(S - R_1) = 20$ psi and $(S - R_2) = 40$ psi.

$$Q_2 = 2200 \frac{(40)^{0.54}}{(20)^{0.54}} = 2200 \frac{7.33}{5.04} = 3200 \text{ gpm}$$

Part 2

From part 1 and the Hazen-Williams equation, we know that

$$\frac{Q_1}{(S - R_1)^{0.54}} = \frac{Q_2}{(S - R_2)^{0.54}} = \text{constant}$$

For this given system the constant is 448.
Therefore, we can rewrite the equation such that

$$(S - R_2)^{0.54} = \frac{Q_2}{\text{constant}}$$

$$S = \left(\frac{Q_2}{\text{constant}}\right)^{1.85} + R_2$$

$$S = \left(\frac{4000}{448}\right)^{1.85} + 20 = 77.4 \text{ psi}$$

Orifice Flow

$$P = (Q/K)^2$$
$$Q = K\sqrt{P}$$

Where

P = pressure (psi)
Q = flow (gpm)
K = the orifice discharge coefficient (from tables in handbooks)

Example 12.22
Problem:
A standard ½ in. orifice automatic sprinkler has a k-factor of 5.6. How much water will the head deliver if the water pressure in the system is equal to 69 psi?

Solution:

$$Q = kp^{1/2}$$
$$Q = ?$$
$$k = 5.6$$
$$p = 69 \text{ psi}$$
$$Q = 5.6(69)^{1/2}$$
$$= 46.5 \text{ gpm}$$

Example 12.23
Problem:
According to NFPA-231 (Standard for General Storage), a certain high-piled storage hazard creates a sprinkler system demand of 0.55 gpm/sq ft over 4000 sq ft. Neglecting friction losses in the system, if 17/32″ orifice sprinklers are used (k-factor of 8.0) and paced at 100 sq ft per head, how much water and at what pressure will be needed?

Solution:

$$Q_{Total} = ? \text{ gpm}$$
$$k = 8.0 \text{ gpm}$$
$$p = ? \text{ in. psi}$$
$$Q_{Total} = (0.55 \text{ gpm/sq ft})(4000 \text{ ft}^2)$$
$$= 2200 \text{ gpm}$$
$$Q_{PerHead} = (0.55 \text{ gpm/ft}^2)(100 \text{ ft}^2/\text{Head})$$
$$= 55 \text{ gpm/Head}$$
$$p = (Q/k)^2 = (55/8.0)^2$$
$$p = 47.27 \text{ psi}$$

Answer: 2200 gpm at 47 psi

Hydraulic Friction Loss

$$P_d = \frac{4.52Q^{1.85}}{C^{1.85}d^{4.87}}$$

Where

Q = flow rate (gpm)
C = the coefficient of friction
d = inside diameter of pipe (in.)
P_d = pressure in pounds per square inch per foot of pipe (psi)

Example 12.24
Problem:
What is the friction loss in psi per foot for a three-inch pipe having a C of 100 with a 400 gpm flow?

Solution:

$$P = 4.52Q^{1.85}/C^{1.85}d^{4.87}$$
$$P = ?$$
$$Q = 400 \text{ gpm}$$
$$C = 100$$
$$D = 3 \text{ inches}$$
$$P = (4.52)(400)^{1.85}/(100)^{1.85}(3)^{4.87}$$
$$= 0.28 \text{ psi per foot}$$

Hydraulic Flow—Pressure Relationship

$$\frac{Q_1}{Q_2} = \frac{\sqrt{P_1}}{\sqrt{P_2}}$$

Where

P = pressure in pounds per square inch (psi)
Q = flow in gallons per minute (gpm)

Example 12.25
Problem:
A sprinkler system was hydraulically designed such that 1500 gpm at 40 psi was needed at the top of the riser to meet a certain storage arrangement demand. The warehouse manager would like to rearrange the storage

in such a way that 1700 gpm will be required over the same area to satisfy the increase in the density of the demand. What will the pressure requirement be at the top of the riser to deliver the new flow requirement?

Solution:

$$P_2 = P_1(Q_2/Q_1)^2$$
$$P_2 = ?$$
$$P_1 = 40 \text{ psi}$$
$$Q_1 = 1500 \text{ gpm}$$
$$Q_2 = 1700 \text{ gpm}$$
$$P_2 = (40 \text{ psi})(1700 \text{ gpm}/1500 \text{ gpm})^2$$
$$\quad = 51.4 \text{ psi}$$

REVIEW QUESTIONS

1. What are the two types of energy?
2. Explain Bernoulli's theorem.
3. In a 3 inch hose (A = 0.049 ft^2), the velocity is 20 ft/sec. What is the water velocity in the 2 1/2 inch (A = 0.034 ft^2) hose that is connected to the end of the 3 inch hose, and how many gallons are flowing?
4. Find the number of gallons in a storage tank that has a volume of 660 ft^3.
5. Suppose a rock weighs 160 pounds in air and 125 pounds under water. What is the specific gravity?
6. A 110 ft diameter cylindrical tank contains 1.6 MG water. What is the water depth?
7. The pressure in a pipeline is 6400 psf. What is the head on the pipe?
8. The pressure on a surface is 35 psi. If the surface area is 1.6 sq ft, what is the force (lb) exerted on the surface?
9. Bernoulli's principle states that the total energy of a hydraulic fluid is _____.
10. What is *pressure head*?
11. What is a *hydraulic grade line*?
12. A flow of 1500 gpm takes place in a 12 inch pipe. Calculate the velocity head.
13. Water flows at 5.00 ml/sec in a 4 inch line under a pressure of 110 psi. What is the pressure head (ft of water)?
14. In question 13, what is the velocity head in the line?
15. What is velocity head in a 6 inch pipe connected to a 1 ft pipe, if the flow in the larger pipe is 1.46 cfs?
16. What is *velocity head*?
17. What is *suction lift*?
18. Explain *energy grade line.*
19. How far above the ground would the level of water in an elevated storage tank need to be in order to create a pressure of 80 psi at the bottom of the tank?
20. What pressure will be shown on a gauge at the base of an elevated water storage tank in which the water level is 140 feet above the pressure gauge?
21. What pressure will be shown on a gauge at the base of a storage tank containing oil with a specific gravity of 0.9, in which the oil level is 90 feet above the pressure gauge?
22. Define *head.*

REFERENCES AND RECOMMENDED READING

Arasmith, S. 1993. *Introduction to small water systems.* Albany, OR: ACR Publications, Inc.
AWWA. 1995. *Basic science concepts and applications: Principles and practices of water supply operations.* 2nd ed. Denver: American Water Works Association.

Barnes, R. G. 1991. Positive displacement flowmeters for liquid measurement. In *Flow measurement*, ed. D. W. Spitzer. Research Triangle Park, NC: Instrument Society of America.

Brauer, R. L. 1994. *Safety and health and engineers.* New York: Van Nostrand Reinhold.

Brown, A. E. 1991. Ultrasonic flowmeters. In *Flow measurement*, ed. D. W. Spitzer. Research Triangle Park, NC: Instrument Society of America.

Cheremisinoff, N. P., and L. Ferrante. 1983. *Pumps, compressors, and fans: Pocket handbook.* Lancaster, PA: Technomic Publishing Company.

Cote, A. E., and P. Bugbee, eds. 1991. *Principles of fire protection.* 17th ed. Quincy, MA: National Fire Protection Association.

Garay, P. N. 1990. *Pump application desk book.* Lilburn, GA: The Fairmont Press (Prentice Hall).

Grant, D. M. 1991. Open channel flow measurement. In *Flow measurement*, ed. D. W. Spitzer. Research Triangle Park, NC: Instrument Society of America.

Hauser, B. A. 1993. *Hydraulics for operators.* Boca Raton, FL: Lewis Publishers.

Hauser, B. A. 1996. *Practical hydraulics handbook.* 2nd ed. Boca Raton, FL: Lewis Publishers.

Hemond, F. H., and E. J. Fechner-Levy. 2000. *Chemical fate and transport in the environment.* 2nd ed. San Diego: Academic Press.

Holman, S. 1998. *A stolen tongue.* New York: Anchor Press, Doubleday.

Husain, Z. D., and M. J. Sergesketter. 1991. Differential pressure flowmeters. In *Flow measurement*, ed. D. W. Spitzer, 119–60. Research Triangle Park, NC: Instrument Society of America.

Kawamura, S. 2000. *Integrated design and operation of water treatment facilities.* 2nd ed. New York: John Wiley & Sons.

Krutzsch, W. C. 1976. Introduction and classification of pumps. In *Pump handbook*, eds. I. J. Karassik et al. New York: McGraw-Hill Inc.

Lindeburg, M. R. 1986. *Civil engineering reference manual.* 4th ed. San Carlos, CA: Professional Publications, Inc.

Magnusson, R. J. 2001. *Water technology in the middle ages.* Baltimore: The John Hopkins University Press.

McGhee, T. J. 1991. *Water supply and sewerage.* 2nd ed. New York: McGraw-Hill.

McKinnon, G. P., and K. Tower. 1997. *Fire protection handbook.* 18th ed. Boston: National Fire Protection Association.

Metcalf, Eddy. 1981. *Wastewater engineering: Collection and pumping of wastewater.* New York: McGraw-Hill Book Company.

Mills, R. C. 1991. Magnetic flowmeters. In *Flow measurement*, ed. D. W. Spitzer. Research Triangle Park, NC: Instrument Society of America.

Nathanson, J. A. 1997. *Basic environmental technology: Water supply, waste management, and pollution control.* 2nd ed. Upper Saddle River, NJ: Prentice Hall.

NFPA. 30. 1969. National Fire Protection Association, Quincy, MA.

Oliver, P. D. 1991. Turbine flowmeters. In *Flow measurement*, ed. D. W. Spitzer. Research Triangle Park, NC: Instrument Society of America.

Spellman, F. R. 1996. *Safe work practices for wastewater operators.* Lancaster, PA: Technomic Publishing Company.

Spellman, F. R. 2007. *The science of water.* 2nd ed. Boca Raton, FL: CRC Press.

Spellman, F. R., and J. Drinan. 2001. *Water hydraulics.* Boca Raton, FL: CRC Press.

Tapley, B., ed. 1990. *Eshbach's handbook of engineering fundamentals.* 4th ed. New York: John Wiley & Sons.

USEPA. 1991. *Flow instrumentation: A practical workshop on making them work.* Sacramento, CA: Water & Wastewater Instrumentation Testing Association.

Viessman, W., Jr., and M. J. Hammer. 1998. *Water supply and pollution control.* 6th ed. Menlo Park, CA: Addison-Wesley.

Wahren, U. 1997. *Practical introduction to pumping technology.* Houston: Gulf Publishing Company.

Water transmission and distribution. 2nd ed. 1996. Denver: American Water Works Association.

Water treatment operators short course. 2008 [1999, 2001]. Blacksburg, VA: Virginia Polytechnic Institute and State University (Virginia Tech).

(Footnote)

[1] Much of the information in this chapter is from Spellman and Drinan, *Water Hydraulics*, 2001; and CDC's *NIOSH: Fire Protection*, 2009.

13

Electrical Safety and Electricity for Engineers

$$R = p \, L/A$$

We will make electricity so cheap that only the rich will burn candles.

—Thomas Edison

ELECTRICITY

Electricity has helped make modern society possible. It not only provides light for employees to work during nondaylight hours but it also powers the machines they use to produce society's products. Today, we would have difficulty imagining ourselves without electricity; it is indispensable to our modern way of life.

For the safety engineer, electricity and electrical circuits and components are areas of major concern. The primary concern, of course, is to ensure that workers do not become part of an electrical circuit (conductors of electricity), and thus become shocked or electrocuted. A secondary (but just as important) concern is the prevention of fires caused by faulty electrical components and/or wiring. No other ignition source comes close to causing the number of industrial fires that electricity has caused. In the sections that follow, we spend some time, not with the goal of trying to turn the safety engineer into an electrician, but rather in providing fundamental information about electricity, electrical circuits, and electrical components that the safety engineer should be familiar with.

What Causes Electric Shock?

Three basic concepts are important to an understanding of electricity: voltage, current, and resistance. We use a common analogy to help us visualize the flow of electrical energy through a wire or conductor: water flowing through a pipe.

The following three factors must be present in every electric circuit. Voltage (V), (also known as electromotive force—emf) measured in volts, can be visualized as the pressure (or pump) of the water in the pipe. Current (I), measured in amperes, is comparable to the total amount of water flowing past a point per unit of time. Resistance (R), measured in ohms, is the friction or any other impediment that tends to retard the flow of the water. Resistance in an electrical conductor is proportional to its length and to the resistivity of the material of which it is made, and it is inversely proportional to its cross-sectional area.

Keep in mind that it is not the voltage in an electrical circuit that shocks and kills or causes fire or arc-blasts. Voltage (like the water pump) does not move or flow; instead, the current (like water) flows—and the current causes electrocutions and ignites fires. In general, the greater the current and the longer it is able to flow through the human body, the greater will be the injury.

How much current is needed to be fatal?

Electric shock is a sudden stimulation of the body's nervous system by an electric current (Hammer 1989), which can cause pain, injury, or death. Experience shows that as little as 100 milliamps (Ma) of 60 Hertz (Hz) alternating current (AC) can be fatal. Note that this level is well below those levels often found in average household electrical circuits, indicating that severe electrical injuries can arise from very ordinary electrical sources. In the industrial setting, however, where much higher energy sources are commonly found, the potential hazards increase dramatically.

NIOSH (1993) points out that the five principal ways that people experience electrical shock are

1. contact with a normally bare energized conductor
2. contact with a normally insulated conductor on which the insulation has deteriorated or been damaged so that it is no longer protected
3. equipment failure that results in an open circuit or short circuit, which, in turn, causes the current to flow in an unexpected manner
4. static electricity discharge
5. lightning strike

Key Terms

Some of the important electrical terms and/or concepts that the safety engineer should be familiar with include:

alternation: A variation, either positive or negative, of a waveform from zero to maximum and back to zero, equaling one half a cycle.

Ohm's Law: Ohm's Law states that the current in a metallic conductor is proportional to the electromotive force of the source if no other emfs are acting in the circuit and the resistance remains constant.

electrolyte: A substance that separates into free ions when dissolved in water is called an electrolyte. The resulting solution is able to conduct electric current (a common use is in wet cell batteries).

battery: A device in which chemical reactions produce an electric current. In a dry cell, the electrolyte is a moist paste rather than a liquid. A storage cell has a liquid electrolyte and can be recharged when passing a current through it in the reverse direction exhausts it; an exhausted dry cell cannot be recharged.

hertz: A unit of frequency; a periodic oscillation has a frequency of n hertz if in one second it goes through n cycles.

frequency: The number of cycles completed by a periodic quantity in a unit time (e.g., the number of complete cycles that occurs in one second).

fuel cell: A device that continuously changes the chemical energy of a fuel directly into electrical energy.

Kirchoff's Laws (for network analysis): 1. The sum of the currents flowing into a junction of three or more wires is equal to the sum of the currents flowing out of the junction. 2. The sum of the emfs around a closed conducting loop is equal to the sum of the IR potential drops around the loop.

conductor: Any substance able to carry electrical charges from place to place.

capacitor: A system of conductors that stores energy in the form of an electric field.

inductance: A part of the total resistance of an electrical circuit, tending to resist the flow of changing current.

self-inductance: The emf induced in a current-carrying conductor.

mutual inductance: Occurs when the magnetic field from one coil cuts through another coil.

insulator (or dielectric substance): Substances that do not have the property to conduct electricity.

charges: Like charges in an electric field always repel; unlike charges always attract.

direct current circuits: In a direct current (DC) circuit, current flows through the circuit in the same direction at all times.

alternating current circuits: In an alternating current (AC) circuit, current flow passes through a regular succession of changing positive and negative values by periodically reversing its direction of flow. Total positive and negative values of current are equal.

wavelength: The distance traveled by a C wave during one complete cycle.

phase angle: Refers to the value of the electrical angle at any point during the cycle.

vector quantity: Has magnitude and direction.

magnetic flux: Similar to current flow in Ohm's Law and comprises the total number of lines of force existing in the magnetic circuit.

magnetomotive force: Force that produces flux in a magnetic circuit.

hysteresis: Condition whereby magnetic flux in an iron core lags behind the increase or decrease of the magnetizing force.

transformers: Allow the transmission of electrical energy at high voltages, which is cheaper than transmitting at low voltages because of the smaller loss of power in the line at high voltages. They are used to step up or step down voltages.

circuit breakers: Perform two major functions in the electrical distribution system: they allow for turning on/off (isolating) a circuit and provide short-circuit current (trip) protection.

switchboards or switchgear: Major control stations containing buses, conductors, control devices, protective devices, circuit-switching interrupting devices, interconnecting wiring, accessories, supporting structures, and enclosures.

protective relays: Consist of an operating element and a set of movable contacts. The operating element receives power from a control power source within the switchgear and obtains input from a sensing element in the circuit. When the relay detects that some present limit has been exceeded, it initiates some action, such as sounding an alarm or tripping a circuit breaker.

fuses: Protect circuits by means of a link that melts from the heat caused by excessive current flow, thereby opening the circuit.

wire and cable: Used for the distribution of electric power and consists of a conducting medium, usually enclosed within an insulating sheath, and sometimes further protected by an outer jacket.

motor controllers and motor control centers: Provide the means for controlling the starting, stopping, reversing, speed, etc., of electric motors.

grounding: An essential aspect of a safe electrical system. Everything that might come in contact with a live system should be maintained at ground potential. Protective grounding must be capable of conducting the maximum fault current that could flow at the point of ground for the time necessary to clear the fault.

ELECTRICAL CONDUCTORS[1]

Electric current moves easily through some materials but with greater difficulty through others. Three good electrical conductors are copper, silver, and aluminum (generally, we can say that most metals are good conductors). At the present time, copper is the material of choice used in electrical conductors. Under special conditions, certain gases are also used as conductors (e.g., neon gas, mercury vapor, and sodium vapor are used in various kinds of lamps).

The function of the wire conductor is to connect a source of applied voltage to a load resistance with a minimum IR voltage drop in the conductor so that most of the applied voltage can produce current in the load resistance. Ideally, a conductor must have a very low resistance (e.g., a typical value for a conductor—copper—is less than 1 Ω per 10 feet).

Because all electrical circuits use conductors of one type or another, in this section we discuss basic features and electrical characteristics of the most common types of conductors. Moreover, because conductor splices and connections (and insulation of such connections) are also an essential part of any electric circuit, they are also discussed. In addition, it has been our experience that many electrical fires (the number one cause of fire in industrial workplaces) are caused by faulty conductors (e.g., short-circuit wiring; conductor insulation breakdown; incorrectly sized conductors; loose connections causing high-resistance connections that produce excessive heat). Thus, the information contained in this section on electrical conductors, connections, and splices is an important part of any safety engineer's toolbox.

Unit Size of Conductors

A standard (or unit) size of a conductor has been established to compare the resistance and size of one conductor with another. The unit of linear measurement used is (in regard to diameter of a piece of wire) the **mil** (0.001 of

---

[1] Note: In order for safety engineers to fully appreciate the benefits and dangers associated with electricity, it is important to have a foundational knowledge of electricity and a full appreciation for electrical safety. Thus, in the following sections we provide an introduction to electrical safety and to the fundamental principles of electricity, circuits, and components—all of which are important in maintaining safe workplaces. Much of the material provided is from F. R. Spellman, *Physics for Nonphysicists*, Rockville, MD: Government Institute Press, 2009.

an inch). A convenient unit of wire length used is the **foot**. Thus, the standard unit of size in most cases is the **mil-foot** (i.e., a wire will have unit size if it has diameter of 1 mil and a length of 1 foot). The resistance in ohms of a unit conductor or a given substance is called the **resistivity** (or specific resistance) of the substance.

As a further convenience, **gage** numbers are also used in comparing the diameter of wires. The B and S (Browne and Sharpe) gage was used in the past; now the most commonly used gage is the **American Wire Gage** (AWG).

Square Mil

Figure 13.1A shows a square mil. The *square mil* is a convenient unit of cross-sectional area for square or rectangular conductors. As shown in figure 13.1A, a square mil is the area of a square, the sides of which are 1 mil. To obtain the cross-sectional area in square mils of a square conductor, square one side measured in mils. To obtain the cross-sectional area in square mils of a rectangular conductor, multiply the length of one side by that of the other, each length being expressed in mils.

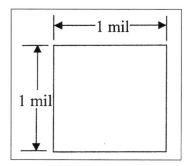

Figure 13.1A   Square mil

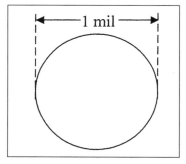

Figure 13.1B   Circular mil.

Example 13.1
Problem:
Find the cross-sectional area of a large rectangular conductor 5/8 inch thick and 5 inches wide.

Solution:
The thickness may be expressed in mils as 0.625 x 1,000 = 625 mils and the width as 5 x 1,000 = 5,000 mils. The cross-sectional area is 625 x 5,000, or 3,125,000 square mils.

Circular Mil

The *circular mil* is the standard unit of wire cross-sectional area used in most wire tables. To avoid the use of decimals (because most wires used to conduct electricity may be only a small fraction of an inch), it is convenient to express these diameters in mils. For example, the diameter of a wire is expressed as 25 mils instead of 0.025 inch. A circular mil is the area of a circle having a diameter of 1 mil, as shown in figure 13.1B. The area in circular mils of a round conductor is obtained by squaring the diameter measured in mils. Thus, a wire having a diameter of 25 mils has an area of 25^2 or 625 circular mils. By way of comparison, the basic formula for the area of a circle is

$$A = \pi R^2 \tag{13.1}$$

and in this example the area in square inches is

$$A = \pi R^2 = 3.14(0.0125)^2 = 0.00049 \text{ sq in.}$$

If D is the diameter of a wire in mils, the area in square mils can be determined using

$$A = \pi(D/2)^2 \tag{13.2}$$

which translates to

$$= 3.14/4 \, D^2$$
$$= 0.785 \, D^2 \text{ sq mils}$$

Thus, a wire 1 mil in diameter has an area of

$$A = 0.785 \times 1^2 = 0.785 \text{ sq mils,}$$

which is equivalent to 1 circular mil. The cross-sectional area of a wire in circular mils is therefore determined as

$$A = \frac{0.785 \, D^2}{0.785} = D^2 \text{ circular mils,}$$

where D is the diameter in mils. Therefore, the constant $\pi/4$ is eliminated from the calculation.

It should be noted that in comparing square and round conductors that the circular mil is a smaller unit of area than the square mil, and therefore there are more circular mils than square mils in any given area. The comparison is shown in figure 13.1C. The area of a circular mil is equal to 0.785 of a square mil.

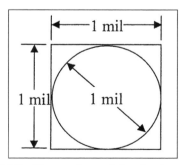

Figure 13.1C   Comparison of circular to square mil.

Did You Know?

To determine the circular-mil area when the square-mil area is given, divide the area in square mils by 0.785. Conversely, to determine the square-mil area when the circular-mil area is given, multiply the area in circular mils by 0.785.

Example 13.2
Problem:
A No. 12 wire has a diameter of 80.81 mils. What is 1. its area in circular mils and 2. its area in square mils?

Solution:

1. $A = D^2 = 80.81^2 = 6,530$ circular mils
2. $A = 0.785 \times 6,530 = 5,126$ square mils

Example 13.3
Problem:
A rectangular conductor is 1.5 inches wide and 0.25 inch thick. 1. What is its area in square mils? 2. What size of round conductor in circular mils is necessary to carry the same current as the rectangular bar?

Solution:

1. $1.5'' = 1.5 \times 1,000 = 1,500$ mils
    $0.25'' = 0.25 \times 1,000 = 250$ mils
    $A = 1,500 \times 250 = 375,000$ sq mils
2. To carry the same current, the cross-sectional area of the rectangular bar and the cross-sectional area of the round conductor must be equal. There are more circular mils than square mils in this area, and therefore

$$A = \frac{375,000}{0.785} = 477,700 \text{ circular mils}$$

Circular-Mil-Foot

As shown in figure 13.2, a *circular-mil-foot* is actually a unit of volume. More specifically, it is a unit conductor 1 foot in length and having a cross-sectional area of 1 circular mil. The circular-mil-foot is useful in making comparisons between wires that are made of different metals because it is considered a unit conductor. Because it is considered a unit conductor, the circular-mil-foot is useful in making comparisons between wires that are made of different metals. For example, a basis of comparison of the **resistivity** of various substances may be made by determining the resistance of a circular-mil-foot of each of the substances.

Note: It is sometimes more convenient to use a different unit of volume when working with certain substances. Accordingly, unit volume may also be taken as the centimeter cube. The inch cube may also be used. The unit of volume employed is given in tables of specific resistances.

Resistivity

All materials differ in their atomic structure and therefore in their ability to resist the flow of an electric current. The measure of the ability of a specific material to resist the flow of electricity is called its *resistivity*,

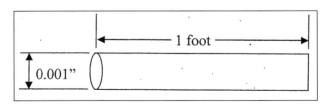

Figure 13.2   Circular-mil-foot

or specific resistance—the resistance in ohms offered by unit volume (the circular-mil-foot) of a substance to the flow of electric current. Resistivity is the reciprocal of conductivity (i.e., the ease by which current flows in a conductor). A substance that has a high resistivity will have a low conductivity, and vice versa.

The resistance of a given length, for any conductor, depends upon the resistivity of the material, the length of the wire, and the cross-sectional area of the wire according to the equation

$$R = \rho \frac{L}{A} \tag{13.3}$$

Where

R = resistance of the conductor, Ω
L = length of the wire, ft
A = cross-sectional area of the wire, CM
ρ = specific resistance or resistivity, CM x Ω/ft

The factor ρ (Greek letter rho, pronounced "roe") permits different materials to be compared for resistance according to their nature without regard to different lengths or areas. Higher values of ρ mean more resistance.

Note: The resistivity of a substance is the resistance of a unit volume of that substance.

Many tables of resistivity are based on the resistance in ohms of a volume of the substance 1 foot long and 1 circular mil in a cross-sectional area. The temperature at which the resistance measurement is made is also specified. If the kind of metal of which the conductor is made is known, the resistivity of the metal may be obtained from a table. The resistivity, or specific resistance, of some common substances are given in table 13.1.

Note: Because silver, copper, gold, and aluminum have the lowest values of resistivity, they are the best conductors. Tungsten and iron have a much higher resistivity.

Example 13.4
Problem:
What is the resistance of 1,000 feet of copper wire having a cross-sectional area of 10,400 circular mils (No. 10 wire), the wire temperature being 20°C?

Solution:
The resistivity (specific resistance), from table 13.1, is 10.37. Substituting the known values in the preceding equation (13.3), the resistance, R, is determined as

$$R = \rho \frac{L}{A} = 10.37 \times \frac{1,000}{10,400} = 1 \text{ ohm, approximately}$$

Table 13.1   Resistivity (Specific Resistance)

Substance	Specific Resistance at 20° (CM ft Ω)
Silver	9.8
Copper (drawn)	10.37
Gold	14.7
Aluminum	17.02
Tungsten	33.2
Brass	42.1
Steel (soft)	95.8
Nichrome	660.0

Wire Measurement

Wires are manufactured in sizes numbered according to a table known as the American wire gage (AWG). Table 13.2 lists the standard wire sizes that correspond to the AWG. The gage numbers specify the size of round wire in terms of its diameter and cross-sectional area. Note the following:

1. As the gage numbers increase from 1 to 40, the diameter and circular area decrease. Higher gage numbers mean smaller wire sizes. Thus, No. 12 is a smaller wire than No. 4.
2. The circular area doubles for every three gage sizes. For example, No. 12 wire has about twice the area of No. 15 wire.
3. The higher the gage number and the smaller the wire, the greater the resistance of the wire for any given length. Therefore, for 1000 ft of wire, No. 12 has a resistance of 1.62 Ω while No. 4 has 0.253 Ω.

Table 13.2   Copper Wire Table

Gage #	Diameter	Circular mils	Ohms/1,000 ft at 25°C
1	289.0	83,700.0	.126
2	258.0	66,400.0	.159
3	229.0	52,600.0	.201
4	204.0	41,700.0	.253
5	182.0	33,100.0	.319
6	162.0	26,300.0	.403
7	144.0	20,800.0	.508
8	128.0	16,500.0	.641
9	114.0	13,100.0	.808
10	102.0	10,400.0	1.02
11	91.0	8,230.0	1.28
12	81.0	6,530.0	1.62
13	72.0	5,180.0	2.04
14	64.0	4,110.0	2.58
15	57.0	3,260.0	3.25
16	51.0	2,580.0	4.09
17	45.0	2,050.0	5.16
18	40.0	1,620.0	6.51
19	36.0	1,290.0	8.21
20	32.0	1,020.0	10.4
21	28.5	810.0	13.1
22	25.3	642.0	16.5
23	22.6	509.0	20.8
24	20.1	404.0	26.4
25	17.9	320.0	33.0
26	15.9	254.0	41.6
27	14.2	202.0	52.5
28	12.6	160.0	66.2
29	11.3	127.0	83.4
30	10.0	101.0	105.0
31	8.9	79.7	133.0
32	8.0	63.2	167.0
33	7.1	50.1	211.0
34	6.3	39.8	266.0
35	5.6	31.5	335.0
36	5.0	25.0	423.0
37	4.5	19.8	533.0
38	4.0	15.7	673.0
39	3.5	12.5	848.0
40	3.1	9.9	1,070.0

Factors Governing the Selection of Wire Size

Several factors must be considered in selecting the size of wire to be used for transmitting and distributing electric power. These factors include: allowable power loss in the line; the permissible voltage drop in the line; the current-carrying capacity of the line; and the ambient temperatures in which the wire is to be used.

1. **Allowable power loss (I^2R) in the line.** This loss represents electrical energy converted into heat. The use of large conductors will reduce the resistance and therefore the I^2R loss. However, large conductors are heavier and require more substantial supports; thus, they are more expensive initially than small ones.

2. **Permissible voltage drop (IR drop) in the line.** If the source maintains a constant voltage at the input to the line, any variation in the load on the line will cause a variation in line current and a consequent variation in the IR drop in the line. A wide variation in the IR drop in the line causes poor voltage regulation at the load.

3. **The current-carrying capacity of the line.** When current is draw through the line, heat is generated. The temperature of the line will rise until the heat radiated, or otherwise dissipated, is equal to the heat generated by the passage of current through the line. If the conductor is insulated, the heat generated in the conductor is not so readily removed as it would be if the conductor were not insulated.

4. **Conductors installed where ambient temperature is relatively high.** When installed in such surroundings, the heat generated by external sources constitutes an appreciable part of the total conductor heating. Due allowance must be made for the influence of external heating on the allowable conductor current, and each case has its own specific limitations.

Copper vs. Other Metal Conductors

If it were not cost prohibitive, silver, the best conductor of electron flow (electricity), would be the conductor of choice in electrical systems. Instead, silver is used only in special circuits where a substance with high conductivity is required.

The two most generally used conductors are copper and aluminum. Each has characteristics that make its use advantageous under certain circumstances. Likewise, each has certain disadvantages, or limitations.

In regards to copper, it has a higher conductivity; it is more ductile (can be drawn out into wire), has relatively high tensile strength, and can be easily soldered. It is more expensive and heavier than aluminum.

Aluminum has only about 60 percent of the conductivity of copper, but its lightness makes possible long spans, and its relatively large diameter for a given conductivity reduces corona (i.e., the discharge of electricity from the wire when it has a high potential). The discharge is greater when smaller diameter wire is used than when larger diameter wire is used. However, aluminum conductors are not easily soldered, and aluminum's relatively large size for a given conductance does not permit the economical use of an insulation covering.

Note: Recent practice involves using copper wiring (instead of aluminum wiring) in house and some industrial applications. This is the case because aluminum connections are not as easily made as they are with copper. In addition, over the years, many fires have been started because of improperly connected aluminum wiring (i.e., poor connections = high resistance connections, resulting in excessive heat generation).

A comparison of some of the characteristics of copper and aluminum is given in table 13.3.

Table 13.3 Characteristics of Copper and Aluminum

Characteristic	Copper	Aluminum
Tensile strength (lb/in²)	55,000	25,000
Tensile strength for same conductivity (lb)	55,000	40,000
Weight for same conductivity (lb)	100	48
Cross-section for same conductivity (CM)	100	160
Specific resistance (Ω/mil. ft)	10.6	17

Table 13.4 Properties of Conducting Materials (approximate)

Material	Temperature Coefficient, $\Omega/°C$
Aluminum	0.004
Carbon	-0.0003
Constantan	0 (average)
Copper	0.004
Gold	0.004
Iron	0.006
Nichrome	0.0002
Nickel	0.005
Silver	0.004
Tungsten	0.005

Temperature Coefficient

The resistance of pure metals—such as silver, copper, and aluminum—increases as the temperature increases. The *temperature coefficient* of resistance, α (Greek letter alpha), indicates how much the resistance changes for a change in temperature. A positive value for α means R increases with temperature; a negative α means R decreases; and a zero α means R is constant, not varying with changes in temperature. Typical values of α are listed in table 13.4.

The amount of increase in the resistance of a 1 ohm sample of the copper conductor per degree rise in temperature (i.e., the temperature coefficient of resistance) is approximately 0.004. For pure metals, the temperature coefficient of resistance ranges between 0.004 and 0.006 ohm.

Thus, a copper wire having a resistance of 50 ohms at an initial temperature of 0°C will have an increase in resistance of 50 x 0.004, or 0.2 ohms (approximate) for the entire length of wire for each degree of temperature rise above 0°C. At 20°C the increase in resistance is approximately 20 x 0.2, or 4 ohms. The total resistance at 20°C is 50 + 4, or 54 ohms.

Note: As shown in table 13.4, carbon has a negative temperature coefficient. In general, α is negative for all semiconductors such as germanium and silicon. A negative value for α means less resistance at higher temperatures. Therefore, the resistance of semiconductor diodes and transistors can be reduced considerably when they become hot with normal load current. Observe, also, that constantan has a value of zero for α (table 13.4). Thus, it can be used for precision wire-wound resistors, which do not change resistance when the temperature increases.

Conductor Insulation

Electric current must be contained; it must be channeled from the power source to a useful load—safely. To accomplish this, electric current must be forced to flow only where it is needed. Moreover, current-carrying conductors must not be allowed (generally) to come in contact with one another, their supporting hardware, or personnel working near them. To accomplish this, conductors are coated or wrapped with various materials. These materials have such a high resistance that they are, for all practical purposes, nonconductors. They are generally referred to as *insulators* or *insulating materials*.

There are a wide variety of insulated conductors available to meet the requirements of any job. However, only the necessary minimum of insulation is applied for any particular type of cable designed to do a specific job. This is the case because insulation is expensive and has a stiffening effect and is required to meet a great variety of physical and electrical conditions.

Two fundamental but distinctly different properties of insulation materials (e.g., rubber, glass, asbestos, and plastics) are insulation resistance and dielectric strength.

1. *Insulation resistance* is the resistance to current leakage through and over the surface of insulation materials.
2. *Dielectric strength* is the ability of the insulator to withstand potential difference and is usually expressed in terms of the voltage at which the insulation fails because of the electrostatic stress.

Various types of materials are used to provide insulation for electric conductors, including rubber, plastics, varnished cloth, paper, silk, cotton, and enamel.

Conductor Splices and Terminal Connections

When conductors join each other, or connect to a load, *splices* or *terminals* must be used. It is important that they be properly made, since any electric circuit is only as good as its weakest connection. The basic requirement of any splice or connection is that it be both mechanically and electrically as strong as the conductor or device with which it is used. High-quality workmanship and materials must be employed to ensure lasting electrical contact, physical strength, and insulation (if required).

Note: Conductor splices and connections are an essential part of any electric circuit.

Soldering Operations

Soldering operations are a vital part of electrical and/or electronics maintenance procedures. Soldering is a manual skill that must be learned by all personnel who work in the field of electricity. Obviously, practice is required to develop proficiency in the techniques of soldering.

In performing a soldering operation both the solder and the material to be soldered (e.g., electric wire and/or terminal lugs) must be heated to a temperature that allows the solder to flow. If either is heated inadequately, cold solder joints result (i.e., high resistance connections are created). Such joints do not provide either the physical strength or the electrical conductivity required. Moreover, in soldering operations it is necessary to select a solder that will flow at a temperature low enough to avoid damage to the part being soldered or to any other part or material in the immediate vicinity.

Solderless Connections

Generally, terminal lugs and splicers that do not require solder are more widely used (because they are easier to mount correctly) than those that do require solder. Solderless connectors—made in a wide variety of sizes and shapes—are attached to their conductors by means of several different devices, but the principle of each is essentially the same. They are all crimped (squeezed) tightly onto their conductors. They afford adequate electrical contact, plus great mechanical strength.

Insulation Tape

The carpenter has his saw, the dentist his pliers, the plumber his wrench, and the electrician his insulating tape. Accordingly, one of the first things the rookie maintenance operator learns (a rookie who is also learning proper and safe techniques for performing electrical work) is the value of electrical insulation tape. Normally, the use of electrical insulating tape comes into play as the final step in completing a splice or joint to place insulation over the bare wire at the connection point.

Typically, insulation tape used should be the same basic substance as the original insulation, usually a rubber-splicing compound. When using rubber (latex) tape as the splicing compound where the original insulation was rubber, it should be applied to the splice with a light tension so that each layer presses tightly against the one underneath it. In addition to the rubber tape application (which restores the insulation to original form), restoring with friction tape is also often necessary.

In recent years, plastic electrical tape has come into wide use. It has certain advantages over rubber and friction tape. For example, it will withstand higher voltages for a given thickness. Single, thin layers of certain commercially available plastic tape will stand several thousand volts without breaking down.

SIMPLE ELECTRIC CIRCUIT

A simple electric circuit includes: an energy source (difference of potential; source of electromotive force, emf; voltage), a conductor (wire), a load, and a means of control (see figure 13.3). The energy source could be a battery, as in figure 13.3, or some other means of producing a voltage. The load that dissipates the energy could be a lamp, a resistor, or some other device that does useful work, such as an electric toaster, a computer, a power drill, radio, or a soldering iron. Conductors are wires that offer low resistance to current; they connect all the loads in the circuit to the voltage source. No electrical device dissipates energy unless current (electrons) flows through it. Because conductors, or wires, are not perfect conductors, they heat up (dissipate energy), so they are actually part of the load. For simplicity, however, we usually think of the connecting wiring as having no resistance, since it would be tedious to assign a very low resistance value to the wires every time we wanted to calculate the solution to a problem. Control devices might be switches, variable resistors, circuit breakers, fuses, or relays.

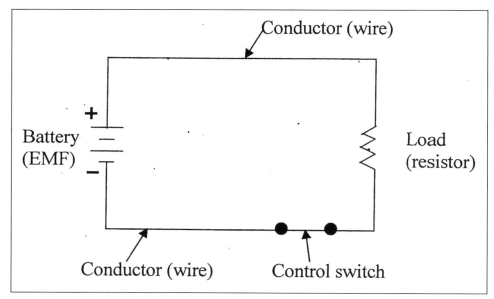

Figure 13.3   Sample DC closed circuit.

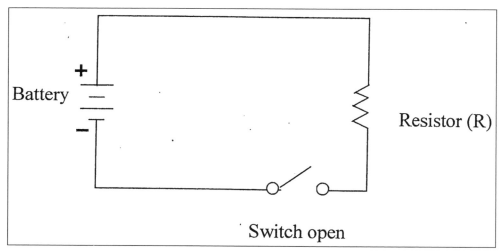

Figure 13.4   Open circuit.

A complete pathway for current flow, or closed circuit (figure 13.3), is an unbroken path for current from the emf, through a load, and back to the source. A circuit is called *open* (see figure 13.4) if a break in the circuit (e.g., open switch) does not provide a complete path for current flow.

To protect a circuit, a fuse (or circuit breaker) is placed directly into the circuit (see figure 13.5). A fuse will open the circuit whenever a dangerous, large current starts to flow (i.e., a short-circuit condition occurs, caused by an accidental connection between two points in a circuit that offers very little resistance). A fuse will permit currents smaller than the fuse value to flow but will melt and therefore break or open the circuit if a larger current flows.

Schematic Representation

The simple circuits shown in figures 13.3, 13.4, and 13.5 are displayed in schematic form. A *schematic diagram* (usually shortened to "schematic") is a simplified drawing that represents the electrical, not the physical, situation in a circuit. The symbols used in schematic diagrams are the electrician's "shorthand"; they make the diagrams easier to draw and easier to understand. Consider the symbol in figure 13.6 used to represent a battery power supply. The symbol is rather simple, straightforward—but is also very important. For example, by convention, the shorter line in the symbol for a battery represents the negative terminal. It is important to remember this because it is sometimes necessary to note the direction of current flow, which is from negative to positive, when you examine the schematic. The battery symbol shown in figure 13.6 has a single cell, so only one short and one long line are used. The number of lines used to represent a battery vary (and they are not necessarily equivalent to the number of cells), but they are always in pairs, with long and short lines alternating. In the circuit shown in figure 13.5, the current would flow in a *counterclockwise* direction; that is, in the opposite direction that a clock's hands move. If the long and short lines of the battery symbol (symbol shown in figure 13.5) were reversed, the current in the circuit shown in figure 13.5 would flow *clockwise*; that is, in the direction of a clock's hands.

Note: In studies of electricity and electronics, many circuits are analyzed that consist mainly of specially designed resistive components. As previously stated, these components are called *resistors*. Throughout the remaining analysis of the basic circuit, the resistive component will be a physical resistor. However, the resistive component could be any one of several electrical devices (e.g., light bulbs, coffee pots, toasters, hair dryers, various other appliances, etc.).

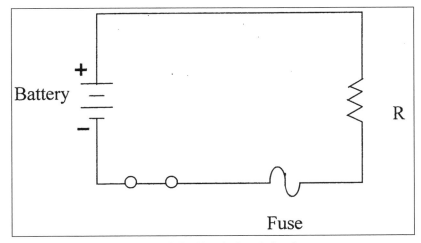

Figure 13.5   Simple fused circuit.

Figure 13.6   Schematic symbol for a battery.

Keep in mind that in the simple circuits shown in the figures to this point we have only illustrated and discussed a few of the many symbols used in schematics to represent circuit components. (Other symbols will be introduced as we need them.)

It is also important to keep in mind that a closed loop of wire (conductor) is not necessarily a circuit. A source of voltage must be included to make it an electric circuit. In any electric circuit where electrons move around a closed loop, current, voltage, and resistance are present. The physical pathway for current flow is actually the circuit. By knowing any two of the three quantities, such as voltage and current, the third (resistance) may be determined. This is done mathematically using Ohm's Law, which is the foundation on which electrical theory is based.

OHM'S LAW

Simply put, Ohm's Law defines the relationship between current, voltage, and resistance in electric circuits. Ohm's Law can be expressed mathematically in three ways.

1. The **current** (I) in a circuit is equal to the voltage applied to the circuit divided by the resistance of the circuit. Stated another way, the current in a circuit is *directly* proportional to the applied voltage and *inversely* proportional to the circuit resistance. Ohm's Law may be expressed as an equation:

$$I = \frac{E}{R} \tag{13.4}$$

Where

I = current in amperes (amps)
E = voltage in volts
R = resistance in ohms

2. The **resistance** (R) of a circuit is equal to the voltage applied to the circuit divided by the current in the circuit:

$$R = \frac{E}{I} \tag{13.5}$$

3. The applied **voltage** (E) to a circuit is equal to the product of the current and the resistance of the circuit:

$$E = I \times R = IR \tag{13.6}$$

If any two of the quantities in equations (13.4–13.6) are known, the third may be easily found. Let's look at an example.

Example 13.5
Problem:
Figure 13.7 shows a circuit containing a resistance of 6 ohms and a source of voltage of 3 volts. How much current flows in the circuit?

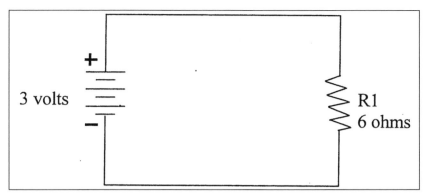

Figure 13.7   Determining current in a simple circuit.

Given:

        E = 3 volts
        R = 6 ohms
        I = ?

Solution:

$$I = \frac{E}{R}$$

$$I = \frac{3}{6}$$

I = 0.5 ampere

To observe the effect of source voltage on circuit current, we use the circuit shown in figure 13.7 but double the voltage to 6 volts.

Example 13.6
Problem:

Given:

        E = 6 volts
        R = 6 ohms
        I = ?

Solution:

$$I = \frac{E}{R}$$

$$I = \frac{6}{6}$$

I = 1 ampere

Notice that as the source of voltage doubles, the circuit current also doubles.

Note: Circuit current is directly proportional to applied voltage and will change by the same factor that the voltage changes.

To verify that current is inversely proportional to resistance, assume the resistor in figure 13.7 to have a value of 12 ohms.

Example 13.7
Problem:

Given:

$$E = 3 \text{ volts}$$
$$R = 12 \text{ ohms}$$
$$I = ?$$

Solution:

$$I = \frac{E}{R}$$

$$I = \frac{3}{12}$$

$$I = 0.25 \text{ ampere}$$

Comparing this current of 0.25 ampere for the 12 ohm resistor to the 0.5 ampere of current obtained with the 6 ohm resistor, it shows that doubling the resistance will reduce the current to one half the original value. The point: Circuit current is inversely proportional to the circuit resistance.

Recall that if you know any two quantities E, I, and R, you can calculate the third. In many circuit applications current is known and either the voltage or the resistance will be the unknown quantity. To solve a problem in which current and resistance are known, the basic formula for Ohm's Law must be transposed to solve for E, for I, or for R.

However, the Ohm's Law equations can be memorized and practiced effectively by using an Ohm's Law pie circle (see figure 13.8).

Example 13.8
Problem:
An electric light bulb draws 0.5 A when operating on a 120 V DC circuit. What is the resistance of the bulb?

Solution:
The first step in solving a circuit problem is to sketch a schematic diagram of the circuit itself, labeling each of the parts and showing the known values (see figure 13.9).

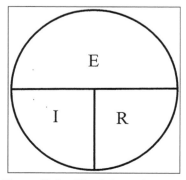

Figure 13.8   Ohm's Law pie circle.

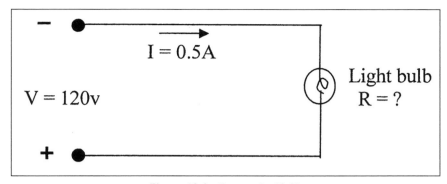

Figure 13.9   [example 13.8].

Because I and E are known, we use

$$R = \frac{E}{I} = \frac{120}{0.5} = 240 \ \Omega$$

ELECTRICAL POWER

Power, whether electrical or mechanical, pertains to the rate at which work is being done, so the power consumption in your plant is related to current flow. A large electric motor or air dryer consumes more power (and draws more current) in a given length of time than, for example, an indicating light on a motor controller. *Work* is done whenever a force causes motion. If a mechanical force is used to lift or move a weight, work is done. However, force exerted *without* causing motion, such as the force of a compressed spring acting between two fixed objects, does not constitute work.

Note: Power is the rate at which work is done.

Electrical Power Calculations

The electric power P used in any part of a circuit is equal to the current I in that part multiplied by the V across that part of the circuit. In equation form,

$$P = I \ x \ E \tag{13.7}$$

Where

P = power, Watts (W)
E = voltage, V
I = current, A

If we know the current I and the resistance R but not the voltage V, we can find the power P by using Ohm's Law for voltage, so we substitute

$$E = IR$$

into

$$P = IR \ x \ I = I^2R \tag{13.8}$$

In the same manner, if we know the voltage V and the resistance R but not the current I, we can find P by using Ohm's Law for current, so we substitute

$$I = \frac{E}{R}$$

into

$$P = E\frac{E}{R} = \frac{E^2}{R} \tag{13.9}$$

Note: If we know any two quantities, we can calculate the third.

Example 13.9
Problem:
The current through a 200 Ω resistor to be used in a circuit is 0.25 A. Find the power rating of the resistor.

Solution:
Since I and R are known, use equation 13.9 to find P.

$$P = I^2R = (0.25)^2(200) = 0.0625(200) = 12.5 \text{ W}$$

Note: The power rating of any resistor used in a circuit should be twice the wattage calculated by the power equation to prevent the resistor from burning out. Thus, the resistor used in example 13.9 should have a power rating of 25 watts.

Example 13.10
Problem:
How many kilowatts of power are delivered to a circuit by a 220 V generator that supplies 30 A to the circuit?

Solution:
Since V and I are given, use

$$P = EI = 220(30) = 6600 \text{ W} = 6.6 \text{ Kw}$$

Example 13.11
Problem:
If the voltage across a 30,000 Ω resistor is 450 V, what is the power dissipated in the resistor?

Solution:
Since R and E are known, use

$$P = \frac{E^2}{R} = \frac{450^2}{30,000} = \frac{202,500}{30,000} = 6.75 \text{ W}$$

In this section, P was expressed in terms of alternate pairs of the other three basic quantities E, I, and R. In practice, you should be able to express any one of the three basic quantities, as well as P, in terms of any two of the others. Figure 13.10 is a summary of twelve basic Ohm's Law formulas. The four quantities E, I, R, and P are at the center of the figure. Adjacent to each quantity are three segments. Note that in each segment, the basic quantity is expressed in terms of two other basic quantities, and no two segments are alike.

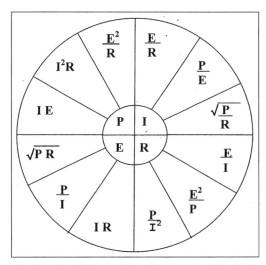

Figure 13.10   Ohm's Law circle: Summary of basic formulas.

ELECTRICAL ENERGY

Energy (the mechanical definition) is defined as the ability to do work (energy and time are essentially the same and are expressed in identical units). Energy is expended when work is done because it takes energy to maintain a force when that force acts through a distance. The total energy expended to do a certain amount of work is equal to the working force multiplied by the distance through which the force moved to do the work.

In electricity, total energy expended is equal to the *rate* at which work is done, multiplied by the length of time the rate is measured. Essentially, energy W is equal to power P times time t.

The kilowatt-hour (Kwh) is a unit commonly used for large amounts of electric energy or work. The amount of kilowatt-hours is calculated as the product of the power in kilowatts (Kw) and the time in hours (h) during which the power is used.

$$Kwh = Kw \times h \tag{13.10}$$

Example 13.12
Problem:
How much energy is delivered in four hours by a generator supplying 12 Kw?

Solution:

$$Kwh = Kw \times h$$
$$= 12(4) = 48$$
$$\text{Energy delivered} = 48 \text{ Kwh}$$

SERIES DC CIRCUITS

As previously mentioned, an electric circuit is made up of a voltage source, the necessary connecting conductors, and the effective load.

If the circuit is arranged so that the electrons have only *one* possible path, the circuit is called a *series circuit*. Therefore, a series circuit is defined as a circuit that contains only one path for current flow. Figure 13.11 shows a series circuit having several loads (resistors).

Note: A *series circuit* is a circuit in which there is only one path for current to flow along.

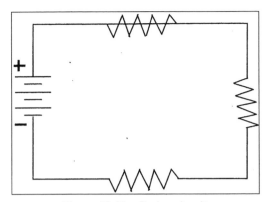

Figure 13.11   Series circuit.

Resistance in Series Circuit

Referring to figure 13.11, the current in a series circuit, in completing its electrical path, must flow through each resistor inserted into the circuit. Thus, each additional resistor offers added resistance. In a series circuit, the total circuit resistance (R_T) is equal to the sum of the individual resistances. As an equation:

$$R_T = R_1 + R_2 + R_3 \dots R_n \tag{13.11}$$

Where

R_T = total resistance, Ω
R_1, R_2, R_3 = resistance in series, Ω
R_n = any number of additional resistors in equation

Example 13.13
Problem:
Three resistors of 10 ohms, 12 ohms, and 25 ohms are connected in series across a battery whose emf is 110 volts (figure 13.12). What is the total resistance?
Solution:

Given: R_1 = 10 ohms
R_2 = 12 ohms
R_3 = 25 ohms
R_T = ?

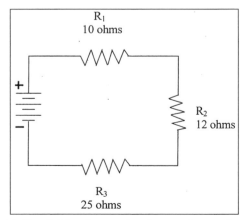

Figure 13.12   Solving for total resistance in a series circuit.

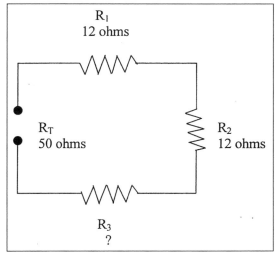

Figure 13.13   Calculating the value of one resistance in a series circuit.

$$R_T = R_1 + R_2 + R_3$$
$$R_T = 10 + 12 + 25$$
$$R_T = 47 \text{ ohms}$$

Example 13.14
Problem:
The total resistance of a circuit containing three resistors is 50 ohms (see figure 13.13). Two of the circuit resistors are 12 ohms each. Calculate the value of the third resistor.

Solution:

Given: R_T = 50 ohms
R_1 = 12 ohms
R_2 = 12 ohms
R_3 = ?
$$R_T = R_1 + R_2 + R_3$$

Subtracting $(R_1 + R_2)$ from both sides of the equation

$$R_3 = R_T - R_1 - R_2$$
$$R_3 = 50 - 12 - 12$$
$$R_3 = 50 - 24$$
$$R_3 = 26 \text{ ohms}$$

Note: When resistances are connected in series, the total resistance in the circuit is equal to the sum of the resistances of all the parts of the circuit.

Current in Series Circuit

Because there is but one path for current in a series circuit, the same *current* (I) must flow through each part of the circuit. Thus, to determine the current throughout a series circuit, only the current through one of the parts need be known.

The fact that the same current flows through each part of a series circuit can be verified by inserting ammeters into the circuit at various points as shown in figure 13.14. As indicated in figure 13.14, each meter indicates the same value of current.

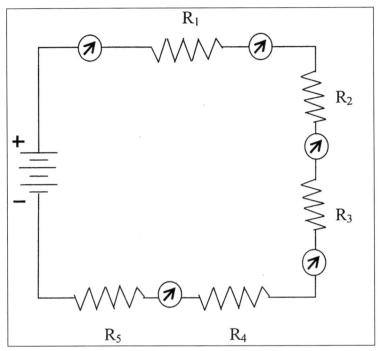

Figure 13.14   Current in a series circuit.

Note: In a series circuit the same current flows in every part of the circuit. *Do not* add the currents in each part of the circuit to obtain current (I).

Voltage in Series Circuit

The *voltage* drop across the resistor in the basic circuit is the total voltage across the circuit and is equal to the applied voltage. The total voltage across a series circuit is also equal to the applied voltage but consists of the sum of two or more individual voltage drops. This statement can be proven by an examination of the circuit shown in figure 13.15.

In this circuit a source potential (E_T) of 30 volts is impressed across a series circuit consisting of two 6 ohm resistors. The total resistance of the circuit is equal to the sum of the two individual resistances, or 12 ohms. Using Ohm's Law, the circuit current may be calculated as follows:

$$I = \frac{E_T}{R_T}$$

$$I = \frac{30}{12}$$

$$I = 2.5 \text{ amperes}$$

Knowing the value of the resistors to be 6 ohms each, and the current through the resistors to be 2.5 amperes, the voltage drops across the resistors can be calculated. The voltage (E_1) across R_1 is therefore:

$E_1 = IR_1$
$E_1 = 2.5$ amps x 6 ohms
$E_1 = 15$ volts

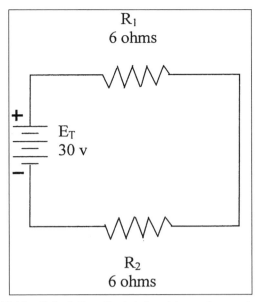

Figure 13.15   Calculating the total resistance in a series circuit.

Since R_2 is the same ohmic value as R_1 and carries the same current, the voltage drop across R_2 is also equal to 15 volts. Adding these two 15 volt drops together gives a total drop of 30 volts exactly equal to the applied voltage. For a series circuit, then:

$$E_T = E_1 + E_2 + E_3 \ldots E_n \qquad\qquad (13.12)$$

Where

E_T = total voltage, V
E_1 = voltage across resistance R_1, V
E_2 = voltage across resistance R_2, V
E_3 = voltage across resistance R_3, V

Example 13.15
Problem:
A series circuit consists of three resistors having values of 10 ohms, 20 ohms, and 40 ohms respectively. Find the applied voltage if the current through the 20 ohm resistor is 2.5 amperes.

Solution:
To solve this problem, a circuit diagram is first drawn and labeled, as shown in figure 13.16.

Given:
R_1 = 10 ohms
R_2 = 20 ohms
R_3 = 40 ohms
I   = 2.5 amps

Since the circuit involved is a series circuit, the same 2.5 amperes of current flows through each resistor. Using Ohm's Law, the voltage drops across each of the three resistors can be calculated and are
E_1 = 25 volts
E_2 = 50 volts
E_3 = 100 volts

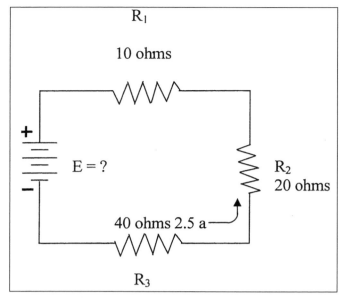

Figure 13.16   Solving for applied voltage in a series circuit.

Once the individual drops are known they can be added to find the total or applied voltage:

$$E_T = E_1 + E_2 + E_3$$
$$E_T = 25 \text{ v} + 50 \text{ v} + 100 \text{ v}$$
$$E_T = 175 \text{ volts}$$

Note 1: The total voltage (E_T) across a series circuit is equal to the sum of the voltages across each resistance of the circuit.

Note 2: The voltage drops that occur in a series circuit are in direct proportions to the resistance across which they appear. This is the result of having the same current flow through each resistor. Thus, the larger the resistor the larger will be the voltage drop across it.

Power in Series Circuit

Each resistor in a series circuit consumes *power*. This power is dissipated in the form of heat. Since this power must come from the source, the total power must be equal in amount to the power consumed by the circuit resistances. In a series circuit the total power is equal to the *sum* of the powers dissipated by the individual resistors. Total power (P_T) is thus equal to:

$$P_T = P_1 + P_2 + P_3 \dots P_n \tag{13.13}$$

Where

P_T = total power, W
P_1 = power used in first part, W
P_2 = power used in second part, W
P_3 = power used in third part, W
P_n = power used in nth part, W

Example 13.16
Problem:
A series circuit consists of three resistors having values of 5 ohms, 15 ohms, and 20 ohms. Find the total power dissipation when 120 volts is applied to the circuit. (See figure 13.17.)

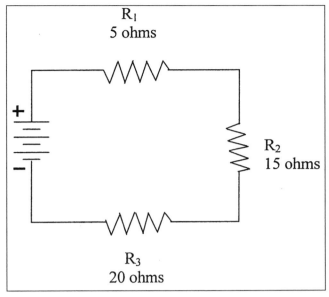

Figure 13.17   Solving for total power in a series circuit.

Solution:
Given:

$$R_1 = 5 \text{ ohms}$$
$$R_2 = 15 \text{ ohms}$$
$$R_3 = 20 \text{ ohms}$$
$$E\ \ = 120 \text{ volts}$$

The total resistance is found first.

$$R_T = R_1 + R_2 + R_3$$
$$R_T = 5 + 15 + 20$$
$$R_T = 40 \text{ ohms}$$

Using total resistance and the applied voltage, the circuit current is calculated.

$$I = \frac{E_T}{R_T}$$

$$I = \frac{120}{40}$$

$$I = 3 \text{ amps}$$

Using the power formula, the individual power dissipations can be calculated.
For resistor R_1:

$$P_1 = I^2 R_1$$
$$P_1 = (3)^2 5$$
$$P_1 = 45 \text{ watts}$$

For R$_2$:

$$P_2 = I^2R_2$$
$$P_2 = (3)^215$$
$$P_2 = 135 \text{ watts}$$

For R$_3$:

$$P_3 = I^2R_3$$
$$P_3 = (3)^220$$
$$P_3 = 180 \text{ watts}$$

To obtain total power:

$$P_T = P_1 + P_2 + P_3$$
$$P_T = 45 + 135 + 180$$
$$P_T = 360 \text{ watts}$$

To check the answer, the total power delivered by the source can be calculated:

$$P = E \times I$$
$$P = 3 \text{ a} \times 120 \text{ v}$$
$$P = 360 \text{ watts}$$

Thus the total power is equal to the sum of the individual power dissipations.

Note: We found that Ohm's Law can be used for total values in a series circuit as well as for individual parts of the circuit. Similarly, the formula for power may be used for total values.

$$P_T = IE_T \tag{13.14}$$

Summary of the Rules for Series DC Circuits

To this point we have covered many of the important factors governing the operation of basic series circuits. In essence, what we have really done is to lay a strong foundation to build upon in preparation for more advanced circuit theory that follows. A summary of the important factors governing the operation of a series circuit is listed as follows:

1. The same current flows through each part of a series circuit.
2. The total resistance of a series circuit is equal to the sum of the individual resistances.
3. The total voltage across a series circuit is equal to the sum of the individual voltage drops.
4. The voltage drop across a resistor in a series circuit is proportional to the size of the resistor.
5. The total power dissipated in a series circuit is equal to the sum of the individual dissipations.

PARALLEL DC CIRCUITS

The principles we applied to solving simple series circuit calculations for determining the reactions of such quantities as voltage, current, and resistance also can be used in parallel and series-parallel circuits.

A *parallel circuit* is defined as one having two or more components connected across the same voltage source (see figure 13.18). Recall that in a series circuit there is only one path for current flow. As additional loads (resistors, etc.) are added to the circuit, the total resistance increases and the total current decreases. This is *not the case* in a parallel circuit. In a parallel circuit, each load (or branch) is connected directly across the voltage source. In figure 13.18, commencing at the voltage source (E$_b$) and tracing counterclockwise around the circuit, two complete and separate paths can be identified in which current can flow. One path

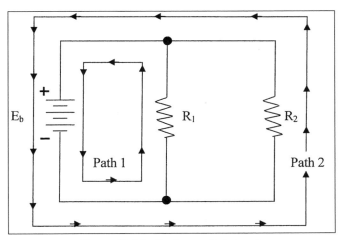

Figure 13.18   Basic parallel circuit.

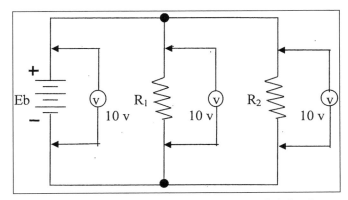

Figure 13.19   Voltage comparison in a parallel circuit.

is traced from the source through resistance R_1 and back to the source; the other, from the source through resistance R_2 and back to the source.

Voltage in Parallel Circuits

Recall that in a series circuit the source voltage divides proportionately across each resistor in the circuit. In a parallel circuit (see figure 13.18), the same voltage is present across all the resistors of a parallel group. This voltage is equal to the applied voltage (E_b) and can be expressed in equation form as:

$$E_b = E_{R1} = E_{R2} = E_{Rn} \tag{13.15}$$

We can verify equation 13.15 by taking voltage measurements across the resistors of a parallel circuit, as illustrated in figure 13.19. Notice that each voltmeter indicates the same amount of voltage; that is, the voltage across each resistor is the same as the applied voltage.

Note: In a parallel circuit the voltage remains the same throughout the circuit.

Example 13.17
Problem:
Assume that the current through a resistor of a parallel circuit is known to be 4.0 milliamperes (ma) and the value of the resistor is 40,000 ohms. Determine the potential (voltage) across the resistor. The circuit is shown in figure 13.20.

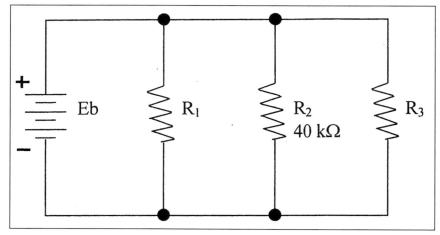

Figure 13.20   [example 13.17].

Solution:
  Given:

$$R_2 = 40 \text{ k}\Omega$$
$$I_{R2} = 4.0 \text{ ma}$$

Find:

$$E_{R2} = ?$$
$$E_b = ?$$

Select the proper equation:

$$E = IR$$

Substitute known values:

$$E_{R2} = I_{R2} \times R_2$$
$$E_{R2} = 4.0 \text{ ma} \times 40,000 \text{ ohms}$$
[use power of tens]
$$E_{R2} = (4.0 \times 10^{-3}) \times (40 \times 10^3)$$
$$E_{R2} = 4.0 \times 40$$

Resultant:

$$E_{R2} = 160 \text{ v}$$

Therefore:

$$E_b = 160 \text{ v}$$

Current in Parallel Circuits

In a series circuit, a single current flows. Its value is determined in part by the total resistance of the circuit. However, the source current in a parallel circuit divides among the available paths in relation to the value of the resistors in the circuit. Ohm's Law remains unchanged. For a given voltage, current varies inversely with resistance.

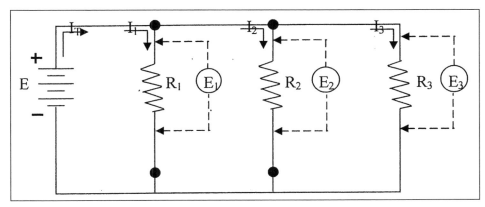

Figure 13.21   Parallel circuit.

The behavior of current in a parallel circuit is best illustrated by example. The example we use is figure 13.21. The resistors R_1, R_2, and R_3 are in parallel with each other and with the battery. Each parallel path is then a branch with its own individual current. When the total current I_T leaves the voltage source E, part I_1 of the current I_T will flow through R_1, part I_2 will flow through R_2, and the remainder I_3 through R_3. The branch current I_1, I_2, and I_3 can be different. However, if a voltmeter (used for measuring the voltage of a circuit) is connected across R_1, R_2, and R_3, the respective voltages E_1, E_2, and E_3 will be equal. Therefore,

$$E = E_1 = E_2 = E_3 \qquad (13.16)$$

The total current I_T is equal to the sum of all branch currents.

$$I_T = I_1 = I_2 = I_3 \qquad (13.17)$$

This formula applies for any number of parallel branches whether the resistances are equal or unequal.

By Ohm's Law, each branch current equals the applied voltage divided by the resistance between the two points where the voltage is applied. Hence (figure 13.21) for each branch we have the following equations:

$$\text{Branch 1:} \qquad I_1 = \frac{E_1}{R_1} = \frac{E}{R_1}$$

$$\text{Branch 2:} \qquad I_2 = \frac{E_2}{R_2} = \frac{E}{R_2}$$

$$\text{Branch 3:} \qquad I_3 = \frac{E_3}{R_3} = \frac{V}{R_3}$$

With the same applied voltage, any branch that has less resistance allows more current through it than a branch with higher resistance.

Example 13.18
Problem:
Two resistors, each drawing 2 A, and a third resistor drawing 1 A are connected in parallel across a 100 V line (see figure 13.22). What is the total current?

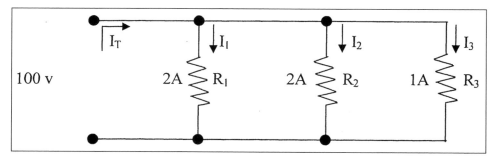

Figure 13.22   [example 13.18].

Parallel Resistance

Unlike series circuits where total resistance (R_T) is the sum of the individual resistances, in a parallel circuit the total resistance is *not* the sum of the individual resistances.

In a parallel circuit we can use Ohm's Law to find total resistance. We use the equation

$$R = \frac{E}{I}$$

or

$$R_T = \frac{E_s}{I_t}$$

R_T is the total resistance of all the parallel branches across the voltage source E_s, and I_T is the sum of all the branch currents.

Example 13.19

Problem:
What is the total resistance of the circuit shown in figure 13.23?

Given:

$$E_s = 120v$$
$$I_T = 26A$$

Solution:
In figure 13.23 the line voltage is 120 V and the total line current is 26 A. Therefore,

$$R_T = \frac{E}{I_T} = \frac{120}{26} = 4.62 \text{ ohms}$$

Note: Notice that R_T is smaller than any of the three resistances in figure 13.23. This fact may surprise you—it may seem strange that the total circuit resistance is *less* than that of the smallest resistor (R_3-12 ohms). However, if we refer back to the water analogy we have used previously, it makes sense. Consider water pressure and water pipes—and assume there is some way to keep the water pressure constant. A small pipe offers more resistance to flow of water than a larger pipe; but if we add another pipe in parallel, one of even smaller diameter, the total resistance to water flow is *decreased*. In an electrical circuit, even a larger resistor in another parallel branch provides an *additional path for current flow*, so the *total* resistance is less.

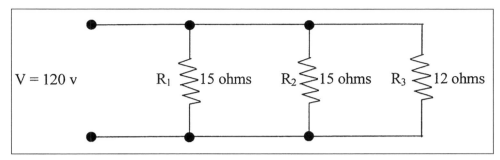

Figure 13.23   [example 13.19].

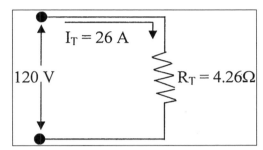

Figure 13.24   Equivalent circuit to that of figure 13.22.

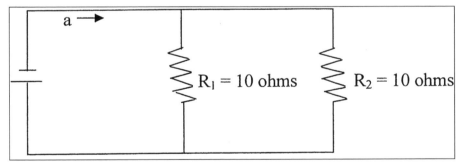

Figure 13.25   Two equal resistors connected in parallel.

Remember, if we add one more branch to a parallel circuit, the total resistance decreases and the total current increases.

The "equivalent resistance" is illustrated in the equivalent circuit shown in figure 13.24.

There are other methods used to determine the equivalent resistance of parallel circuits. The most appropriate method for a particular circuit depends on the number and value of the resistors. For example, consider the parallel circuit shown in figure 13.25.

For this circuit, the following simple equation is used:

$$R_{eq} = \frac{R}{N} \qquad (13.18)$$

Where

R_{eq} = equivalent parallel resistance
R = ohmic value of one resistor
N = number of resistors

Thus,

$$R_{eq} = \frac{10 \text{ ohms}}{2}$$

$$R_{eq} = 5 \text{ ohms}$$

Note 1: Equation 13.18 is valid for any number of equal value parallel resistors.

Note 2: When two equal value resistors are connected in parallel, they present a total resistance equivalent to a single resistor of one half the value of either of the original resistors.

Example 13.20
Problem:
Five 50 ohm resistors are connected in parallel. What is the equivalent circuit resistance?

Solution:

$$R_{eq} = \frac{R}{N} = \frac{50}{5} = 10 \text{ ohms}$$

What about parallel circuits containing resistance of unequal value? How is equivalent resistance determined?

Example 13.21 demonstrates how this is accomplished.

Example 13.21
Problem:
Refer to figure 13.26.

Solution:
Given:
$$R_1 = 3\Omega$$
$$R_2 = 6\Omega$$
$$E_a = 30v$$

Known:
$$I_1 = 10 \text{ amps}$$
$$I_2 = 5 \text{ amps}$$
$$I_t = 15 \text{ amps}$$

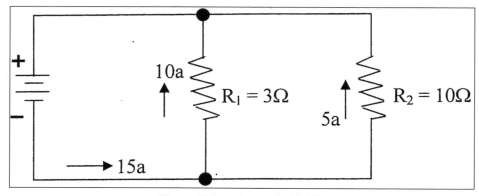

Figure 13.26   [example 13.21].

Determine:

$$R_{eq} = ?$$

$$R_{eq} = \frac{E_a}{I_t}$$

$$R_{eq} = \frac{30}{15} = 2 \text{ ohms}$$

Reciprocal Method

When circuits are encountered in which resistors of unequal value are connected in parallel, the equivalent resistance may be computed by using the *reciprocal method*.

Note: A *reciprocal* is an inverted fraction; the reciprocal of the fraction 3/4, for example is 4/3. We consider a whole number to be a fraction with 1 as the denominator, so the reciprocal of a whole number is that number divided into 1. For example, the reciprocal of R_t is $1/R_t$.

The equivalent resistance in parallel is given by the formula

$$\frac{1}{R_T} = \frac{1}{R_1} + \frac{1}{R_2} + \frac{1}{R_3} + \dots + \frac{1}{R_n} \qquad (13.19)$$

Where

R_T is the total resistance in parallel and R_1, R_2, R_3, and R_n are the branch resistances.

Example 13.22
Problem:
Find the total resistance of a 2 ohm, a 4 ohm, and an 8 ohm resistor in parallel (figure 13.27).
Solution:

Write the formula for three resistors in parallel.

$$\frac{1}{R_T} = \frac{1}{R_1} + \frac{1}{R_2} + \frac{1}{R_3}$$

Substitute the resistance values.

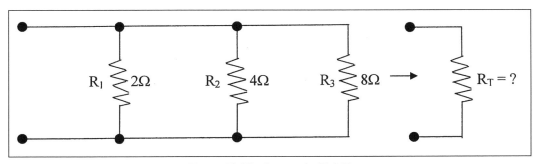

Figure 13.27 [example 13.22].

$$\frac{1}{R_T} = \frac{1}{2} + \frac{1}{4} + \frac{1}{8}$$

Add fractions.

$$\frac{1}{R_T} = \frac{4}{8} + \frac{2}{8} + \frac{1}{8} = \frac{7}{8}$$

Invert both sides of the equation to solve for R_T.

$$R_T = \frac{8}{7} = 1.14\ \Omega$$

Note: When resistances are connected in parallel, the total resistance is always *less* than the smallest resistance of any single branch.

Product over the Sum Method

When any two unequal resistors are in parallel, it is often easier to calculate the total resistance by multiplying the two resistances and then dividing the product by the sum of the resistances.

$$R_T = \frac{R_1 \times R_2}{R_1 + R_2} \qquad\qquad (13.20)$$

where

R_T is the total resistance in parallel and R_1 and R_2 are the two resistors in parallel.

Example 13.23
Problem:
What is the equivalent resistance of a 20 ohm resistor and a 30 ohm resistor connected in parallel?

Solution:
   Given:

$$R_1 = 20$$
$$R_2 = 30$$

$$R_T = \frac{R_1 \times R_2}{R_1 + R_2}$$

$$R_T = \frac{20 \times 30}{20 + 30}$$

$$R_T = 12\ ohms$$

Power in Parallel Circuits

As in the series circuit, the total *power* consumed in a parallel circuit is equal to the sum of the power consumed in the individual resistors.

Note: Because power dissipation in resistors consists of a heat loss, power dissipations are additive regardless of how the resistors are connected in the circuit.

$$P_T = P_1 + P_2 + P_3 + \dots + P_n \qquad (13.21)$$

Where

P_T is the total power and P_1, P_2, P_3, and P_n are the branch powers.

Total power can also be calculated by the equation

$$P_T = EI_T \qquad (13.22)$$

Where

P_T is the total power, E is the voltage source across all parallel branches, and I_T is the total current.

The power dissipated in each branch is equal to EI and equal to V^2/R.

Note: In both parallel and series arrangements the sum of the individual values of power dissipated in the circuit equals the total power generated by the source. The circuit arrangements cannot change the fact that all power in the circuit comes from the source.

Example 13.24
Problem:
Find the total power consumed by the circuit in figure 13.28.

Solution:

$P_{R1} = E_b \times I_{R1}$
$P_{R1} = 50 \times 5$
$P_{R1} = 250 \text{ watts}$
$P_{R2} = E_b \times I_{R2}$
$P_{R2} = 50 \times 2$
$P_{R2} = 100 \text{ watts}$
$P_{R3} = E_b \times I_{R3}$
$P_{R3} = 50 \times 1$
$P_{R3} = 50 \text{ watts}$
$P_T = P_1 + P_2 + P_3$
$P_T = 250 + 100 + 50$

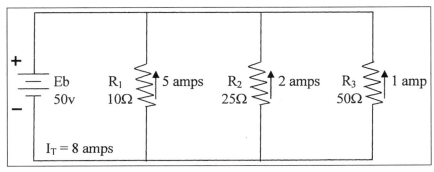

Figure 13.28 [example 13.24].

$$P_T = 400 \text{ watts}$$

Note: The power dissipated in the branch circuits in figure 13.28 is determined in the same manner as the power dissipated by individual resistors in a series circuit. The total power (P_T) is then obtained by summing up the power dissipated in the branch resistors using equation 13.22.

Because, in the example shown in figure 13.28, the total current is known, we could determine the total power by the following method:

$$P_T = E_b \times I_T$$
$$P_T = 50 \text{ v} \times 8a$$
$$P_T = 400 \text{ watts}$$

Rules for Solving Parallel DC Circuits

Problems involving the determination of resistance, voltage, current, and power in a parallel circuit are solved as simply as in a series circuit. The procedure is basically the same—1. draw a circuit diagram, 2. state the values given and the values to be found, 3. state the applicable equations, and 4. substitute the given values and solve for the unknown.

Along with following the problem-solving procedure above, it is also important to remember and apply the rules for solving parallel DC circuits. These rules are

1. The same voltage exists across each branch of a parallel circuit and is equal to the source voltage.
2. The current through a branch of a parallel network is inversely proportional to the amount of resistance of the branch.
3. The total current of a parallel circuit is equal to the sum of the currents of the individual branches of the circuit.
4. The total resistance of a parallel circuit is equal to the reciprocal of the sum of the reciprocals of the individual resistances of the circuit.
5. The total power consumed in a parallel circuit is equal to the sum of the power consumption of the individual resistances.

SERIES-PARALLEL CIRCUITS

To this point we have discussed series and parallel DC circuits. However, the maintenance operator will seldom encounter a circuit that consists solely of either type of circuit. Most circuits consist of both series and parallel elements. A circuit of this type is referred to as a *series-parallel circuit* (see figure 13.29), or as a combination circuit. The solution of a series-parallel (combination) circuit is simply a matter of application of the laws and rules discussed prior to this point.

ELECTROMAGNETISM

In medicine, anatomy and physiology are so closely related that the medical student cannot study one at length without involving the other. A similar relationship holds for the electrical field; that is, magnetism and basic electricity are so closely related that one cannot be studied at length without involving the other. This close fundamental relationship is continually borne out in subsequent chapters of this manual, such as in the study of generators, transformers, and motors. To be proficient in electricity, the maintenance operator must become familiar with such general relationships that exist between magnetism and electricity as follows:

1. Electric current flow will always produce some form of magnetism.
2. Magnetism is by far the most commonly used means for producing or using electricity.
3. The peculiar behavior of electricity under certain conditions is caused by magnetic influences.

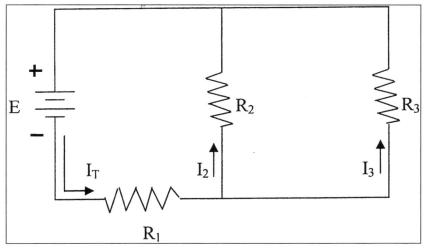

Figure 13.29   A series-parallel circuit.

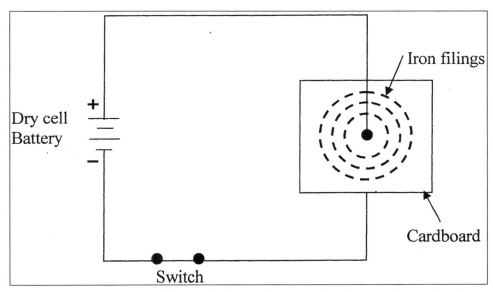

Figure 13.30   A circular pattern of magnetic force exists around a wire carrying an electric current.

Magnetic Field around a Single Conductor

In 1819, Hans Christian Oersted, a Danish scientist, discovered that a field of magnetic force exists around a single wire conductor carrying an electric current. In figure 13.30, a wire is passed through a piece of cardboard and connected through a switch to a dry cell. With the switch open (no current flowing), if we sprinkle iron filings on the cardboard then tap it gently, the filings will fall back haphazardly. Now, if we close the switch, current will begin to flow in the wire. If we tap the cardboard again, the magnetic effect of the current in the wire will cause the filings to fall back into a definite pattern of concentric circles with the wire as the center of the circles. Every section of the wire has this field of force around it in a plane perpendicular to the wire, as shown in figure 13.31.

The ability of the magnetic field to attract bits of iron (as demonstrated in figure 13.30) depends on the number of lines of force present. The strength of the magnetic field around a wire carrying a current depends on the current, since it is the current that produces the field. The greater the current, the greater the strength of the field. A large current will produce many lines of force extending far from the wire, while a small current will produce only a few lines close to the wire, as shown in figure 13.32.

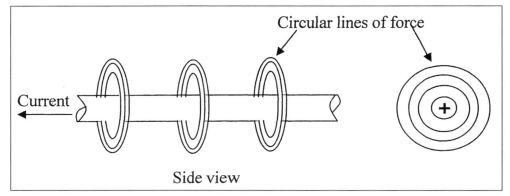

Figure 13.31   The circular fields of force around a wire carrying a current are in planes that are perpendicular to the wire.

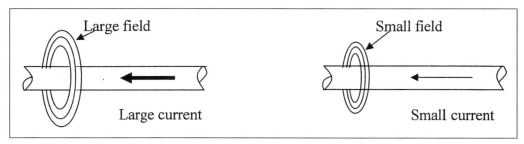

Figure 13.32   The strength of the magnetic field around a wire carrying a current depends on the amount of current.

Polarity of a Single Conductor

The relation between the direction of the magnetic lines of force around a conductor and the direction of current flow along the conductor may be determined by means of the *left-hand rule for a conductor*. If the conductor is grasped in the left hand with the thumb extended in the direction of electron flow (- to +), the fingers will point in the direction of the magnetic lines of force. This is the same direction that the north pole of a compass would point if the compass were placed in the magnetic field.

Note: Arrows are generally used in electric diagrams to denote the direction of current flow along the length of wire. Where cross sections of wire are shown, a special view of the arrow is used. A cross-sectional view of a conductor that is carrying current toward the observer is illustrated in figure 13.33A. The direction of current is indicated by a dot, which represents the head of the arrow. A conductor that is carrying current away from the observer is illustrated in figure 13.33B. The direction of current is indicated by a cross, which represents the tail of the arrow.

Field around Two Parallel Conductors

When two parallel conductors carry current in the same direction, the magnetic fields tend to encircle both conductors, drawing them together with a force of attraction, as shown in figure 13.34A. Two parallel conductors carrying currents in opposite directions are shown in figure 13.34B. The field around one conductor is opposite in direction to the field around the other conductor. The resulting lines of force are crowded together in the space between the wires and tend to push the wires apart. Therefore, two parallel, adjacent conductors carrying currents in the same direction attract each other, and two parallel conductors carrying currents in opposite directions repel each other.

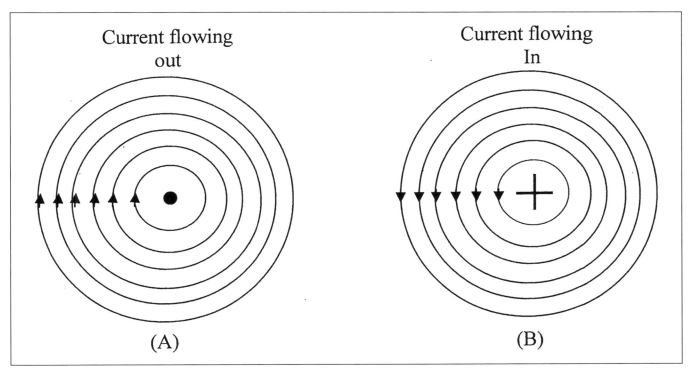

Figure 13.33   Magnetic field around a current-carrying conductor.

Magnetic Field of a Coil

The magnetic field around a current-carrying wire exists at all points along its length. Bending the current-carrying wire into the form of a single loop has two results. First, the magnetic field consists of more dense concentric circles in a plane perpendicular to the wire, although the total number of lines is the same as for the straight conductor. Second, all the lines inside the loop are in the same direction. When this straight wire is wound around a core, as is shown in figure 13.35, it becomes a coil and the magnetic field assumes a different shape. When current is passed through the coiled conductor, the magnetic field of each turn of wire links with the fields of adjacent turns. The combined influence of all the turns produces a two-pole field similar to that of a simple bar magnet. One end of the coil will be a north pole, and the other end will be a south pole.

Polarity of an Electromagnetic Coil

Earlier it was shown that the direction of the magnetic field around a straight conductor depends on the direction of current flow through that conductor. Thus, a reversal of current flow through a conductor causes a reversal in the direction of the magnetic field that is produced. It follows that a reversal of the current flow through a coil also causes a reversal of its two-pole field. This is true because that field is the product of the linkage between the individual turns of wire on the coil. Therefore, if the field of each turn is reversed, it follows that the total field (coils' field) is also reversed.

When the direction of electron flow through a coil is known, its polarity may be determined by use of the *left-hand rule for coils*. This rule is illustrated in figure 13.35 and is stated as follows: Grasping the coil in the left hand, with the fingers "wrapped around" in the direction of electron flow, the thumb will point toward the north pole.

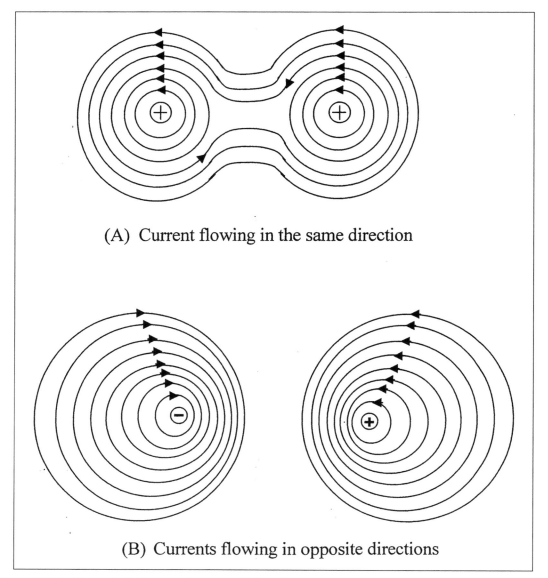

(A) Current flowing in the same direction

(B) Currents flowing in opposite directions

Figure 13.34   Magnetic fields around two parallel conductors: (A) currents flowing in the same direction; (B) currents flowing in opposite directions.

Strength of an Electromagnetic Field

The strength, or intensity, of the magnetic field of a coil depends on a number of factors.

- The *number of turns* of conductor.
- The *amount of current flow* through the coil.
- The *ratio of the coil's length to its width*.
- The *type of material in the core*.

MAGNETIC UNITS

The law of current flow in the electric circuit is similar to the law for the establishing of flux in the magnetic circuit.

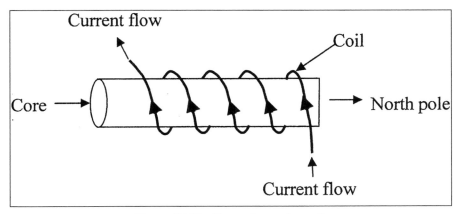

Figure 13.35 Current-carrying coil.

The *magnetic flux*, φ (phi) is similar to current in the Ohm's Law formula and comprises the total number of lines of force existing in the magnetic circuit. The **maxwell** is the unit of flux—that is, 1 line of force is equal to 1 maxwell.

Note: The maxwell is often referred to as simply a line of force, line of induction, or line.

The *strength* of a magnetic field in a coil of wire depends on how much current flows in the turns of the coil. The more current, the stronger the magnetic field. Also, the more turns, the more concentrated are the lines of force. The *force* that produces the flux in the magnetic circuit (comparable to electromotive force in Ohm's Law) is known as *magnetomotive force*, or mmf. The practical unit of magnetomotive force is the **ampere-turn** (At). In equation form,

$$F = \text{ampere-turns} = NI \tag{13.23}$$

Where

F = magnetomotive force, At
N = number of turns
I = current, A

Example 13.25
Problem:
Calculate the ampere-turns for a coil with 2,000 turns and a 5 Ma current.

Solution:
Use equation 13.23 and substitute N = 2,000 and I = 5 x 10⁻³ A.

$$NI = 2,000(5 \times 10^{-3}) = 10 \text{ At}$$

The unit of *intensity* of magnetizing force per unit of length is designated as H and is sometimes expressed as Gilberts per centimeter of length. Expressed as an equation,

$$H = \frac{NI}{L} \tag{13.24}$$

Where

H = magnetic field intensity, ampere-turns per meter (At/m)
NI = ampere-turns, At

L = length between poles of the coil, m

Note: Equation 13.24 is for a solenoid. *H* is the intensity of an air core. With an iron core, *H* is the intensity through the entire core and *L* is the length or distance between poles of the iron core.

PROPERTIES OF MAGNETIC MATERIALS

In this section we discuss two important properties of magnetic materials: permeability and hysteresis.

Permeability

When the core of an electromagnet is made of annealed sheet steel, it produces a stronger magnet than if a cast iron core is used. This is the case because annealed sheet steel is more readily acted upon by the magnetizing force of the coil than is the hard cast iron. Simply put, soft sheet steel is said to have greater *permeability* because of the greater ease with which magnetic lines are established in it.

Recall that permeability is the relative ease with which a substance conducts magnetic lines of force. The permeability of air is arbitrarily set at 1. The permeability of other substances is the ratio of their ability to conduct magnetic lines compared to that of air. The permeability of nonmagnetic materials, such as aluminum, copper, wood, and brass, is essentially unity, or the same as for air.

Note 1: The permeability of magnetic materials varies with the degree of magnetization, being smaller for high values of flux density.

Note 2: *Reluctance*—which is analogous to resistance, is the opposition to the production of flux in a material—is inversely proportional to permeability. Iron has high permeability and, therefore, low reluctance. Air has low permeability and hence high reluctance.

Hysteresis

When the current in a coil of wire reverses thousands of times per second, a considerable loss of energy can occur. This loss of energy is caused by *hysteresis*. Hysteresis means "a lagging behind"; that is, the magnetic flux in an iron core lags behind the increases or decreases of the magnetizing force.

The simplest method of illustrating the property of hysteresis is by graphical means, such as the hysteresis loop shown in figure 13.36.

The hysteresis loop (figure 13.36) is a series of curves that show the characteristics of a magnetic material. Opposite directions of current result are in the opposite directions of +H and -H for field intensity. Similarly, opposite polarities are shown for flux density as +B and -B. The current starts at the center 0 (zero) when the material is unmagnetized. Positive H values increase B to saturation at +B$_{max}$. Next H decreases to zero, but B drops to the value of B, because of hysteresis. The current that produced the original magnetization now is reversed so that H becomes negative. B drops to zero and continues to -B$_{max}$. As the -H values decrease, B is reduced to -B, when H is zero. Now with a positive swing of current, H becomes positive, producing saturation at +B$_{max}$ again. The hysteresis loop is now complete. The curve doesn't return to zero at the center because of hysteresis.

ELECTROMAGNETS

An *electromagnet* is composed of a coil of wire wound around a core that is normally soft iron, because of its high permeability and low hysteresis. When direct current flows through the coil, the core will become magnetized with the same polarity that the coil would have without the core. If the current is reversed, the polarity of the coil and core are reversed.

The electromagnet is of great importance in electricity simply because the magnetism can be "turned on" or "turned off" at will. The starter solenoid (an electromagnet) in automobiles and power boats is a good example. In an automobile or boat, an electromagnet is part of a relay that connects the battery to the induction coil, which generates the very high voltage needed to start the engine. The starter solenoid iso-

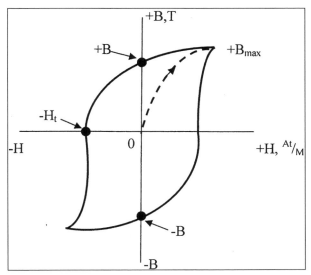

Figure 13.36   Hysteresis loop.

lates this high voltage from the ignition switch. When no current flows in the coil, it is an "air core," but when the coil is energized, a movable soft-iron core does two things. First, the magnetic flux is increased because the soft-iron core is more permeable than the air core. Second, the flux is more highly concentrated. All this concentration of magnetic lines of force in the soft-iron core results in a very good magnet when current flows in the coil. But soft-iron loses its magnetism quickly when the current is shut off. The effect of the soft iron is, of course, the same whether it is movable, as in some solenoids, or permanently installed in the coil. An electromagnet consists basically of a coil and a core; it becomes a magnet when current flows through the coil.

The ability to control the action of magnetic force makes an electromagnet very useful in many circuit applications. Many of the applications of electromagnets are discussed throughout this manual.

AC THEORY

Since voltage is induced in a conductor when lines of force are cut, the amount of the induced emf depends on the number of lines cut in a unit time. To induce an emf of 1 volt, a conductor must cut 100,000,000 lines of force per second. To obtain this great number of "cuttings," the conductor is formed into a loop and rotated on an axis at great speed (see figure 13.37). The two sides of the loop become individual conductors in series, each side of the loop cutting lines of force and inducing twice the voltage that a single conductor would induce. In commercial generators, the number of "cuttings" and the resulting emf are increased by: 1. increasing the number of lines of force by using more magnets or stronger electromagnets, 2. using more conductors or loops, and 3. rotating the loops faster.

How an AC generator operates to produce an AC voltage and current is a basic concept today, taught in elementary and middle school science classes. Of course, we accept technological advances as commonplace today—we surf the Internet, watch cable television, use our cell phones, and take space flight as a given—and consider producing the electricity that makes all these technologies possible as our right. These technologies are bottom shelf to us today—we have them available to us so we simply use them—no big deal, right? Not worth thinking about.

This point of view surely was not held initially—especially by those who broke ground in developing technology and electricity.

In groundbreaking years of electric technology development, the geniuses of the science of electricity (including Georg Simon Ohm) performed their technological breakthroughs in faltering steps. We tend to forget that those first faltering steps of scientific achievement in the field of electricity were performed with crude, and for the most part, homemade apparatus. (Sounds something like the more contemporary young garage-and-basement inventors who came up with the first basic user-friendly microcomputer and software

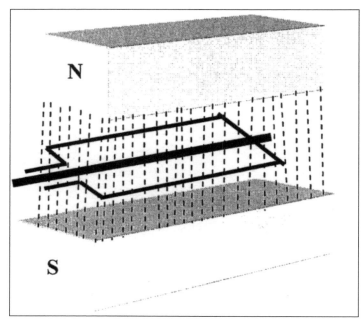

Figure 13.37   Loop rotating in magnetic field produces an AC voltage.

packages, doesn't it?)

Indeed, the innovators of electricity had to make nearly all the laboratory equipment used in their experiments. At the time, the only convenient source of electrical energy available to these early scientists was the voltaic cell, invented some years earlier. Because of the fact that cells and batteries were the only sources of power available, some of the early electrical devices were designed to operate from **direct current**.

Thus, initially, direct current was used extensively. However, when the use of electricity became widespread, certain disadvantages in the use of direct current became apparent. In a direct-current system, the supply voltage must be generated at the level required by the load. To operate a 240 volt lamp, for example, the generator must deliver 240 volts. A 120 volt lamp could not be operated from this generator by any convenient means. A resistor could be placed in series with the 120 volt lamp to drop the extra 120 volts, but the resistor would waste an amount of power equal to that consumed by the lamp.

Another disadvantage of direct-current systems is the large amount of power lost due to the resistance of the transmission wires used to carry current from the generating station to the consumer. This loss could be greatly reduced by operating the transmission line at very high voltage and low current. This is not a practical solution in a DC system, however, since the load would also have to operate at high voltage. As a result of the difficulties encountered with direct current, practically all modern power distribution systems use **alternating current (AC).**

Unlike DC voltage, AC voltage can be stepped up or down by a device called a **transformer**. Transformers permit the transmission lines to be operated at high voltage and low current for maximum efficiency. Then at the consumer end the voltage is stepped down to whatever value the load requires by using a transformer. Due to its inherent advantages and versatility, alternating current has replaced direct current in all but a few commercial power distribution systems.

The purpose of this section is to provide the basic information required to understand basic AC theory. Accordingly, we explain how an alternator generates AC voltage; explain frequency and state the factors that affect it; analyze the sine waveform of AC voltage or current; make the mathematical conversions between maximum and effective values of voltage or current; and explain vectors, or phasors.

BASIC AC GENERATOR

As shown in figure 13.37, AC voltage and current can be produced when a conductor loop rotates through a

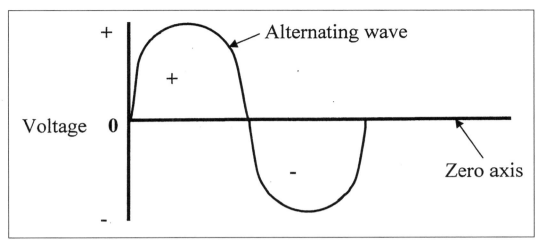

Figure 13.38   An AC voltage waveform.

magnetic field and cuts lines of force to generate an induced AC voltage across its terminals. This describes the basic principle of operation of an alternating current generator, or **alternator**. An alternator converts mechanical energy into electrical energy. It does this by using the principle of **electromagnetic induction**. The basic components of an alternator are an armature, about which many turns of conductor are wound and which rotates in a magnetic field, and some means of delivering the resulting alternating current to an external circuit.

Cycle

An AC voltage is one that continually changes in magnitude and periodically reverses in polarity (see figure 13.38). The zero axis is a horizontal line across the center. The vertical variations on the voltage wave show the changes in magnitude. The voltages above the horizontal axis have positive (+) polarity, while voltages below the horizontal axis have negative (-) polarity.

Note: To bring the important points presented up to this point into finer focus and to expand the presentation, figure 13.39 is provided with accompanying explanation.

Figure 13.39 shows a suspended loop of wire (conductor or armature) being rotated (moved) in a counterclockwise direction through the magnetic field between the poles of a permanent magnet. For ease of explanation, the loop has been divided into a thick and thin half. Notice that in part (A), the thick half is moving along (parallel to) the lines of force. Consequently, it is cutting none of these lines. The same is true of the thin half, moving in the opposite direction. Because the conductors are not cutting any lines of force, no emf is induced. As the loop rotates toward the position shown in part (B), it cuts more and more lines of force per second because it is cutting more directly across the field (lines of force) as it approaches the position shown in (B). At position (B) the induced voltage is greatest because the conductor is cutting directly across the field.

As the loop continues to be rotated toward the position shown in part (C), it cuts fewer and fewer lines of force per second. The induced voltage decreases from its peak value. Eventually, the loop is once again moving in a plane parallel to the magnetic field, and no voltage (zero voltage) is induced. The loop has now been rotated through half a circle (one alternation, or 180°). The sine curve shown in the lower part of figure 13.39 shows the induced voltage at every instant of rotation of the loop. Notice that this curve contains 360°, or two alternations. **Two alternations represent one complete circle of rotation.**

Note: Two complete alternations in a period of time is called a *cycle*.

In figure 13.39, if the loop is rotated at a steady rate, and if the strength of the magnetic field is uniform,

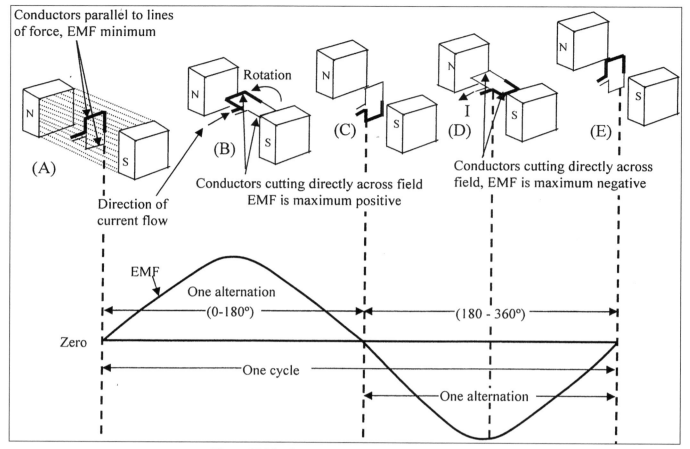

Figure 13.39 Basic alternating current generator.

the number of cycles per second (cps), or **hertz**, and the voltage will remain at fixed values. Continuous rotation will produce a series of sine-wave voltage cycles, or, in other words, AC voltage. In this way mechanical energy is converted into electrical energy.

Frequency, Period, and Wavelength

The *frequency* of an alternating voltage or current is the number of complete cycles occurring in each second of time. It is indicated by the symbol f and is expressed in hertz (Hz). One cycle per second equals one hertz. Thus 60 cycles per second (cps) equals 60 Hz. A frequency of 2 Hz (figure 13.40 (A) is twice the frequency of 1 Hz [figure 13.40 (B)]).

The amount of time for the completion of one cycle is the *period*. It is indicated by the symbol T for time and is expressed in second(s). Frequency and period are reciprocals of each other.

$$f = \frac{1}{T} \tag{13.25}$$

$$T = \frac{1}{f} \tag{13.26}$$

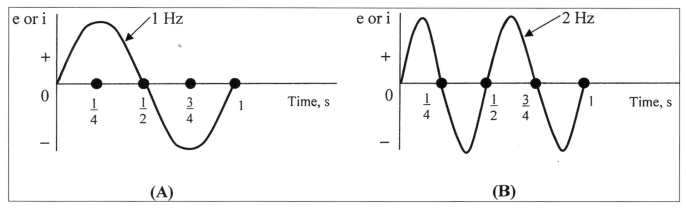

Figure 13.40 Comparison of frequencies.

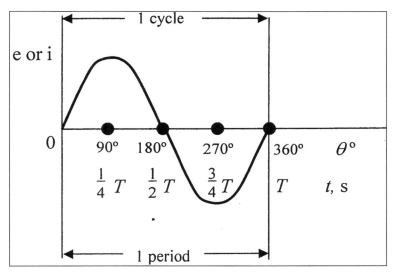

Figure 13.41 Relationship between electrical degrees and time.

Note: The higher the frequency, the shorter the period.

The angle of 360° represents the time for one cycle, or the period T. So we can show the horizontal axis of the sine wave in units of either electrical degrees or seconds (see figure 13.41).

The *wavelength* is the length of one complete wave or cycle. It depends upon the frequency of the periodic variation and its velocity of transmission. It is indicated by the symbol λ (Greek lowercase lambda). Expressed as a formula:

$$\lambda = \frac{\text{velocity}}{\text{frequency}} \tag{13.27}$$

CHARACTERISTIC VALUES OF AC VOLTAGE AND CURRENT

Because AC sine wave voltage or current has many instantaneous values throughout the cycle, it is convenient to specify magnitudes for comparing one wave with another. The peak, average, or root-mean-square (rms) value can be specified (see figure 13.42). These values apply to current or voltage.

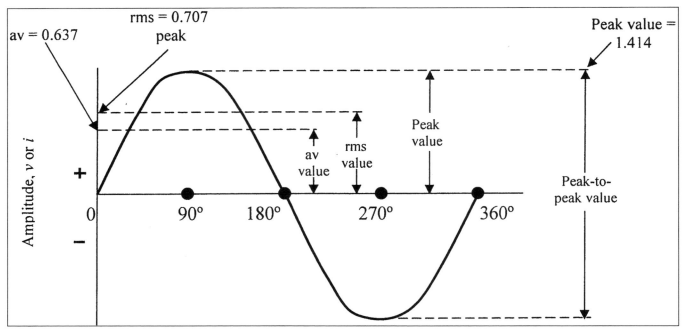

Figure 13.42   Amplitude values for AC sine wave.

Peak Amplitude

One of the most frequently measured characteristics of a sine wave is its amplitude. Unlike DC measurement, the amount of alternating current or voltage present in a circuit can be measured in various ways. In one method of measurement, the maximum amplitude of either the positive or the negative alternation is measured. The value of current or voltage obtained is called the *peak voltage* or the *peak current*. To measure the peak value of current or voltage, an oscilloscope must be used. The peak value is illustrated in figure 13.42.

Peak-to-Peak Amplitude

A second method of indicating the amplitude of a sine wave consists of determining the total voltage or current between the positive and negative peaks. This value of current or voltage is called the *peak-to-peak value* (see figure 13.42). Because both alternations of a pure sine wave are identical, the peak-to-peak value is twice the peak value. Peak-to-peak voltage is usually measured with an oscilloscope, although some voltmeters have a special scale calibrated in peak-to-peak volts.

Instantaneous Amplitude

The *instantaneous value* of a sine wave of voltage for any angle of rotation is expressed by the formula:

$$e = E_m \text{ x sin } \theta \tag{13.28}$$

where

e = the instantaneous voltage
E_m = the maximum or peak voltage
$\sin\theta$ = the sine of angle at which e is desired

Similarly, the equation for the instantaneous value of a sine wave of current would be:

$$i = I_m \text{ x sin } \theta \tag{13.29}$$

Where

> i = the instantaneous current
> I_m = the maximum or peak current
> $\sin\theta$ = the sine of the angle at which i desired

Note: The instantaneous value of voltage constantly changes as the armature of an alternator moves through a complete rotation. Because current varies directly with voltage, according to Ohm's Law, the instantaneous changes in current also result in a sine wave whose positive and negative peaks and intermediate values can be plotted exactly as we plotted the voltage sine wave. However, instantaneous values are not useful in solving most AC problems, so an **effective** value is used.

Effective or RMS Value

The *effective value* of an AC voltage or current of sine waveform is defined in terms of an equivalent heating effect of a direct current. Heating effect is independent of the direction of current flow.

Note: Because all instantaneous values of induced voltage are somewhere between zero and E_M (maximum, or peak voltage), the effective value of a sine wave voltage or current must be greater than zero and less than E_M (the maximum or peak voltage).

The alternating current of sine waveform having a maximum value of 14.14 amps produces the same amount of heat in a circuit having a resistance of one ohm as a direct current of 10 amps. Because this is true, we can work out a constant value for converting any peak value to a corresponding effective value. This constant is represented by X in the simple equation below. Solve for X to three decimal places.

$$14.14X = 10$$
$$X = 0.707$$

The effective value is also called the *root-mean-square (rms)* value because it is the square root of the average of the squared values between zero and maximum. The effective value of AC current is stated in terms of an equivalent DC current. The phenomenon used as the standard comparison is the heating effect of the current.

Note: Anytime AC voltage or current is stated without any qualifications, it is assumed to be an effective value.

In many instances it is necessary to convert from effective to peak or vice versa using a standard equation. Figure 13.42 shows that the peak value of a sine wave is 1.414 times the effective value; therefore, the equation we use is

$$E_m = E \times 1.414 \tag{13.30}$$

Where

> E_m = maximum or peak voltage
> E = effective or RMS voltage

and

$$I_m = I \times 1.414 \tag{13.31}$$

Where

> I_m = maximum or peak current
> I = effective or RMS current

Upon occasion it is necessary to convert a peak value of current or voltage to an effective value. This is accomplished by using the following equations:

$$E = E_m \times 0.707 \tag{13.32}$$

Where

E = effective voltage
E_m = the maximum or peak voltage

$$I = I_m \times 0.707 \tag{13.33}$$

Where

I = the effective current
I_m = the maximum or peak current

Average Value

Because the positive alternation is identical to the negative alternation, the *average value* of a complete cycle of a sine wave is zero. In certain types of circuits, however, it is necessary to compute the average value of one alternation. Figure 13.42 shows that the average value of a sine wave is 0.637 x peak value and therefore:

$$\text{Average Value} = 0.637 \times \text{peak value} \tag{13.34}$$

or

$$E_{avg} = E_m \times 0.637$$

Where

E_{avg} = the average voltage of one alternation
E_m = the maximum or peak voltage

Similarly

$$I_{avg} = I_m \times 0.637 \tag{13.35}$$

Where

I_{avg} = the average current in one alternation
I_m = the maximum or peak current

Table 13.5 lists the various values of sine wave amplitude used to multiply in the conversion of AC sine wave voltage and current.

RESISTANCE IN AC CIRCUITS

If a sine wave of voltage is applied to a resistance, the resulting current will also be a sine wave. This follows Ohm's Law that states that the current is directly proportional to the applied voltage. Figure 13.43 shows a

Table 13.5   AC Sine Wave Conversion Table

Multiply the Value	By	To Get the Value
Peak	2	Peak-to-peak
Peak-to-peak	0.5	Peak
Peak	0.637	Average
Average	1.637	Peak
Peak	0.707	RMS (effective)
RMS (effective)	1.414	Peak
Average	1.110	RMS (effective)
RMS (effective)	0.901	Average

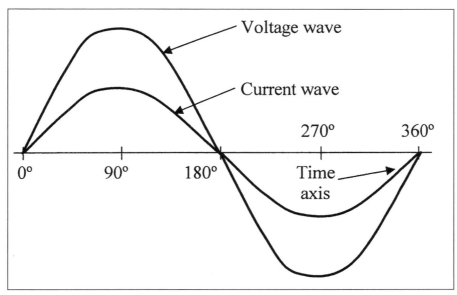

Figure 13.43 Voltage and current waves in phase.

sine wave of voltage and the resulting sine wave of current superimposed on the same time axis. Notice that as the voltage increases in a positive direction, the current increases along with it. When the voltage reverses direction, the current reverses direction. At all times the voltage and current pass through the same relative parts of their respective cycles at the same time. When two waves, such as those shown in figure 13.43, are precisely in step with one another, they are said to be **in phase**. To be in phase, the two waves must go through their maximum and minimum points at the same time and in the same direction.

In some circuits, several sine waves can be in phase with each other. Thus, it is possible to have two or more voltage drops in phase with each other and also in phase with the circuit current.

Note: It is important to remember that Ohm's Law for DC circuits is applicable to AC circuits **with resistance only**.

Voltage waves are not always in phase. For example, figure 13.44 shows a voltage wave E_1 considered to start at 0° (time 1). As voltage wave E_1 reaches its positive peak, a second voltage wave E_2 starts to rise (time 2). Because these waves do not go through their maximum and minimum points at the same instant of time, a **phase difference** exists between the two waves. The two waves are said to be out of phase. For the two waves in figure 13.44 this phase difference is 90°.

Phase Relationships

In the preceding section we discussed the important concepts of **in phase** and **phase difference**. Another important phase concept is **phase angle**. The phase angle between two waveforms of the same frequency is the angular difference at a given instant of time. As an example, the phase angle between waves B and A (see figure 13.45) is 90°. Take the instant of time at 90°. The horizontal axis is shown in angular units of time. Wave B starts at maximum value and reduces to zero value at 90°, while wave A starts at zero and increases to maximum value at 90°. Wave B reaches its maximum value 90° ahead of wave A, so wave B **leads** wave A by 90° (and wave A **lags** wave B by 90°). This 90° phase angle between waves B and A is maintained throughout the complete cycle and all successive cycles. At any instant of time, wave B has the value that wave A will have 90° later. Wave B is a cosine wave because it is displaced 90° from wave A, which is a sine wave.

Note: The amount by which one wave leads or lags another is measured in degrees.

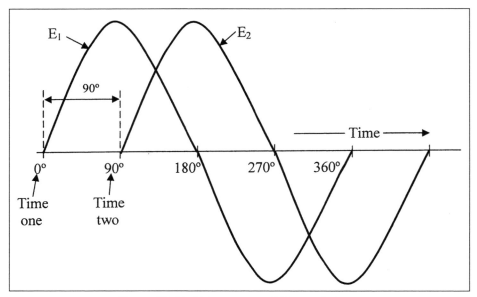

Figure 13.44 Voltage waves 90° out of phase.

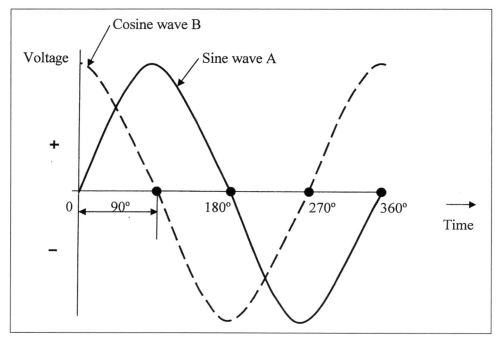

Figure 13.45 Wave B leads wave A by a phase angle of 90°.

To compare phase angles or phases of alternating voltages or currents, it is more convenient to use vector diagrams corresponding to the voltage and current waveforms. A *vector* is a straight line used to denote the magnitude and direction of a given quantity. Magnitude is denoted by the length of the line drawn to scale, and the direction is indicated by the arrow at one end of the line, together with the angle that the vector makes with a horizontal reference vector.

Note: In electricity, since different directions really represent **time** expressed as a phase relationship, an electrical vector is called a **phasor**. In an AC circuit containing only resistance, the voltage and current oc-

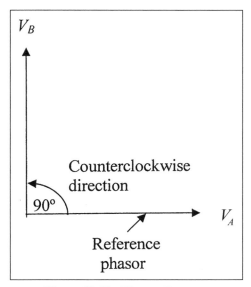

Figure 13.46   Phasor diagram.

cur at the **same time**, or are in phase. To indicate this condition by means of phasors, all that is necessary is to draw the phasors for the voltage and current in the same direction. The value of each is indicated by the **length** of the phasor.

A vector, or phasor, diagram is shown in figure 13.46, where vector V_B is vertical to show the phase angle of 90° with respect to vector V_A, which is the reference. Since lead angles are shown in the counterclockwise direction from the reference vector, V_B leads V_A by 90°.

INDUCTANCE

To this point we have learned the following key points about magnetic fields:

- A field of force exists around a wire carrying a current.
- This field has the form of concentric circles around the wire, in planes perpendicular to the wire, and with the wire at the center of the circles.
- The strength of the field depends on the current. Large currents produce large fields; small currents produce small fields.
- When lines of force cut across a conductor, a voltage is induced in the conductor.

Moreover, to this point we have studied circuits that have been **resistive** (i.e., resistors presented the only opposition to current flow). Two other phenomena—inductance and capacitance—exist in DC circuits to some extent, but they are major players in AC circuits. Both inductance and capacitance present a kind of opposition to current flow that is called **reactance**.

Inductance: What Is It?

Inductance is the characteristic of an electrical circuit that makes itself evident by opposing (acts like a resistance) the starting, stopping, or changing of current flow. A simple analogy can be used to explain inductance. We are all familiar with how difficult it is to push a heavy load (a cart full of heavy materials, etc.). It takes more work to start the load moving than it does to keep it moving. This is because the load possesses the property of **inertia**. Inertia is the characteristic of mass that opposes a *change* in velocity. Therefore, inertia can hinder us in some ways and help us in others. Inductance exhibits the same effect on current in an electric circuit as inertia does on velocity of a mechanical object. The effects of inductance are sometimes desirable—sometimes undesirable.

Note: Simply put, inductance is the characteristic of an electrical conductor that opposes a *change* in current flow.

What does all this mean? Good question.

What it means is that since inductance is the property of an electric circuit that opposes any change in the current through that circuit, if the current increases, a self-induced voltage opposes this change and delays the increase. On the other hand, if the current decreases, a self-induced voltage tends to aid (or prolong) the current flow, delaying the decrease. Thus, current can neither increase nor decrease as fast in an inductive circuit as it can in a purely resistive circuit.

In AC circuits, this effect becomes very important because it affects the phase relationships between voltage and current. We learned earlier that voltages (or currents) can be out of phase if they are induced in separate armatures of an alternator. In that case, the voltage and current generated by each armature were in phase. When inductance is a factor in a circuit, the voltage and current generated by the same armature are out of phase.

Unit of Inductance

The unit for measuring inductance, L, is the *henry* (named for the American physicist, Joseph Henry), abbreviated *h* and normally written in lowercase. Figure 13.47 shows the schematic symbol for an inductor. An inductor has an inductance of 1 henry if an emf of 1 volt is induced in the inductor when the current through the inductor is changing at the rate of 1 ampere per second. The relation between the induced voltage, inductance, and rate of change of current with respect to time is stated mathematically as

$$E = L \frac{\Delta I}{\Delta t} \tag{13.36}$$

Where

E = the induced emf in volts
L = the inductance in henrys
ΔI = is the change in amperes occurring in Δt seconds

Note: The symbol Δ (delta) means "a change in …".

The henry is a large unit of inductance and is used with relatively large inductors. The unit employed with small inductors is the millihenry (mh). For still smaller inductors the unit of inductance is the microhenry (μh).

SELF-INDUCTANCE

As previously explained, current flow in a conductor always produces a magnetic field surrounding, or linking with, the conductor. When current changes, the magnetic field changes, and an emf is induced in the conductor. This emf is called a *self-induced emf* because it is induced in the conductor carrying the current.

Note: Even a perfectly straight length of conductor has some inductance.

Figure 13.47 Schematic symbol for an inductor.

The direction of the induced emf has a definite relation to the direction in which the field that induces the emf varies. When the current in a circuit is increasing, the flux linking with the circuit is increasing. This flux cuts across the conductor and induces an emf in the conductor in such a direction to oppose the increase in current and flux. This emf is sometimes referred to as **counterelectromotive force** (cemf). The two terms are used synonymously throughout this manual. Likewise, when the current is decreasing, an emf is induced in the opposite direction and opposes the decrease in current.

Note: The effects just described are summarized by **Lenz's Law**, which states that the induced emf in any circuit is always in a direction opposed to the effect that produced it.

Shaping a conductor so that the electromagnetic field around each portion of the conductor cuts across some other portion of the same conductor increases the inductance. This is shown in its simplest form in figure 13.48 (A). A loop of conductor is looped so that two portions of the conductor lie adjacent and parallel to one another. These portions are labeled Conductor 1 and Conductor 2. When the switch is closed, electron flow through the conductor establishes a typical concentric field around *all* portions of the conductor. The field is shown in a single plane (for simplicity) that is perpendicular to both conductors. Although the field originates simultaneously in both conductors, it is considered as originating in Conductor 1, and its effect on Conductor 2 will be noted. With increasing current, the field expands outward, cutting across a portion of Conductor 2. The resultant induced emf in Conductor 2 is shown by the dashed arrow. Note that it is in *opposition* to the battery current and voltage, according to Lenz's Law.

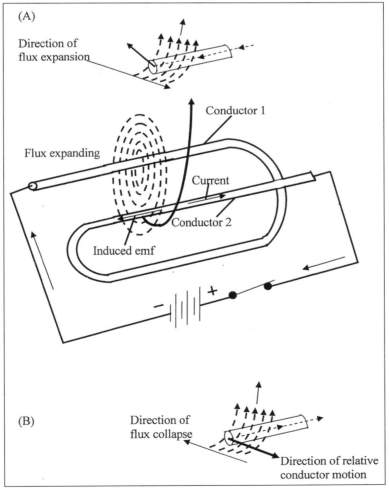

Figure 13.48 Self-inductance.

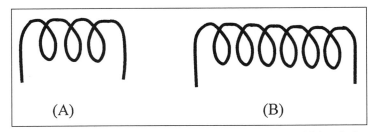

Figure 13.49   (A) Few turns, low inductance; (B) More turns, higher inductance.

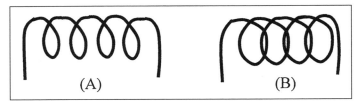

Figure 13.50   (A) Wide spacing between turns, low inductance; (B) Close spacing between turns, higher inductance.

Figure 13.51   (A) Small diameter, low inductance; (B) Larger diameter, higher inductance.

In Figure 13.48 (B), the same section of Conductor 2 is shown, but with the switch opened and the flux collapsing.

Note: From figure 13.48, the important point to note is that the voltage of self-induction opposes both changes in current. It delays the initial buildup of current by opposing the battery voltage and delays the breakdown of current by exerting an induced voltage in the same direction that the battery voltage acted.

Four major factors affect the self-inductance of a conductor, or circuit.

1. **Number of turns:** Inductance depends on the number of wire turns. Wind more turns to increase inductance. Take turns off to decrease the inductance. Figure 13.49 compares the inductance of two coils made with different numbers of turns.
2. **Spacing between turns:** Inductance depends on the spacing between turns, or the inductor's length. Figure 13.50 shows two inductors with the same number of turns. The first inductor's turns have a wide spacing. The second inductor's turns are close together. The second coil, though shorter, has a larger inductance value because of its close spacing between turns.
3. **Coil diameter:** Coil diameter, or cross-sectional area, is highlighted in figure 13.51. The larger-diameter inductor has more inductance. Both coils shown have the same number of turns, and the spacing between turns is the same. The first inductor has a small diameter, and the second one has a larger diameter. The second inductor has more inductance than the first one.
4. **Type of core material:** Permeability, as pointed out earlier, is a measure of how easily a magnetic field goes through a material. Permeability also tells us how much stronger the magnetic field will be with the material inside the coil.

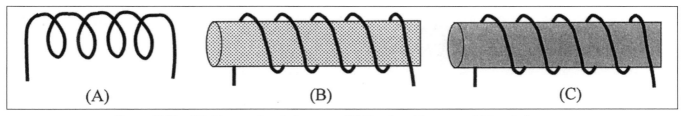

Figure 13.52   (A) Air core, low inductance; (B) Powdered-iron core, higher inductance; (C) Soft-iron core, highest inductance.

Figure 13.52 shows three identical coils. One has an air core, one has a powdered-iron core in the center, and the other has a soft-iron core. This figure illustrates the effects of core material on inductance. The inductance of a coil is affected by the magnitude of current when the core is a magnetic material. When the core is air, the inductance is independent of the current.

Note: The inductance of a coil increases very rapidly as the number of turns is increased. It also increases as the coil is made shorter, the cross-sectional area is made larger, or the permeability of the core is increased.

Growth and Decay of Current in an RL Series Circuit

If a battery is connected across a pure inductance, the current builds up to its final value at a rate that is determined by the battery voltage and the internal resistance of the battery. The current buildup is gradual because of the counter emf (cemf) generated by the self-inductance of the coil. When the current starts to flow, the magnetic lines of force move out, cut the turns of wire on the inductor, and build up a cemf that opposes the emf of the battery. This opposition causes a delay in the time it takes the current to build up to steady value. When the battery is disconnected, the lines of force collapse, again cutting the turns of the inductor and building up an emf that tends to prolong the current flow.

Although the analogy is not exact, electrical inductance is somewhat like mechanical inertia. A boat begins to move on the surface of water at the instant a constant force is applied to it. At this instant its rate of change of speed (acceleration) is greatest, and all the applied force is used to overcome the inertia of the boat. After a while the speed of the boat increases (its acceleration decreases) and the applied force is used up in overcoming the friction of the water against the hull. As the speed levels off and the acceleration becomes zero, the applied force equals the opposing friction force at this speed and the inertia effect disappears. In the case of inductance, it is electrical inertia that must be overcome.

Figure 13.53 shows a circuit that includes two switches, a battery, and a voltage divider containing a resistor (R) and an inductor (L). Switches S_1 and S_2 are mechanically interlocked ("ganged"), as indicated by the dashed line, so that when one is closed, the other is opened at exactly the same instant. Such an arrangement is called an **R-L** (resistive-inductive) series circuit. The source voltage of the battery is applied across the R-L combination (the resistor and inductor) when S_1 is closed. As S_1 is closed, as shown in figure 13.53, a voltage (E) appears across the R-L combination. A current attempts to flow, but the inductor opposes this current by building up a cemf. At the first instant S_1 is closed, the cemf *exactly equals* the battery emf, and its polarity is opposite. Under this condition, no current will flow in resistor R. Because no current can flow when the cemf is exactly equal to the battery voltage, no voltage is dropped across R. As time goes on, more of the battery voltage appears across the resistor and less across the inductor. The rate of change of current is less, and the induced emf is less. As the steady-state condition of the current flow is approached, the drop across the inductor approaches zero and all the battery voltage is used to overcome the resistance of the circuit.

When S_2 is closed (source voltage E removed from the circuit), the flux that has been established around L collapses through the windings and induces a voltage, ^eL, in L that has a polarity opposite to E and essentially equal to it in magnitude. The induced voltage, ^eL, causes i_d to flow through R in the same direction that it was flowing when S_1 was closed. A voltage, ^eR, that is initially equal to E, is developed across R. It rapidly falls to zero as the voltage, ^eL, across L, due to the collapsing flux, falls to zero.

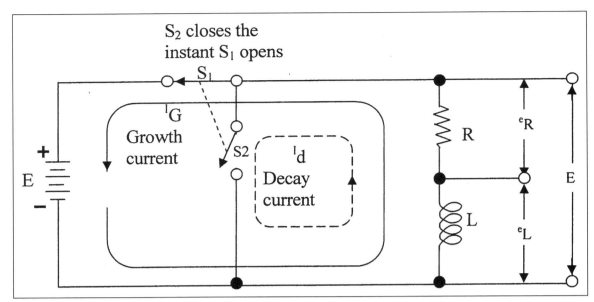

Figure 13.53   Growth and decay of current in an RL circuit.

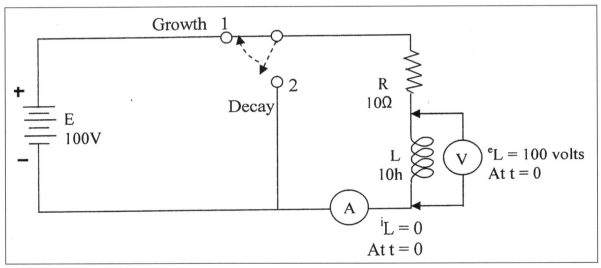

Figure 13.54   L/R time constant.

Note 1: When a switch is first closed to complete a circuit, inductance opposes the buildup of current in the circuit.

Note 2: In many electronic circuits, the time required for the growth or decay of current is important. However, these applications are beyond the scope of this manual, but you need to learn the "fundamentals" of the L/R time constant, which are discussed in the following section.

L/R Time Constant

The time required for the current through an inductor to increase to 63.2 percent of the maximum current or to decrease to 36.7 percent is known as the *time constant* of the circuit. An RL circuit is shown in figure 13.54.

The value of the time constant in seconds is equal to the inductance in henrys divided by the circuit resistance in ohms. One set of values is given in figure 13.54. L/R is the symbol used for this time constant. If L is in henrys and R is in ohms, t (time) is in seconds. If L is in microhenrys and R is in ohms, t is in microseconds. If L is in millihenrys and R is in ohms, t is in milliseconds.

R in the L/R equation is always in ohms, and the time constant is on the same order of magnitude as L. Two useful relations used in calculating L/R time constants are as follows:

$$\frac{\text{L (in henrys)}}{\text{R (in ohms)}} = t \text{ (in seconds)} \tag{13.37}$$

$$\frac{\text{L (microhenrys)}}{\text{R (in ohms)}} = t \text{ (in milliseconds)} \tag{13.38}$$

Note: The time constant of an L/R circuit is always expressed as a ratio between inductance (or L) and resistance (or R).

MUTUAL INDUCTANCE

When the current in a conductor or coil changes, the varying flux can cut across any other conductor or coil located nearby, thus inducing voltages in both. A varying current in L_1, therefore, induces voltage across L_1 and across L_2 (see figure 13.55; see figure 13.56 for the schematic symbol for two coils with mutual inductance). When the induced voltage e_{L2} produces current in L_2, its varying magnetic field induces voltage in L_1. Hence, the two coils L_1 and L_2 have *mutual inductance* because current change in one coil can induce voltage in the other. The unit of mutual inductance is the henry, and the symbol is L_M. Two coils have L_M of 1 H when a current change of 1 A/s in one coil induces 1 E in the other coil.

The factors affecting the mutual inductance of two adjacent coils is dependent upon

- physical dimensions of the two coils
- number of turns in each coil
- distance between the two coils
- relative positions of the axes of the two coils
- the permeability of the cores

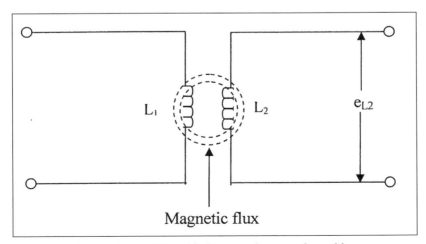

Figure 13.55   Mutual inductance between L$_1$ and L$_2$.

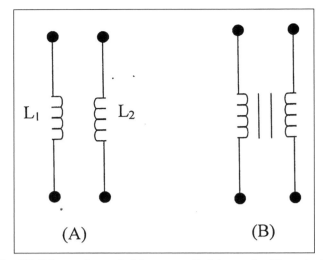

**Figure 13.56   (A) Schematic symbol for two coils (air core) with mutual inductance;
(B) Two coils (iron core) with mutual inductance.**

Note: The amount of mutual inductance depends on the relative position of the two coils. If the coils are separated by a considerable distance, the amount of flux common to both coils is small and the mutual inductance is low. Conversely, if the coils are close together so that nearly all the flow of one coil links the turns of the other, mutual inductance is high. The mutual inductance can be increased greatly by mounting the coils on a common iron core.

CALCULATION OF TOTAL INDUCTANCE

Note: In the study of advanced electrical theory, it is necessary to know the effect of mutual inductance in solving for total inductance in both series and parallel circuits. However, for our purposes, in this text we do not attempt to make these calculations. Instead, we discuss the basic total inductance calculations that the safety engineer should be familiar with.

Inductors in Series

If inductors in series are located far enough apart, or well shielded to make the effects of mutual inductance negligible, the total inductance is calculated in the same manner as for resistances in series; we merely add them.

$$L_t = L_1 + L_2 + L_3 \text{ ... (etc.)} \tag{13.39}$$

Example 13.26
Problem:
If a series circuit contains three inductors whose values are 40 µh, 50 µh, and 20 µh, what is the total inductance?

Solution:

$$L_t = 40\mu h + 50\mu h + 20\mu$$
$$= 110 \ \mu h$$

Example 13.27
Problem:
What is the equivalent inductance of a circuit that has three inductances within a series if

$L_1 = 0.8$ H; $L_2 = 110$ mH; and $L_3 = 55$ mH?
Note: mH = millihenries

Solution:

$$\begin{aligned}
L_{Series} &= L_1 + L_2 + L_3 \\
&= (0.8 + 0.11 + 0.055) \text{ H} \\
&= 0.965 \text{ henries}
\end{aligned}$$

Inductors in Parallel

In a parallel circuit containing inductors (without mutual inductance), the total inductance is calculated in the same manner as for resistances in parallel:

$$\frac{1}{L_t} = \frac{1}{L_1} + \frac{1}{L_2} + \frac{1}{L_3} + \dots \text{ (etc.)} \tag{13.40}$$

Example 13.28
Problem:
A circuit contains three totally shielded inductors in parallel. The values of the three inductances are: 4 mh, 5 mh, and 10 mh. What is the total inductance?

Solution:

$$\frac{1}{L_t} = \frac{1}{4} + \frac{1}{5} + \frac{1}{10}$$

$$= 0.25 + 0.2 + 0.1$$

$$= 0.55$$

$$L_t = \frac{1}{0.55}$$

$$= 1.8 \text{ mh}$$

CAPACITANCE

No matter how complex the electrical circuit, it is composed of no more than three basic electrical properties: resistance, inductance, and capacitance. Accordingly, gaining a thorough understanding of each of these three basic properties is a necessary step toward the understanding of electrical equipment. Because resistance and inductance have been covered, the last of the basic three, capacitance, is covered in this section.

Capacitance: What Is It?

Earlier, we learned that inductance opposes any change in current. *Capacitance* is the property of an electric circuit that opposes any change of *voltage* in a circuit. That is, if applied voltage is increased, capacitance opposes the change and delays the voltage increase across the circuit. If applied voltage is decreased, capacitance tends to maintain the higher original voltage across the circuit, thus delaying the decrease.

Capacitance is also defined as that property of a circuit that enables energy to be stored in an electric field. Natural capacitance exists in many electric circuits. However, in this manual, we are concerned only with the capacitance that is designed into the circuit by means of devices called *capacitors*.

Note: The most noticeable effect of capacitance in a circuit is that voltage can neither increase nor decrease rapidly in a capacitive circuit as it can in a circuit that does not include capacitance.

THE CAPACITOR

A *capacitor*, or condenser, is a manufactured electrical device that consists of two conducting plates of metal separated by an insulating material called a *dielectric* (see figure 13.57). (Note: The prefix *di-* means "through" or "across.")

Schematic symbol for a capacitor is shown in figure 13.58.

When a capacitor is connected to a voltage source, there is a short current pulse. A capacitor stores this electric charge in the dielectric (it can be charged and discharged, as we shall see later). To form a capacitor of any appreciable value, however, the area of the metal pieces must be quite large and the thickness of the dielectric must be quite small.

Note: A capacitor is essentially a device that stores electrical energy.

Capacitors are used in a number of ways in electrical circuits. They may block DC portions of a circuit because they are effectively barriers to direct current (but not to AC current). A capacitor may be part of a tuned circuit—one such application is in the tuning of a radio to a particular station. It may be used to filter AC out of a DC circuit. Most of these are advanced applications that are beyond the scope of this manual; however, a basic understanding of capacitance is necessary to the fundamentals of AC theory.

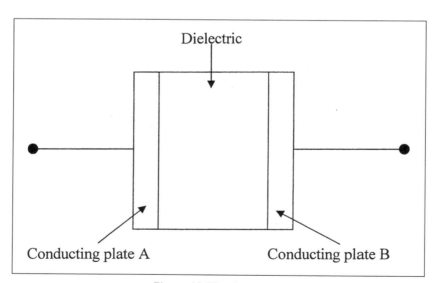

Figure 13.57   Capacitor.

Figure 13.58   (A) Schematic for a fixed capacitor; (B) Variable capacitor.

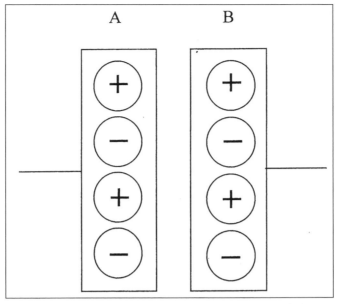

Figure 13.59 Two plates of a capacitor with a neutral charge.

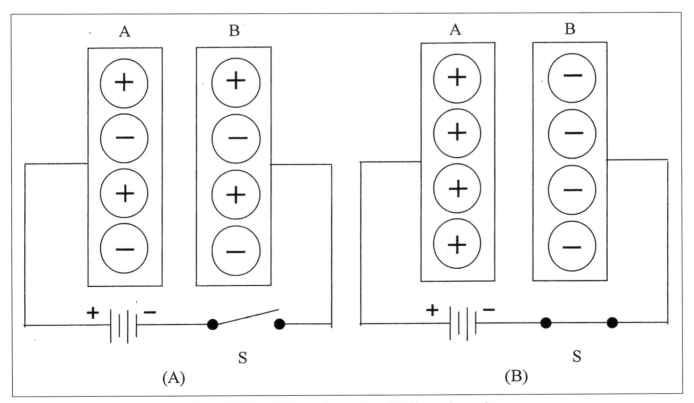

Figure 13.60 (A) Neutral capacitor; (B) Charged capacitor.

Note: A capacitor does not conduct DC current. The insulation between the capacitor plates blocks the flow of electrons. We learned earlier there is a short current pulse when we first connect the capacitor to a voltage source. The capacitor quickly charges to the supply voltage, and then the current stops.

The two plates of the capacitor shown in figure 13.59 are electrically neutral since there are as many protons (positive charge) as electrons (negative charge) on each plate. Thus the capacitor has *no charge*.

Now a battery is connected across the plates (see figure 13.60 (A)). When the switch is closed (see figure 13.60 (B)), the negative charge on plate A is attracted to the positive terminal of the battery. This movement of charges will continue until the difference in charge between plates A and B is equal to the electromotive force (voltage) of the battery.

The capacitor is now *charged*. Because almost none of the charge can cross the space between plates, the capacitor will remain in this condition even if the battery is removed (see figure 13.61 (A)). However, if a conductor is placed across the plates (see figure 13.61 (B), the electrons find a path back to plate A and the charges on each plate are again neutralized. The capacitor is now *discharged*.

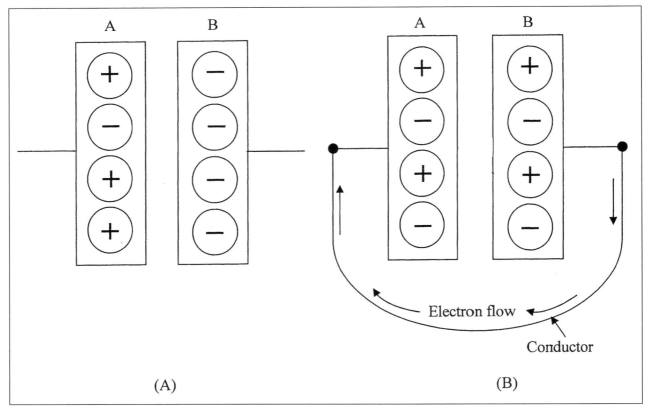

Figure 13.61   (A) Charged capacitor; (B) Discharging a capacitor.

Table 13.6 Dielectric Constants

Material	Constant
Vacuum	1.0000
Air	1.0006
Paraffin paper	3.5
Glass	5–10
Quartz	3.8
Mica	3–6
Rubber	2.5–35
Wood	2.5–8
Porcelain	5.1–5.9
Glycerine (15°C)	56
Petroleum	2
Pure water	81

Note: In a capacitor, electrons cannot flow through the dielectric, because it is an insulator. Because it takes a definite quantity of electrons to charge ("fill up") a capacitor, it is said to have *capacity*. This characteristic is referred to as *capacitance*.

DIELECTRIC MATERIALS

Somewhat similar to the phenomenon of permeability in magnetic circuits, various materials differ in their ability to support electric flux (lines of force) or to serve as dielectric material for capacitors. Materials are rated in their ability to support electric flux in terms of a number called a *dielectric constant*. Other factors being equal, the higher the value of the dielectric constant, the better is the dielectric material. Dry air is the standard (the reference) by which other materials are rated.

Dielectric constants for some common materials are given in table 13.6.

Note: From table 13.6 it is obvious that "pure" water is the best dielectric. Keep in mind that the key word is "pure." Water capacitors are used today in some high-energy applications, in which differences in potential are measured in thousands of volts.

UNIT OF CAPACITANCE

Capacitance is equal to the amount of charge that can be stored in a capacitor divided by the voltage applied across the plates.

$$C = \frac{Q}{E}$$ (13.41)

Where

C = capacitance, F (farads)
Q = amount of charge, C (coulombs)
E = voltage, V

Example 13.29
Problem:
What is the capacitance of two metal plates separated by one centimeter of air, if 0.001 coulomb of charge is stored when a potential of 200 volts is applied to the capacitor?

Solution:
Given:

Q = 0.001 coulomb
F = 200 volts

$$C = \frac{Q}{E}$$

Converting to power of ten

$$C = \frac{10 \times 10^{-4}}{2 \times 10^{2}}$$

$C = 5 \times 10^{-6}$
$C = 0.000005$ farads

Note: Although the capacitance value obtained in example 13.29 appears small, many electronic circuits require capacitors of much smaller value. Consequently, the farad is a cumbersome unit, far too large for most applications. The **microfarad,** which is one millionth of a farad (1 x 10⁻⁶ farad), is a more convenient unit. The symbols used to designated microfarad are μf.

Equation 13.41 can be rewritten as follows:

$$Q = CE \qquad\qquad (13.42)$$

$$E = \frac{Q}{C} \qquad\qquad (13.43)$$

Note: From equation 13.42, do not deduce the mistaken idea that capacitance is dependent upon charge and voltage. Capacitance is determined entirely by physical factors.

The symbol used to designate a capacitor is (C). The unit of capacitance is the farad (F). The farad is that capacitance that will store one coulomb of charge in the dielectric when the voltage applied across the capacitor terminals is one volt.

Factors Affecting the Value of Capacitance

The capacitance of a capacitor depends on three main factors: plate surface area; distance between plates; and dielectric constant of the insulating material.

- **Plate surface area:** Capacitance varies directly with place surface area. We can double the capacitance value by doubling the capacitor's plate surface area. Figure 13.62 shows a capacitor with a small surface area and another one with a large surface area. Adding more capacitor plates can increase the plate surface area. Figure 13.63 shows alternate plates connecting to opposite capacitor terminals.
- **Distance between plates:** Capacitance varies inversely with the distance between plate surfaces. The capacitance increases when the plates are closer together. Figure 13.64 shows capacitors with the same plate surface area, but with different spacing.
- **Dielectric constant of the insulating material:** An insulating material with a higher dielectric constant produces a higher capacitance rating. Figure 13.65 shows two capacitors. Both have the same plate surface area and spacing. Air is the dielectric in the first capacitor, and mica is the dielectric in the second

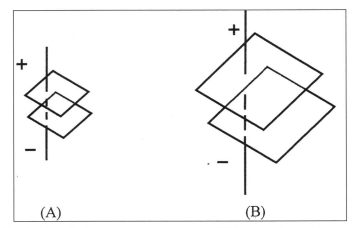

Figure 13.62 (A) Small plates, small capacitance; (B) Larger plates, higher capacitance.

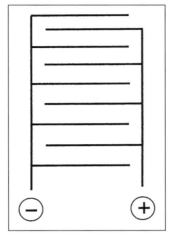

Figure 13.63   Several sets of plates connected to produce capacitor with more surface area.

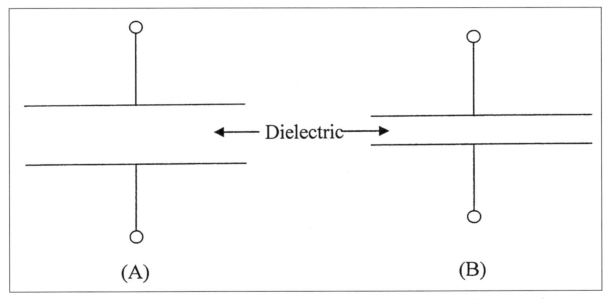

Figure 13.64   (A) Wide plate spacing, small capacitance; (B) Narrow plate spacing, larger capacitance.

one. Mica's dielectric constant is 5.4 times greater than air's dielectric constant. The mica capacitor has 5.4 times more capacitance than the air-dielectric capacitor.

Voltage Rating of Capacitors

There is a limit to the voltage that may be applied across any capacitor. If too large a voltage is applied, it will overcome the resistance of the dielectric and a current will be forced through it from one plate to the other, sometimes burning a hole in the dielectric. In this event, a short circuit exists and the capacitor must be discarded. The maximum voltage that may be applied to a capacitor is known as the **working voltage** and must never be exceeded.

The working voltage of a capacitor depends on 1. the type of material used as the dielectric, and 2. the thickness of the dielectric. As a margin of safety, the capacitor should be selected so that its working voltage

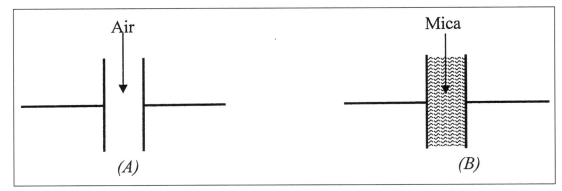

Figure 13.65   (A) Low capacitance; (B) Higher capacitance.

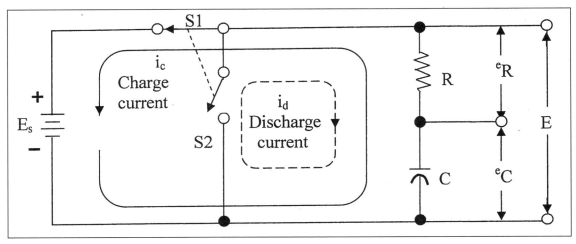

Figure 13.66   Charge and discharge of an RC series circuit.

is at least 50 percent greater than the highest voltage to be applied to it. For example, if a capacitor is expected to have a maximum of 200 volts applied to it, its working voltage should be at least 300 volts.

Charge and Discharge of an RC Series Circuit

According to Ohm's Law, the voltage across a resistance is equal to the current through it times the value of the resistance. This means that a voltage will be developed across a resistance **only when current flows through it**.

As previously stated, a capacitor is capable of storing or holding a charge of electrons. When uncharged, both plates contain the same number of free electrons. When charged, one plate contains more free electrons than the other. The difference in the number of electrons is a measure of the charge on the capacitor. The accumulation of this charge builds up a voltage across the terminals of the capacitor, and the charge continues to increase until this voltage equals the applied voltage. The greater the voltage, the greater the charge on the capacitor. Unless a discharge path is provided, a capacitor keeps its charge indefinitely. Any practical capacitor, however, has some leakage through the dielectric so that the voltage will gradually leak off.

A voltage divider containing resistance and capacitance may be connected in a circuit by means of a switch, as shown in figure 13.66. Such a series arrangement is called an **RC series circuit**.

If S1 is closed, electrons flow counterclockwise around the circuit containing the battery, capacitor, and resistor. This flow of electrons ceases when C is charged to the battery voltage. At the instant current begins to flow, there is no voltage on the capacitor and the drop across R is equal to the battery voltage. The initial charging current, I, is therefore equal to E_s/R.

The current flowing in the circuit soon charges the capacitor. Because the voltage on the capacitor is proportional to its charge, a voltage, e_c, will appear across the capacitor. This voltage opposes the battery voltage—that is, these two voltages buck each other. As a result, the voltage e_r across the resistor is $E_s - e_C$, and this is equal to the voltage drop $(i_C R)$ across the resistor. Because E_s is fixed, i_C decreases as e_C increases.

The charging process continues until the capacitor is fully charged and the voltage across it is equal to the battery voltage. At this instant, the voltage across is zero and no current flows through it.

If S2 is closed (S1 opened) in figure 13.66, a discharge current, i_d, will discharge the capacitor. Because i_d is opposite in direction to i_c, the voltage across the resistor will have a polarity opposite to the polarity during the charging time. However, this voltage will have the same magnitude and will vary in the same manner. During discharge the voltage across the capacitor is equal and opposite of the drop across the resistor. The voltage drops rapidly from its initial value and then approaches zero slowly.

The actual time it takes to charge or discharge is important in advanced electricity and electronics. Because the charge or discharge time depends on the values of resistance and capacitance, an RC circuit can be designed for the proper timing of certain electrical events. RC time constant is covered in the next section.

RC Time Constant

The time required to charge a capacitor to 63 percent of maximum voltage or to discharge it to 37 percent of its final voltage is known as the *time constant* of the current. An RC circuit is shown in figure 13.67.

The time constant T for a RC circuit is

$$T = RC \hspace{4cm} (13.44)$$

The time constant of an RC circuit is usually very short because the capacitance of a circuit may be only a few microfarads or even picofarads.

Note: An RC time constant expresses the charge and discharge times for a capacitor.

CAPACITORS IN SERIES AND PARALLEL

Like resistors or inductors, capacitors may be connected in series, in parallel, or in a series-parallel combination. Unlike resistors or inductors, however, total capacitance in series, in parallel, or in a series-parallel combination is found in a different manner. Simply put, the rules are not the same for the calculation of total capacitance. This difference is explained as follows:

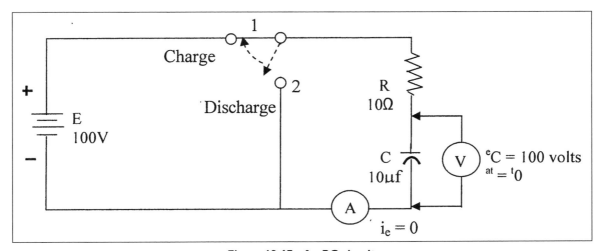

Figure 13.67 An RC circuit.

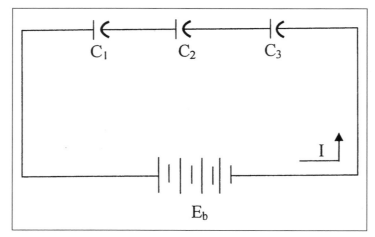

Figure 13.68   Series capacitive circuit.

Parallel capacitance is calculated like series resistance, and series capacitance is calculated like parallel resistance.

Capacitors in Series

When capacitors are connected in **series** (see figure 13.68), the total capacitance C_T is

$$\text{Series:}\quad \frac{1}{C_T} = \frac{1}{C_1} + \frac{1}{C_2} + \frac{1}{C_3} + \ldots + \frac{1}{C_n} \tag{13.45}$$

Example 13.30
Problem:
Find the total capacitance of a 3 µF, a 5 µF, and a 15 µF capacitor in series.

Solution:
   Write equation 13.45 for three capacitors in series.

$$\frac{1}{C_T} = \frac{1}{C_1} + \frac{1}{C_2} + \frac{1}{C_3}$$

$$= \frac{1}{3} + \frac{1}{5} + \frac{1}{15} \qquad = \frac{9}{15} = \frac{3}{5} = \frac{5}{3} = 1.7 \; \mu F$$

When capacitors are connected in *parallel* (see figure 13.69), the total capacitance C_T is the sum of the individual capacitances.

Capacitors in Parallel

$$C_T = C_1 + C_2 + C_3 + \ldots + C_n \tag{13.46}$$

Example 13.31
Problem:
Determine the total capacitance in a parallel capacitive circuit:

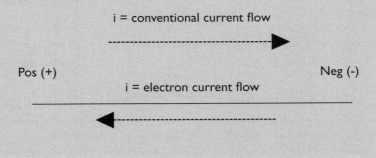

Did You Know?

Benjamin Franklin (1706–1790), without the benefit of our modern atomic theories, surmised that electricity traveled from positive to negative, setting a convention that we still use today. Based on modern theories, however, we note that the path of electron flow in a material is actually from negative to positive. Therefore, conventional current, which we use in circuit analysis, is the movement of positive charges. The movement of electrons in the material (called electron current) is actually in the opposite direction (NIOSH 1993).

i = conventional current flow

Pos (+)   Neg (-)

i = electron current flow

Given:

$$C_1 = 2 \ \mu F$$
$$C_2 = 3 \ \mu F$$
$$C_3 = 0.25 \ \mu F$$

Solution:

Write equation 13.46 for three capacitors in parallel:

$$C_T = C_1 + C_2 + C_3$$
$$= 2 + 3 + 0.25$$
$$= 5.25 \ \mu F$$

Capacitors in Series-Parallel

If capacitors are connected in a combination of *series and parallel* (see figure 13.70), the total capacitance is found by applying equations 13.45 and 13.46 to the individual branches.

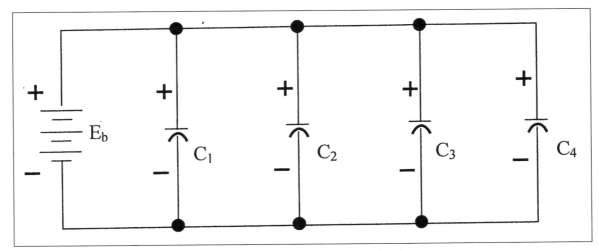

Figure 13.69 Parallel capacitive circuit.

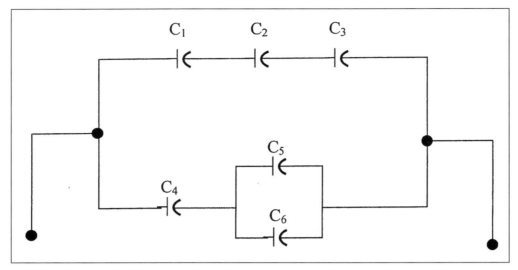

Figure 13.70   Series-parallel capacitance configuration.

Table 13.7 Comparison of Capacitor Types

Dielectric	Construction	Capacitance Range
Air	Meshed plates	10–400 pF
Mica	Stacked plates	10–5000 pF
Paper	Rolled foil	0.001–1 μF
Ceramic	Tubular	0.5–1600 pF
	disk	0.002–0.1 μF
Electrolytic	Aluminum	5–1000 μF
	tantalum	0.01–300 μF

TYPES OF CAPACITORS

Capacitors used for commercial applications are divided into two major groups—fixed and variable—and are named according to their dielectric. Most common are air, mica, paper, and ceramic capacitors, plus the electrolytic type. These types are compared in table 13.7.

The fixed capacitor has a set value of capacitances that is determined by its construction. The construction of the variable capacitor allows a range of capacitances. Within this range, the desired value of capacitance is obtained by some mechanical means, such as by turning a shaft (as in turning a radio tuner knob, for example) or adjusting a screw to adjust the distance between the plates.

The electrolytic capacitor consists of two metal plates separated by an electrolyte. The electrolyte, either paste or liquid, is in contact with the negative terminal, and this combination forms the negative electrode. The dielectric is a very thin film of oxide deposited on the positive electrode, which is aluminum sheet. Electrolytic capacitors are polarity sensitive (i.e., they must be connected in a circuit according to their polarity markings) used where a large amount of capacitance is required.

REVIEW QUESTIONS

1. What is the function of the wire conductor?
2. The resistance in ohms of a unit conductor or a given substance is called the _____.
3. The _____ _____ is the standard unit of wire cross-sectional area used in most wire tables.
4. Resistivity is the reciprocal of _____.
5. List the four factors that govern the selection of wire size.
6. Carbon has a _____ temperature coefficient.

7. What is the relationship between flux density and magnetic field strength?
8. Permeability depends on what two factors?
9. What is hysteresis?
10. Lines of force flow from _____ _____.
11. If the peak AC voltage across a resistor is 200 volts, what is the rms voltage?
12. The rms voltage generated by an alternator is 600 volts. What is the peak voltage?
13. List four factors affecting the inductance of a coil.
14. An increase in the number of turns in a coil will _____ inductance.
15. The inductance of an R/L circuit is 20 mh and the resistance is 5 ohms. What is the L/R time constant of the circuit?
16. Name the three factors that affect capacitance.
17. What is the total capacitance of two 60 microfarad connected in parallel?
18. The sum of all voltages in a series circuit is equal to:
19. What is the total resistance of four 30 Ω resistors connected in series?
20. There is (are) only _____ voltage across all components in parallel.

REFERENCES AND RECOMMENDED READING

Cooper, R. 1986. *Electrical safety engineering.* 2nd ed. London, England: Butterworth.

Hammer, W. 1989. *Occupational safety management and Engineering.* 4th ed. Englewood Cliffs, NJ: Prentice-Hall.

National Fire Protection Association. 1993. *The national electrical code handbook.* 6th ed. Boston: NFPA.

NIOSH. 1993. *An introduction to electrical safety for engineers.* Cincinnati, OH: National Institute for Occupational Safety and Health.

Ridley, J. 1996. *Safety at work.* 2nd ed. Boston: Butterworths.

Spellman, F. R. 2009. *Physics for nonphysicists.* Rockville, MD: Government Institutes Press.

Tapley, B., ed. 1990. *Eshbach's handbook of engineering fundamentals.* 4th ed. New York: John Wiley & Sons.

Engineering Site Security

You may say Homeland Security is a Y2K problem that doesn't end Jan. 1 of any given year.

—Governor Tom Ridge

The greatest of faults, I should say, is to be conscious of none.

—Thomas Carlyle

INTRODUCTION

One consequence of the events of 9/11 is a heightened concern among citizens in the United States over their personal security as well as the security of their vital infrastructure (transportation, manufacturing, energy, water, wastewater, electrical, telecommunications, and so forth). Governor Tom Ridge points out the security role for the public professional:

> Americans should find comfort in knowing that millions of their fellow citizens are working every day to ensure our security at every level—federal, state, county, municipal. These are dedicated professionals who are good at what they do. I've seen it up close, as Governor of Pennsylvania. . . . But there may be gaps in the system. The job of the Office of Homeland Security will be to identify those gaps and work to close them (Henry 2002).

Many of us are driven to increase security—to shore up the "gaps in the system."

SAFETY AND UPGRADING SECURITY

Safety and *security* are words that often belong together. 9/11 has brought the need for this pairing to the forefront. Moreover, it is not an overstatement when we say that since the phrase *homeland security* entered our national lexicon, safety and security have become fundamental to the design, construction, and management of buildings and other structures (bridges, roads, etc.).

Recent catastrophic events, most notably the 9/11 terrorist attacks on the World Trade Center and the Pentagon, have brought about fundamental changes in our lives and lifestyles. These events made all of us examine our lives, both at home and at work, in terms of safety and security, and security engineering has quickly become a rapidly growing field. People who have lived out their lives in peace and comfort take a degree of personal safety for granted and being jarred out of that complacency has advantages and disadvantages. One of the hardest lessons to learn, for most people, is that none of us can really control how safe we are. We can, however, take some sensible steps in that direction, and at work, for many different types of facilities, some added security is necessary.

Critical infrastructure operations such as utilities, public works, transportation systems, and local businesses hold the lives of their community in their hands. Think about how difficult living in a large metropolitan area becomes with flooding, for example, which damages property, disrupts electrical service and mobility, fouls the potable water supply, and releases wastewater into the floodwaters, contaminating that water. Now, think about the possibilities for disaster with deliberate sabotage of potable water and wastewater systems, roads and bridges, nuclear power sites, economic infrastructure (ATM machines), personal privacy (medical record systems), electrical power grids, communication systems, and others. Not a pleasant thought, is it? And, of course, many facilities store large quantities of potentially hazardous chemicals, too. A natural disaster can easily be terribly damaging and disruptive, as Hurricane Isabel demonstrated to the East Coast in September 2003 and as the even more devastating one-two punch provided by Hurricanes Katrina and Rita to the Gulf Coast in 2005. Deliberate sabotage adds to those problems the emotional damage caused by what the victims can only consider betrayal and treachery on the deepest human levels. In the United States of America, where certain rights and freedoms are safeguarded by our Constitution, such betrayal strikes at the foundation of our faith in our government and our fellow human creatures. By ignoring our own foundational belief in the value of our freedom, we may gain a marginal degree of improved safety, but we lose more.

SECURITY ENGINEERING

As mentioned, the catastrophic events of 9/11 brought about fundamental change to our lives and lifestyle. Again, due to these tragic events, security engineering has quickly become a rapidly growing field.

Ross J. Anderson (2008), one of the pioneers of security engineering as a formal field of study, points out that security engineering is a specialized field that deals with the development of detailed engineering plans and designs for security features, controls, and systems all intended to prevent misuse and malicious behavior. A sound security system is composed of the following elements:

Deter—detect—alarm—delay—respond

SECURITY ENGINEERING TRAINING

Recently, various U.S. universities have incorporated security engineering into their engineering schools course of study (curriculum). Security engineers, like safety engineers, must be generalists—Jacks or Jills of many trades and disciplines. Anderson (2008) points out that security engineers should be trained in aspects of social science, psychology (such as designing a system to "fail well" instead of trying to eliminate all sources of error—error elimination is an admirable goal but impossible so long as humans are involved), and economics as well as physics, chemistry, mathematics, architecture, and landscaping. Fault tree analysis, primarily used in safety engineering (and cryptography, primarily used in military applications) is also used in security engineering.

Based on our experience, the problem with university-level security engineering degree-granting programs, especially graduate-level programs, is that security engineering is thought of as an extension or subset of computer science concerned with the design and development of secure systems and applications. The emphasis is on training that covers security of computer networks and systems, with physical security usually placed on a back shelf somewhere. In addition to cryptography, protocols, and access control, the curriculum deals with policy and management issues, integration, and logistics and budgeting. Typical courses provided in these programs include:

- Computer System Security
- Operating Systems and Systems Software
- Data and Network Security
- Software Security
- Software Testing and Quality Assurance
- Logistics Systems Engineering
- User Interface Design
- Data Mining
- Software Reliability and Safety
- Fault Tolerant Computing

Based on experience, we have found that the above training is a plus for the safety engineer or anyone else concerned with security engineering. However, physical security is also an important (we feel it is the most important) part of engineering for security and cannot be overlooked. Besides, most companies go to great lengths, such as creating firewalls and network monitors, to protect against virtual attack. We have found, however, in many of the facilities we have inspected, physical security is nonexistent or an afterthought not yet enacted. This always confounds us; physical security prevents unauthorized access to information, property, and assets. Thus, physical security makes good sense—a no-brainer—and is absolutely necessary for survival in the post-9/11 environment. Even before 9/11, physical security was important in preventing destruction of property and disablement of communications as well as compromising propriety information. At the bare minimum, physical security should include:

- Surveillance
- Intrusion detection
- Access control
- Fire alarm
- Mass notifications and emergency communications
- Command and control

In order to ensure that safety engineers have the proper formal training in security engineering, we strongly recommend classroom coverage of all the topics discussed to this point. To ensure that physical security aspects are covered, we also recommend course study that includes:

- Introduction to Homeland Security
- Transportation Security
- Occupational Security/Physical Security
- Aviation Security
- Law and Policy
- Emergency Management
- Risk Assessment
- Introduction to Occupational Safety and Health
- Design of Engineering Hazard Controls
- Control and Command Structures

PHYSICAL SECURITY

Physical security is composed of elements dealing with explosion protection; obstacles to frustrate trivial attackers and delay serious ones; alarms, security lighting, security guard patrols, or closed-circuit television cameras to make it likely that attacks will be noticed; and security response to repel, catch, or frustrate attackers when an attack is detected. Ross (2008) points out that in a well-designed system, these features must complement each other. There are at least four layers of physical security:

- Environmental design
- Mechanical and electronic access control
- Intrusion detection
- Video monitoring

The goal of physical security implementation is to convince potential attackers that the likely costs of attack exceed the value of making the attack.

HAZARD ASSESSMENT

As with every other hazard to safety, health, and/or life, the first step toward increasing security is hazard analysis—the area or facility to be made secure must be identified. A threat assessment team can be created to evaluate the facility's vulnerability on several different levels. These include recommending and implementing employee training programs to improve employee response to threats to security, implementing facility-wide plans for dealing with threats to security, and creating communication channels for employees to bring concerns to effective attention. The threat assessment team should employ a three-step process for security analysis: hazard assessment, workplace security analysis, and workplace surveys.

Hazard Assessment Procedure

In industry, a security hazard assessment is a scientific approach to determine the security profile of an industrial facility or complex. In order to perform a hazard assessment, an assessment team should be formed. Because threats to a workplace's security can come from many different areas, assessment teams should include representatives from diverse groups within the facility. Management, operations, security, finance, legal, and human resources should all be represented, as should general employees. The hazard assessment should begin with a review of pertinent records.

A records review allows the development of a baseline, which will help the team analyze trends in past security threats. Records to examine include OSHA 200 and 300 logs, incidence reports, and records or information relating to assault or attempted assault incidents. Insurance, medical, and workers' compensation records should be examined. Police reports and accident investigations may also be useful. Other records that may show trends include grievance records, training records, and other miscellaneous records—minutes of pertinent meetings, for example. Communicating with other similar local businesses or community and civic associations to discuss their own experiences and concerns with security may also prove useful.

Workplace Security Analysis

When the threat assessment team inspects the workplace in a workplace security analysis, they are looking at both facility and work tasks to determine the presence of hazards, conditions, operations, and situations that might place workers at risk from either outside or inside threats. Follow-up inspections should be scheduled with some regularity, for continuous improvement.

Workplace Survey

As in any other type of hazard assessment, frequently the employees who face the hazards every day know more about potential problems or threats than anyone else. Pinpointing areas that make employees feel uncomfortable or insecure are key areas to examine closely and to implement changes. The team may wish to interview people who work in higher risk areas and will often receive more complete responses through questioning.

WHAT TO IMPLEMENT: HAZARD CONTROL AND PREVENTION

Control methods can be used to improve, eliminate, or minimize the risks involved with a breach of security by examining the general workplace design, workstations, area designs, and the existing security measures and by looking at the existing security equipment, work practice controls and procedures, and the workplace violence prevention program.

Buildings, Workstations, and Areas

- Review all new or renovated facilities designs to ensure safe and secure conditions. Design facilities to allow employees to communicate with other staff in emergency situations via clear partitions, video cameras, speakers or alarms, or other measures appropriate to the workplace.
- Prevent entrapment of the employees and/or minimize potential for assault incidents by the design of work areas and furniture placement.
- Control access to employee work areas by use of keyed entrances, buzzers or keycard accesses, and security badges.

- Adequate lighting systems increase safety and security, both in indoor areas and the grounds around the facility, especially parking areas. Lighting should meet the requirements of nationally recognized standards (ANSI A-85, ANSI/IES RP-7 1983, ANSI/IES RP-1 1993) and local building codes.

Security Equipment

- Use electronic alarm systems that are activated visually or audibly and that identify the location of the room or employee by an alarm, lighted indicator, or other effective system. Make sure such systems are adequately manned.
- Use closed-circuit televisions to monitor high-risk areas inside and outside the building.
- Use metal detection systems to identify persons with weapons.
- Use cell phones, beepers, CB radios, handheld alarms, or noise devices for personnel in the field.
- Inspect and repair security equipment regularly to ensure effectiveness.

Work Practice Controls and Procedures

- Provide identification cards for all employees and require employees to wear them.
- Establish sign-in and sign-out books and an escort policy for nonemployees.
- Base staffing consideration on safety and security assessment for both fixed-site and field locations.
- Develop internal communication systems to respond to emergencies.
- Develop policy on responding to emergency or hostage situations.
- Develop and implement security procedures for employees who work late or off hours; accounting for field staff; when to involve in-house security or local law enforcement in an assault incident; enforcing weapons ban in facilities; and employer response to assault incidents.
- Develop written procedures for employees who must enter locations where they feel threatened or unsafe.
- Provide information and give assistance to employees who are victims of domestic violence.
- Develop procedures to ensure confidentiality and safety for affected employees.
- Train employees on awareness, avoidance, and action to take to prevent mugging, robbery, rapes, and other assaults.

Don't Be a Stranger

Here's something to think about. The best on-the-job security for personnel is to know the people you're working with.

Sounds too simple, doesn't it? Yet this is how small towns used to function, and in some places, still do. As an example, middle schools and high schools small enough that the school administrators know all of the students tend to have fewer problems with student violence, suicide, and murder—by knowing their students better, teachers, administrators, and staff are better able to evaluate potential problems. As your home is more secure when you know your neighbors (they watch your property and carry in your paper when you go on vacation and vice versa), exchanging the ordinary information of polite conversation and getting to know the people you work with increases everyone's safety. When we pay attention to those around us, when we learn something about them, we have a better idea of their normal state of mind and condition—and would be more apt to know if something goes wrong for them. This kind of knowledge, though, must not happen just to be nosy but because we care about those we work with on the most basic human terms.

For safety engineers this may mean that your visible presence is an ordinary part of the worker's day in your facility—that you don't just hole up in your office, writing safety programs, doing paperwork, and reading CFRs. Remember, our goal is to create a good security program and not just a feel-good security program. A very effective school management philosophy is called "School Management by Walking Around" and is abbreviated as SMWA—among other elements, this methodology involves school principals and other administrators to visit classrooms, walk the hallways when students are present, chat with staff, faculty, and students during lunch and other activities—in short, to be an active, visible school presence. As safety engineer, you might be wise to adopt a level of Safety Management by Walking Around—your own version of SMWA.

Why is this important? Strangers are much easier to hate and to attack than those we know. Kidnappers and terrorists, as well as others, sometimes including people who open fire on small children in school yards or at crowds in public places, often begin by mentally mapping groups of other people as "the enemy," or as "less than human," or as "strangers." Kidnap victim survival rates go up if the victim can establish communication and a personal relationship with the kidnapper. Becoming something other than a face in the crowd to your coworkers reinforces your position as a real person to those around you. Encouraging your coworkers to interact with other people in other departments could improve their safety as well.

Table 14.1   Sample Employee Security Survey

This survey will help detect security problems in your building or at an alternate work site. Please fill out this form and get your coworkers to fill it out and review it to see where the potential for major security problems lie.

NAME:

WORK LOCATION:

(IN BUILDING OR ALTERNATE WORK SITE)

1. Do either of these two conditions exist in your building or at your alternate work site?

___ Work alone during working hours.

___ No notification given to anyone when you finish work.

Are these conditions a problem? If so, when? Please describe. (For example, Mondays, evening, daylight savings time)

2. Do you have any of the following complaints that may be associated with causing an unsafe work site? (Check all that apply.)

___ Does your workplace have a written policy to follow for addressing general problems?

___ Does your workplace have a written policy on how to handle a violent client?

___ When and how to request the assistance of a coworker

___ When and how to request the assistance of police

___ What to do about a verbal threat

___ What to do about a threat of violence

___ What to do about harassment

___ Working alone

___ Alarm system(s)

___ Security in and out of building

___ Security in parking lot

___ Have you been assaulted by a coworker?

___ To your knowledge, have incidents of violence ever occurred between your coworkers?

3. Are violence-related incidents worse during shift work, on the road, or in other situations?

Please specify: _____

4. Where in the building or work site would a violence-related incident be most likely to occur?

___ lounge ___ exits ___ deliveries ___ private offices

___ parking lot ___ bathroom ___ entrance ___ other (specify)_____

5. Have you ever noticed a situation that could lead to a violent incident?

6. Have you missed work because of potential violent act(s) committed during your course of employment?

7. Do you receive workplace violence-related training or assistance of any kind?

8. Has anything happened recently at your work site that could have led to violence?

9. Can you comment about the situation?

10. Has the number of violent clients increased?

Security and the Industrial Setting: What Does Facility Security Entail?

Different facilities have differing security needs. Community placement, the nature of the community, the facility location in the community, and the type of operations and equipment the facility is equipped for are all considerations for designing the security system. What would be essential for a facility that serves a large metropolitan area might be serious overkill for an industrial facility in an isolated rural area. In short, the security should match the possibility of threat and the need for protection caused by the value of what it protects. In many cases, systems put into place for protection from theft serve multiple purposes; keeping intruders out prevents more than theft. Security evaluation is essential for assessing the level of protection needed.

Areas of Concern for all Workplaces

Security provides three-way protection: protecting people from themselves and each other, ensuring the safety and security (in short, the integrity) of the facility, contents, and its environs, and liability protection. Security systems are intended to provide protection of property, operations, equipment, and personnel. Fencing is installed to restrict access to areas for reasons that include safety for both those outside and inside the fence; protection for processes, equipment, and operations; protection against theft and damage; and protection for outsiders from hazardous areas. Outsiders who might decide to come on facility property might just be curious—or they could be intent on casual or more serious damage, malicious mischief, theft, and physical harm to personnel or equipment. If the idea that some kid might break into facility property on a dare and get killed messing around where he had no business being seems far-fetched, think about the graffiti spray-painted onto water towers, and think again.

A serious fence, lock, and gate system (especially with warning signage) is a visible symbol that protects a facility physically, psychologically, and legally. Many people who would willfully ignore a "stay off the grass" sign would never think of deliberately scaling a fence to enter a similarly restricted area—and even if they did, they couldn't pass it off as only a casual mistake. Fences and other outward signs of enforced security work to discourage the attention of unwanted visitors. By presenting the image of a difficult target, visible security bespeaks the idea that breaking in isn't worth the effort it will take.

But fencing isn't the only necessary outside security measure. Of high concern is improving facility visibility, including trimming landscaping so that it doesn't provide cover, and directional lighting systems to illuminate potential problem areas. After all, facility protection is also for protecting personnel. Parking areas should be well lit and access to them limited. Security guards should escort workers who don't feel secure walking to their car at night, for example.

Many industrial facilities have to deal with the problem of security for outlying facilities and equipment as well as security for the principle facility. Off-site work centers, transportable equipment that might need to be left at a site for the duration of a job, access points to utility systems (manholes) and other outlying facilities can all be targets of theft, damage, graffiti, and unwanted and dangerous exploration. Remote security cameras, stout fencing, and other means of making entry difficult, solid locks, lighting, trimmed landscaping, and regular monitoring help to protect these areas. Such measures also provide a certain level of liability protection, as well. Actively preventing idle access by fences and warning signage flags the organization's efforts for safety. Under such circumstances, flagrant disregard of safety measures and warnings cannot easily be blamed on the organization.

Protection from Theft

Materials that have intrinsic value must be protected—no need to go any further into why theft is of concern. However, the definition of "theft" is changing, and many people don't think of activities they do as "stealing," although management might. Blatant theft of tangible goods is hard to argue over, but theft of small items and of time are becoming more and more important to those who must keep track of the bottom line. Both internal and external security systems can help control costly losses.

Expensive equipment (portable-sized power equipment, computer components, and other office machines) may have to be kept in place by physically locking down the equipment. Fully document and register equipment information and model numbers and serial numbers for insurance as well as for tracking of stolen goods.

Tools are another big area of concern. Both small hand tools and power tools as well as larger power equipment can be vulnerable to "walking off the site" without some sort of controlled access system, especially in areas where security cameras are impractical. Guilty parties can include facility workers and outsider theft as well as subcontractor crewmembers or temporary workers. In our experience, tools can vanish overnight, and some member of a trusted contractor crew's home shop was improved. These job "souvenirs"—often very practical souvenirs—can be kept where they belong with the use of a regulated tag-in/tag-out system and lockdown system for larger tools. Security camera tapes can provide evidence.

Equipment that must remain portable presents additional security headaches. For personnel, the tag-in/ tag-out systems may work very well, but for equipment that would be attractive to outsiders (laptop computers, for example, which are being used more and more in the field), workers may need training on techniques for protecting such equipment from theft. Key to keeping such equipment in the inventory is making sure the employee using it keeps the equipment secured at all times.

Losses of small office and shop goods don't seem like much on the surface, but they add up fast. The half-deliberate theft or absentminded disappearance of common office supplies or the absolutely deliberate theft of tools and equipment or coffee service perishables or cleaning supplies has happened at every facility at one time or another. While management doesn't want to feel like "Big Brother" and workers don't want to feel like "Big Brother Is Watching You," supplies are expensive, and the time spent in inventorying and replacing them adds to the cost.

"Theft" can also include time employees spend attending to tasks other than their assigned duties. While employers need to keep a sense of balance about time valued employees spend "on the job" but not on *their* job, excessive use of personal phone calls, email, chatter, or too much time with the solitaire game on the screen costs the company money, just as stolen equipment does. Keeping track of these losses is much more possible now than in the past.

Whether or not employees sign a contract or agreement that stipulates the boundaries of their employment, an implied agreement exists between employer and employee—and this agreement works both ways. Disgruntled workers sometimes feel "entitled" to break that contract; whether their grievances are real or imagined does not mean that they aren't seriously angry or unhappy about some aspect of their employment. The results can be loss of worker productivity, deliberate damage, theft, lawsuits, or in extreme cases, violence. Employers who work various legal or illegal methods to exploit their workforce need to understand that underpaying workers, allowing unsafe conditions, denying them standard benefits, and ignoring their basic needs puts a serious strain on the employer/employee relationship—particularly when jobs are being cut and top management is pulling down big bucks. While employees should give workplace value for their pay, employers should give concrete evidence of concern for their workers as well. The safety engineer and the safety division is a visible sign of employer concern for the employee.

Protection from Harm

In the case of industrial facilities and general businesses, security systems also provide protection for the public from hazardous processes and materials. Layers of security that prevent unauthorized people from going where they have no business being can protect outsiders as well as workers from everything from minor annoyances to gunfire. Key-code, password, or ID-card entry systems that limit access to the building and reception areas, and locked access doors to nonpublic areas (crash-barred on the inside, of course) keep unwelcome guests out. Use common sense in allowing former employees access to former work areas; disgruntled former employees have been known to "go postal" in more facilities than USPS offices.

To go along with entry systems, a visible security presence—guards on duty, security cameras, security firm signage or reception or guard desk entry that requests sign-in and sign-out and proof of identification—provides a disincentive for interested outsiders.

ON-SITE HAZARDOUS MATERIALS SECURITY

All sorts of materials that industrial plants use or manufacture are in and of themselves dangerous; to handle them or work around them, workers must undergo training for their own safety. Measures put into place to ensure that an unprepared worker doesn't walk into a place that might pose an inhalant risk for methane, for example, also protects anyone who passes by, not just the worker. Well-designed security measures provide safety as well.

Did You Know?

OSHA defines a hazardous material as:

Any item or agent (biological, chemical, physical) which has the potential to cause harm to humans, animals, or the environment, either by itself or through interaction with other factors.

—OSHA 2009

Hazardous materials are always an issue under OSHA's jurisdiction, but safety engineers should also be aware of hazardous materials security. While we're accustomed to thinking about hazardous materials in terms of safety for workers and the surrounding community, we have sometimes been less stringent about how securely our chemicals are stored on site. Hazardous materials safety entails more than all the rules and regulations that go along with transporting, storing, using, and disposing of chemicals and the attendant waste products that remain after the chemicals do their job. Proper compliance with process safety management and risk management planning (OSHA's 29 CFR 1910.119) must contain a "worst-case scenario" emergency plan for facilities that take into account all of the hazardous substances on-site. The last several years have enlarged the terrible possibilities of the "worst-case scenario." The scope of our attention to hazardous materials security has been changing over the last several years, and security needs will continue to increase. We can't limit our site protection to keeping out petty theft, malicious mischief, and straightforward accident protection anymore.

Prevent Random Acts of Violence

Companies that handle hazardous materials have long known that casual trespass is risky—even someone who doesn't mean any harm can cause problems tampering with equipment or materials they don't understand. More and more common is deliberate damage. Plant security systems limit physical access to the plant and to the hazardous materials themselves on-site. Many safety engineers are finding that facility security must be more than warning signage and fencing; facilities must now create serious barriers, whether physical or electronic, securing site and personnel, and providing higher levels of community protection.

What's Normal Isn't What's Normal for You

Many hazardous materials used in public utilities and in industry are used in such quantities that removing them from the facility takes specialized equipment and considerable effort. Tank-cars can't just hitch to the back of the average family car and pull away. While this isn't true for all chemicals with the potential for trouble, chlorine tanks, for example, are well-protected, designed with several layers of security in place—the dome tank lid, the cable sealing system, the heavy-duty valve, and the valve seal plugs. Releasing the chlorine takes a level of knowledge and the right tools to breach the tanks. This, however, presumes a reasonably "normal" use of the contents of the tank. Look at the Material Safety Data Sheet (MSDS) for chlorine, and use a little suspense-movie imagination. Someone who wanted to create havoc in a community could cause deadly results in a fairly widespread area, especially if they didn't care about their own survival.

It's in My Backyard and That's Where It Belongs

As an example of why hazardous materials are becoming such a widespread problem, consider the black market use of anhydrous ammonia. Businesses that use anhydrous ammonia and the law enforcement and environmental problems that have attended the growing illicit market for this hazardous material illustrate the widespread growing need for facility security. Bootleg methamphetamine (meth) labs have become common (rural agricultural areas have been especially hard-hit—old moonshine stills have been replaced by meth labs), and because of the related problems, both chemical tank security and facility security had to improve. Small amounts of anhydrous ammonia are part of the process used to manufacture meth, but it's a substance subject to control, although it is not particularly expensive. To acquire it without leaving a paper trail, meth manufacturers either steal it or buy it illicitly. Black market prices are high (sometimes reaching more per gallon on the black market than a legitimate buyer would pay per ton), making theft of this chemical attractive to a wide spectrum of criminals. Thieves cut through fences to reach the tanks and break or saw off valves to rig chemical transfer. They damage equipment and facilities and leave valves open, releasing the remainder of the stored supplies and dangerously exposing themselves. Those who live near such spills, as well as rescue, law enforcement, and clean-up personnel, are put at risk for exposure, and the spills also cause serious environmental problems. To top it off, tanks and meth labs in rural areas end up putting a tremendous financial and human-power strain on local community services—services not ordinarily accustomed to handling the processes for criminal activity of this kind.

Obviously, we have to look at hazardous materials availability and usage more cynically than we did even a few years ago. We can't just think about standard usages any more—safety engineers have to think about much more unpleasantly creative possibilities.

The Protection Dilemma

Planning protection against terrorist attack leads to an examination of the "protection dilemma." Experience has shown us that no one-size-fits-all when it comes to implementing security schemes. How much, in terms of resources, should a business or industrial facility devote to securing the areas where dangerous materials are stored, just in case? Statistically, the odds of someone breaking into a facility to deliberately cause a toxic material release are pretty slim—but we're all now hyperaware that the odds of someone deciding to deliberately direct passenger jets into skyscrapers was pretty low, too, until 9/11. As a safety engineer, your written emergency response plans and actual procedures may need to be altered to include the possibility of your facility as a target for terrorism, as well as vandalism, bomb threats, theft, and malicious mischief. Protection against terrorism presents a difficult problem, simply because the issue is amorphous. What kind of attack? When? Military personnel know too well that stopping an all-out attack from a group bent on destruction, willing to die for their cause in the process—and armed with enough force—can be virtually impossible to stop. As a civilian safety engineer, armed with your emergency response plan, a big part of your task is evaluating the potential value of your facility to a potential terrorist. High-interest industries (does your facility manufacture directional devices for mirv-type missiles?) and facilities whose destruction could cause great havoc are of more potential interest than a plastics widget plant in the rural Midwest. The World Trade Center and the Pentagon were targeted because of their potential for the greatest amount of bang for the terrorist buck, to put it bluntly.

When security companies plan security systems for a building, they base their plans on a number of concerns, which include the value of the contents of the building, the attractiveness of the contents to a potential threat, and the security of the building itself. As mentioned, the goal is to make entry into the facility more difficult than the contents are worth—in short, more costly to the potential entrant than the value of the materials being guarded.

Checklist Questions for Industrial Facilities

- Is the facility protected by fences?
- Do fences or buildings protect critical equipment and chemicals?
- Do in-place systems serve to detect intruders (guards, security cameras)?
- Is the facility protected by an alarm system?
- Do critical areas have controlled access?

Is Your Current System Adequate?

- Chemicals Available: Chemicals stored on-site may be a particularly attractive target because of the potential of greater damage if they are released.
- Site Location: Sites in densely populated areas may need more stringent security measures than those distant from populous areas. Yet remote sites may provide easier, thus more attractive, targets.
- Site Accessibility: Can your existing security systems do the job of repelling intruders?
- Facility Age and Condition: Both older and newer buildings can have security problems—older buildings often have more windows; newer ones are sometimes designed for easy access.

- Hours of Operation: Facilities with 24/7 hours of operation typically need less security than those with limited nighttime and weekend hours. However, if facility traffic is markedly less at certain times, the need for increased security measures at specific hours should be evaluated.

Safe Work Practices

- Bad community or employee relations can increase facility risk of attack.
- Don't forget rapid shutdown procedures in emergency planning, especially for those processes that operate under extreme conditions of pressure or temperature.
- Disconnecting outside or otherwise vulnerable storage tanks or delivery vehicles from connecting pipes, hoses, or distribution systems may decrease the chances of deliberate or accidental release.
- Remember, details of security systems should not be included in the facility emergency plan because that plan must be made available to the general public.

PHYSICAL ASSET MONITORING AND CONTROL DEVICES

Aboveground, Outdoor Equipment Enclosures

Many industrial facilities and systems consist of multiple components spread over a wide area and typically include a centralized operations area as well as satellite offices or building complexes that are typically distributed at multiple locations throughout the community. However, in recent years, system designers have favored placing critical operational equipment—especially assets that require regular use and maintenance—aboveground.

One of the primary reasons for doing so is that locating this equipment aboveground eliminates the safety risks associated with confined space entry, which is often required for the maintenance of equipment located belowground. In addition, space restrictions often limit the amount of equipment that can be located inside, and there are concerns that some types of equipment (chemical processing units) can, under certain circumstances, discharge chemicals and off-gases that could flood pits, vaults, or equipment rooms. Therefore, many pieces of critical equipment are located outdoors and aboveground.

Many different system components can be installed outdoors and aboveground. Examples of these types of components could include:

- Backflow prevention devices
- Air release and control valves
- Pressure vacuum breakers
- Pumps and motors
- Chemical storage and feed equipment
- Meters
- Sampling equipment
- Instrumentation

Much of this equipment is installed in remote locations and/or in areas where the public can access it.

One of the most effective security measures for protecting aboveground equipment is to place it inside a building. When/where this is not possible, enclosing the equipment or parts of the equipment using some sort of commercial or homemade add-on structure may help to prevent tampering with the equipment. These types of add-on structures or enclosures, which are designed to protect the equipment both from the elements and from unauthorized access or tampering, typically consist of a boxlike structure that is placed over the entire component or over critical parts of the component (i.e., valves, etc.), and is then secured to delay or prevent intruders from tampering with the equipment. The enclosures are typically locked or otherwise anchored to a solid foundation, which makes it difficult for unauthorized personnel to remove the enclosure and access the equipment.

Standardized aboveground enclosures are available in a wide variety of materials, sizes, and configurations. Many options and security features are also available for each type of enclosure, and this allows system operators the flexibility to customize an enclosure for a specific application and/or price range. In addition, most manufacturers can custom-design enclosures if standard, off-the-shelf enclosures do not meet a user's needs.

Many of these enclosures are designed to meet certain standards. For example, the American Society of Sanitary Engineers (ASSE) has developed Standard #1060, *Performance Requirements for Outdoor Enclosures for Backflow Prevention Assemblies*. If an enclosure will be used to house a backflow preventer, this standard specifies the acceptable construction materials for the enclosure as well as the performance requirements that the enclosure should meet, including specifications for freeze protection, drainage, air inlets, access for maintenance, and hinge requirements. ASSE #1060 also states that the enclosure should be lockable to enhance security.

Did You Know?

A backflow prevention device is used to protect water supplies from contamination or pollution.

Equipment enclosures can generally be categorized into one of four main configurations, which include:

- One-piece, drop-over enclosures
- Hinged or removable top enclosures
- Sectional enclosures
- Shelters with access locks

All enclosures, including those with integral floors, must be secured to a foundation to prevent them from being moved or removed. Unanchored or poorly anchored enclosures may be blown off the equipment being protected or may be defeated by intruders. In either case, this may result in the equipment beneath the enclosure becoming exposed and damaged. Therefore, ensuring that the enclosure is securely anchored will increase the security of the protected equipment.

The three basic types of foundations that can be used to anchor the aboveground equipment enclosure are concrete footers, concrete slabs-on-grade, or manufactured fiberglass pads. The most common types of foundations used for equipment enclosures are standard or slab-on-grade footers; however, local climate and soil conditions may dictate whether either of these types of foundations can be used. These foundations can be either precast or poured in place at the installation site. Once the foundation is installed and properly cured, the equipment enclosure is bolted or anchored to the foundation to secure it in place.

An alternative foundation, specifically for use with smaller hot-box enclosures, is a manufactured fiberglass pad known as the Glass Pad™. The Glass Pad™ has the center cut out so that it can be dropped directly over the piece of equipment being enclosed. Once the pad is set level on the ground, it is backfilled over a two-inch flange located around its base. The enclosure is then placed on top of the foundation and is locked in place with either a staple anchor or a slotted anchor, depending on the enclosure configuration.

One of the primary attributes of a security enclosure is its strength and resistance to breaking and penetration. Accordingly, the materials from which the enclosure is constructed will be important in determining the strength of the enclosure, and thus its usefulness for security applications. Enclosures are typically manufactured from either fiberglass or aluminum. With the exception of the one-piece, drop-over enclosure, which is typically made of fiberglass, each configuration described above can be constructed from either material. In addition, enclosures can be custom-manufactured from polyurethane, galvanized steel, or stainless steel. Galvanized or stainless steel is often offered as an exterior layer, or "skin," for an aluminum enclosure. Although they are typically used in underground applications, precast concrete structures can also be used as aboveground equipment enclosures. However, precast structures are much heavier and more difficult to maneuver than are their fiberglass and aluminum counterparts. Concrete is also brittle, and that can be a security concern; however, products can be applied to concrete structures to add strength and minimize security risks (i.e., epoxy coating). Because precast concrete structures can be purchased from any concrete producers, this document does not identify specific vendors for these types of products.

In addition to the construction materials, enclosure walls can be configured or reinforced to give them added strength. Adding insulation is one option that can strengthen the structural characteristics of an enclosure; however, some manufacturers offer additional features to add strength to exterior walls. For example, while most enclosures are fabricated with a flat-wall construction, some vendors manufacture fiberglass shelters with ribbed exterior walls. These ribs increase the structural integrity of the wall and allow the fabrication of standard shelters up to twenty feet in length. Another vendor has developed a proprietary process that uses a series of integrated fiberglass beams that are placed throughout a foam inner core to tie together the interior and exterior walls and roof. Yet another vendor constructs aluminum enclosures with horizontal and vertical redwood beams for structural support.

Other security features that can be implemented on aboveground, outdoor equipment enclosures include locks, mounting brackets, tamper-resistant doors, and exterior lighting.

Active Security Barriers (Crash Barriers)

Active security barriers (also known as crash barriers) are large structures that are placed in roadways at entrance and exit points to protected facilities to control vehicle access to these areas. These barriers are placed perpendicular to traffic to block the roadway, so that the only way that traffic can pass the barrier is for the barrier to be moved out of the roadway. These types of barriers are typically constructed from sturdy materials, such as concrete or steel, such that vehicles cannot penetrate through them. They are also designed at a certain height off the roadway so that vehicles cannot go over them.

The key difference between active security barriers, which include wedges, crash beams, gates, retractable bollards, and portable barricades; and passive security barriers, which include nonmoveable bollards, jersey barriers, and planters, is that active security barriers are designed so that they can be raised and lowered or moved out of the roadway easily to allow authorized vehicles to pass them. Many of these types of barriers are designed so that they can be opened and closed automatically (i.e., mechanized gates, hydraulic wedge barriers), while others are easy to open and close manually (swing crash beams, manual gates). In contrast to active barriers, passive barriers are permanent, nonmovable barriers, and thus they are typically used to protect the perimeter of a protected facility, such as sidewalks and other areas that do not require vehicular traffic to pass them. Several of the major types of active security barriers such as wedge barriers, crash beams, gates, bollards, and portable/removable barricades are described below.

Wedge barriers are plated, rectangular, steel buttresses approximately two to three feet high that can be raised and lowered from the roadway. When they are in the open position, they are flush with the roadway and vehicles can pass over them. However, when they are in the closed (armed) position, they project up from the road at a forty-five degree angle, with the upper end pointing toward the oncoming vehicle and the base of the barrier away from the vehicle. Generally, wedge barriers are constructed from heavy gauge steel or concrete that contains an impact-dampening iron rebar core that is strong and resistant to breaking or cracking, thereby allowing them to withstand the impact from a vehicle attempting to crash through them. In addition, both of these materials help to transfer the energy of the impact over the barrier's entire volume, thus helping to prevent the barrier from being sheared off its base. In addition, because the barrier is angled away from traffic, the force of any vehicle impacting the barrier is distributed over the entire surface of the barrier and is not concentrated at the base, which helps prevent the barrier from breaking off at the base. Finally, the angle of the barrier helps hang up any vehicles attempting to drive over it.

Wedge barriers can be fixed or portable. Fixed wedge barriers can be mounted on the surface of the roadway ("surface-mounted wedges") or in a shallow mount in the road's surface, or they can be installed completely below the road surface. Surface-mounted wedge barricades operate by rising from a flat position on the surface of the roadway, while shallow-mount wedge barriers rise from their resting position just below the road surface. In contrast, below-surface wedge barriers operate by rising from beneath the road surface. Both the shallow-mounted and surface-mounted barriers require little or no excavation and thus do not interfere with buried utilities. All three barrier mounting types project above the road surface and block traffic when they are raised into the armed position. Once they are disarmed and lowered, they are flush with the road, thereby allowing traffic to pass. Portable wedge barriers are moved into place on wheels that are removed after the barrier has been set into place.

Installing rising wedge barriers requires preparation of the road surface. Installing surface-mounted wedges does not require that the road be excavated; however, the road surface must be intact and strong enough to allow the bolts anchoring the wedge to the road surface to attach properly. Shallow-mount and below-surface wedge barricades require excavation of a pit that is large enough to accommodate the wedge structure as well

Table 14.2   Pros and Cons of Wedge Barriers

Pros	Cons
Can be surface-mounted or completely installed below the roadway surface.	Installations below the surface of the roadway will require construction that may interfere with buried utilities.
Wedge barriers have a quick response time (normally 3.5 to 10.5 seconds, but can be 1 to 3 seconds in emergency situations. Because emergency activation of the barrier causes more wear and tear on the system than does normal activation, it is recommended for use only in true emergency situations.	Regular maintenance is needed to keep wedge barrier fully operational.
Surface or shallow-mount wedge barricades can be used in locations with a high water table and/or corrosive soils.	Improper use of the system may result in authorized vehicles being hung up by the barrier and damaged. Guards must be trained to use the system properly to ensure that this does not happen. Safety technologies may also be installed to reduce the risk of the wedge activating under an authorized vehicle.
All three wedge barrier designs have a high crash rating, thereby allowing them to be employed for higher security applications.	
These types of barriers are extremely visible, which may deter potential intruders.	

Source: USEPA, 2005.

as any arming/disarming mechanisms. Generally, the bottom of the excavation pit is lined with gravel to allow for drainage. Areas not sheltered from rain or surface runoff can install a gravity drain or self-priming pump. Table 14.2 lists the pros and cons of wedge barriers.

Crash beam barriers consist of aluminum beams that can be opened or closed across the roadway. While there are several different crash beam designs, every crash beam system consists of an aluminum beam that is supported on each side by a solid footing or buttress, which is typically constructed from concrete, steel, or some other strong material. Beams typically contain an interior steel cable (typically at least one inch in diameter) to give the beam added strength and rigidity. The beam is connected by a heavy-duty hinge or other mechanism to one of the footings so that it can swing or rotate out of the roadway when it is open and can swing back across the road when it is in the closed (armed) position, blocking the road and inhibiting access by unauthorized vehicles. The nonhinged end of the beam can be locked into its footing, thus providing anchoring for the beam on both sides of the road and increasing the beam's resistance to any vehicles attempting to penetrate through it. In addition, if the crash beam is hit by a vehicle, the aluminum beam transfers the impact energy to the interior cable, which in turn transfers the impact energy through the footings and into their foundation, thereby minimizing the chance that the impact will snap the beam and allow the intruding vehicle to pass through.

Crash beam barriers can employ drop-arm, cantilever, or swing beam designs. Drop-arm crash beams operate by raising and lowering the beam vertically across the road. Cantilever crash beams are projecting structures that are opened and closed by extending the beam from the hinge buttress to the receiving buttress located on the opposite side of the road. In the swing beam design, the beam is hinged to the buttress such that it swings horizontally across the road. Generally, swing beam and cantilever designs are used at locations where a vertical lift beam is impractical. For example, the swing beam or cantilever designs are used at entrances and exits with overhangs, trees, or buildings that would physically block the operation of the drop-arm beam design.

Installing any of these crash beam barriers involves the excavation of a pit approximately forty-eight inches deep for both the hinge and the receiver footings. Due to the depth of excavation, the site should be inspected for underground utilities before digging begins. Table 14.3 lists the pros and cons of crash beams.

In contrast to wedge barriers and crash beams, which are typically installed separately from a fence line, *gates* are often integrated units of a perimeter fence or wall around a facility.

Gates are basically movable pieces of fencing that can be opened and closed across a road. When the gate is in the closed (armed) position, the leaves of the gate lock into steel buttresses that are embedded in concrete foundation located on both sides of the roadway, thereby blocking access to the roadway. Generally, gate barricades are constructed from a combination of heavy gauge steel and aluminum that can absorb an impact

Table 14.3   Pros and Cons of Crash Beams

Pros	Cons
Requires little maintenance while providing long-term durability.	Crash beams have a slower response time (normally 9.5 to 15.3 seconds, but can be reduced to 7 to 10 seconds in emergency situations) than do other types of active security barriers, such as wedge barriers. Because emergency activation of the barrier causes more wear and tear on the system than does normal activation, it is recommended for use only in true emergency situations.
No excavation is required in the roadway itself to install crash beams.	All three crash beam designs possess a low crash rating relative to other types of barriers, such as wedge barriers, and thus they typically are used for lower security applications.
	Certain crash barriers may not be visible to oncoming traffic and therefore may require additional lighting and/or other warning markings to reduce the potential for traffic to accidentally run into the beam.

Source: USEPA, 2005.

Table 14.4   Pros and Cons of Gates

Pros	Cons
All three gate designs possess an intermediate crash rating, thereby allowing them to be used for medium to higher security applications.	Gates have a slower response time (normally 10 to 15 seconds, but can be reduced to 7 to 10 seconds in emergency situations) than do other types of active security barriers, such as wedge barriers. Because emergency activation of the barrier causes more wear and tear on the system than does normal activation, it is recommended for use only in true emergency situations.
Requires very little maintenance.	
Can be tailored to blend in with perimeter fencing.	
Gate construction requires no roadway excavation.	
Cantilever gates are useful for roads with high crowns or drainage gutters.	
These types of barriers are extremely visible, which may deter intruders.	
Gates can also be used to control pedestrian traffic.	

Source: USEPA, 2005.

from vehicles attempting to ram through them. Any remaining impact energy not absorbed by the gate material is transferred to the steel buttresses and their concrete foundation.

Gates can use a cantilever, linear, or swing design. Cantilever gates are projecting structures that operate by extending the gate from the hinge footing across the roadway to the receiver footing. A linear gate is designed to slide across the road on tracks via a rack and pinion drive mechanism. Swing gates are hinged so that they can swing horizontally across the road.

Installation of the cantilever, linear, or swing gate designs described above involve the excavation of a pit approximately forty-eight inches deep for both the hinge and receiver footings to which the gates are attached. Due to the depth of excavation, the site should be inspected for underground utilities before digging begins. Table 14.4 lists the pros and cons of gates.

Bollards are vertical barriers at least three feet tall and one to two feet in diameter that are typically set four to five feet apart from each other so that they block vehicles from passing between them. Bollards can either be fixed in place, removable, or retractable. Fixed and removable bollards are passive barriers that are typically used along building perimeters or on sidewalks to prevent vehicles from them, while allowing pedestrians to pass them. In contrast to passive bollards, retractable bollards are active security barriers that can easily be raised and lowered to allow vehicles to pass between them. Thus, they can be used in driveways or on roads to control vehicular access. When the bollards are raised, they project above the road surface and block the roadway; when they are lowered, they sit flush with the road surface, and thus allow traffic to pass over them. Retractable bollards are typically constructed from steel or other materials that have a low weight-to-volume ratio so that they require low power to raise and lower. Steel is also more resistant to breaking than is a more brittle material, such as concrete, and is better able to withstand direct vehicular impact without breaking apart.

Retractable bollards are installed in a trench dug across a roadway—typically at an entrance or gate. Installing retractable bollards requires preparing the road surface. Depending on the vendor, bollards can be installed either in a continuous slab of concrete or in individual excavations with concrete poured in place. The required excavation for a bollard is typically slightly wider and slightly deeper than the bollard height when extended aboveground. The bottom of the excavation is typically lined with gravel to allow drainage. The bollards are then connected to a control panel that controls the raising and lowering of the bollards. Installation typically requires mechanical, electrical, and concrete work; if utility personnel with these skills are available, then the utility can install the bollards themselves. Table 14.5 lists the pros and cons of retractable bollards.

Portable/removable barriers, which can include removable crash beams and wedge barriers, are mobile obstacles that can be moved in and out of position on a roadway. For example, a crash beam may be completely removed and stored off-site when it is not needed. An additional example would be wedge barriers that are equipped with wheels that can be removed after the barricade is towed into place.

When portable barricades are needed, they can be moved into position rapidly. To provide them with added strength and stability, they are typically anchored to buttress boxes that are located on either side of the road. These buttress boxes, which may or may not be permanent, are usually filled with sand, water, cement, gravel, or concrete to make them heavy and aid in stabilizing the portable barrier. In addition, these buttresses can help dissipate any impact energy from vehicles crashing into the barrier itself.

Because these barriers are not anchored into the roadway, they do not require excavation or other related construction for installation. In contrast, they can be assembled and made operational in a short period of time. The primary shortcoming to this type of design is that these barriers may move if they are hit by vehicles. Therefore, it is important to carefully assess the placement and anchoring of these types of barriers to

Table 14.5   Pros and Cons of Retractable Bollards

Pros	Cons
Bollards have a quick response time (normally 3 to 10 seconds, but can be reduced to 1 to 3 seconds in emergency situations).	Bollard installations will require construction below the surface of the roadway, which may interfere with buried utilities.
Bollards have an intermediate crash rating, which allows them to be used for medium to higher security applications.	Some maintenance is needed to ensure barrier is free to move up and down.
	The distance between bollards must be decreased (i.e., more bollards must be installed along the same perimeter) to make these systems effective against small vehicles (i.e., motorcycles).

Source: USEPA, 2005.

Table 14.6   Pros and Cons of Portable/Removable Barricades

Pros	Cons
Installing portable barricades requires no foundation or roadway excavation.	Portable barriers may move slightly when hit by a vehicle, resulting in a lower crash resistance.
Can be moved in and out of position in a short period of time.	Portable barricades typically require 7.75 to 16.25 seconds to move into place, and thus they are considered to have a medium response time when compared with other active barriers.
Wedge barriers equipped with wheels can be easily towed into place.	
Minimal maintenance is needed to keep barriers fully operational.	

Source: USEPA, 2005.

ensure that they can withstand the types of impacts that may be anticipated at that location. Table 14.6 lists the pros and cons of portable/removable barricades.

Because the primary threat to active security barriers is that vehicles will attempt to crash through them, their most important attributes are their size, strength, and crash resistance. Other important features for an active security barrier are the mechanisms by which the barrier is raised and lowered to allow authorized vehicle entry, and other factors, such as weather resistance and safety features.

Alarms

An *alarm system* is a type of electronic monitoring system that is used to detect and respond to specific types of events—such as unauthorized access to an asset, or a possible fire. In industrial and business facilities, alarms are also used to alert operators/employees when process operating or monitoring conditions go out of preset parameters (i.e., process alarms). These types of alarms are primarily integrated with process monitoring and reporting systems (i.e., SCADA systems). Note that this discussion does not focus on alarm systems that are not related to a utility's processes.

Alarm systems can be integrated with fire detection systems, IDSs, access control systems, or closed-circuit television (CCTV) systems, such that these systems automatically respond when the alarm is triggered. For example, a smoke detector alarm can be set up to automatically notify the fire department when smoke is detected, or an intrusion alarm can automatically trigger cameras to turn on in a remote location so that personnel can monitor that location.

An alarm system consists of sensors that detect different types of events; an arming station that is used to turn the system on and off; a control panel that receives information, processes it, and transmits the alarm; and an annunciator that generates a visual and/or audible response to the alarm. When a sensor is tripped, it sends a signal to a control panel, which triggers a visual or audible alarm and/or notifies a central monitoring station. A more complete description of each of the components of an alarm system is provided below.

Detection devices (also called *sensors*) are designed to detect a specific type of event (such as smoke, intrusion, etc.). Depending on the type of event they are designed to detect, sensors can be located inside or outside of the facility or other asset. When an event is detected, the sensors use some type of communication method (such as wireless radio transmitters, conductors, or cables) to send signals to the control panel to generate the alarm. For example, a smoke detector sends a signal to a control panel when it detects smoke.

Alarms use either normally closed (NC) or normally open (NO) electric loops, or "circuits," to generate alarm signals. These two types of circuits are discussed separately below.

In NC loops or circuits, all of the system's sensors and switches are connected in series. The contacts are "at rest" in the closed (on) position, and current continually passes through the system. However, when an event triggers the sensor, the loop is opened, breaking the flow of current through the system and triggering the alarm. NC switches are used more often than are NO switches because the alarm will be activated if the loop or circuit is broken or cut, thereby reducing the potential for circumventing the alarm. This is known as a "supervised" system.

In NO loops or circuits, all of the system's sensors and switches are connected in parallel. The contacts are "at rest" in the open (off) position, and no current passes through the system. However, when an event triggers the sensor, the loop is closed. This allows current to flow through the loop, powering the alarm. NO systems are not "supervised" because the alarm will not be activated if the loop or circuit is broken or cut. However, adding an end-of-line resistor to an NO loop will cause the system to alarm if tampering is detected.

An *arming station*, which is the main user interface with the security system, allows the user to arm (turn on), disarm (turn off), and communicate with the system. How a specific system is armed will depend on how it is used. For example, while IDSs can be armed for continuous operation (twenty-four hours/day), they are usually armed and disarmed according to the work schedule at a specific location so that personnel going about their daily activities do not set off the alarms. In contrast, fire protection systems are typically armed twenty-four hours/day.

A *control panel* receives information from the sensors and sends it to an appropriate location, such as to a central operations station or to a twenty-four-hour monitoring facility. Once the alarm signal is received at the central monitoring location, personnel monitoring for alarms can respond (such as by sending security teams to investigate or by dispatching the fire department).

An *annunciator* responds to the detection of an event by emitting a signal. This signal may be visual, audible, electronic, or a combination of these three. For example, fire alarm signals will always be connected to audible annunciators, whereas intrusion alarms may not be.

Alarms can be reported locally, remotely, or both locally and remotely. Local and remotely (centrally) reported alarms are discussed in more detail below.

A *local alarm* emits a signal at the location of the event (typically using a bell or siren). A "local only" alarm emits a signal at the location of the event but does not transmit the alarm signal to any other location (i.e., it does not transmit the alarm to a central monitoring location). Typically, the purpose of a "local only" alarm is to frighten away intruders and possibly to attract the attention of someone who might notify the proper authorities. Because no signal is sent to a central monitoring location, personnel can only respond to a local alarm if they are in the area and can hear and/or see the alarm signal.

Fire alarm systems must have local alarms, including both audible and visual signals. Most fire alarm signal and response requirements are codified in the National Fire Alarm Code, National Fire Protection Association (NFPA) 72. NFPA 72 discusses the application, installation, performance, and maintenance of protective signaling systems and their components. In contrast to fire alarms, which require a local signal when fire is detected, many IDSs do not have a local alert device because monitoring personnel do not wish to inform potential intruders that they have been detected. Instead, these types of systems silently alert monitoring personnel that an intrusion has been detected, thus allowing monitoring personnel to respond.

In contrast to systems that are set up to transmit "local only" alarms when the sensors are triggered, systems can also be set up to transmit signals to a *central location*, such as to a control room or guard post at the utility or to a police or fire station. Most fire/smoke alarms are set up to signal both at the location of the event and at a fire station or central monitoring station. Many insurance companies require that facilities install certified systems that include alarm communication to a central station. For example, systems certified by the Underwriters Laboratory (UL) require that the alarm be reported to a central monitoring station.

Table 14.7 Perimeter and Interior Sensors

Type of Perimeter Sensor	Description
Foil	Foil is a thin, fragile, lead-based metallic tape that is applied to glass windows and doors. The tape is applied to the window or door, and electric wiring connects this tape to a control panel. The tape functions as a conductor and completes the electric circuit with the control panel. When an intruder breaks the door or window, the fragile foil breaks, opening the circuit and triggering an alarm condition.
Magnetic switches (reed switches)	The most widely used perimeter sensor. They are typically used to protect doors and windows that can be opened (windows that cannot be opened are more typically protected by foil alarms).
Glass-break detectors	Placed on glass and senses vibrations in the glass when it is disturbed. The two most common types of glass-break detectors are shock sensors and audio discriminators.

Type of Interior Sensor	Description
Passive infrared (PIR)	Presently the most popular and cost-effective interior sensors. PIR detectors monitor infrared radiation (energy in the form of heat) and detect rapid changes in temperature within a protected area. Because infrared radiation is emitted by all living things, these types of sensors can be very effective.
Quad PIRs	Consists of two dual-element sensors combined in one housing. Each sensor has a separate lens and a separate processing circuitry, which allows each lens to be set up to generate a different protection pattern.
Ultrasonic detectors	Emits high-frequency sound waves and sense movement in a protected area by sensing changes in these waves. The sensor emits sound waves that stabilize and set a baseline condition in the area to be protected. Any subsequent movement within the protected area by a would-be intruder will cause a change in these waves, thus creating an alarm condition.
Microwave detectors	Emits ultra-high frequency radio waves, and the detector senses any changes in these waves as they are reflected throughout the protected space. Microwaves can penetrate through walls, and thus a unit placed in one location may be able to protect multiple rooms.
Dual technology devices	Incorporates two different types of sensor technology (such as PIR and microwave technology) together in one housing. When both technologies sense an intrusion, an alarm is triggered.

Source: USEPA, 2005.

The main differences between alarm systems lie in the types of event detection devices used. **Intrusion sensors**, for example, consist of two main categories: perimeter sensors and interior (space) sensors. *Perimeter intrusion sensors* are typically applied on fences, doors, walls, windows, etc., and are designed to detect an intruder before he or she accesses a protected asset (i.e., perimeter intrusion sensors are used to detect intruders attempting to enter through a door, window, etc.). In contrast, *interior intrusion sensors* are designed to detect an intruder who has already accessed the protected asset (i.e., interior intrusion sensors are used to detect intruders once they are already within a protected room or building). These two types of detection devices can be complementary, and they are often used together to enhance security for an asset. For example, a typical intrusion alarm system might use a perimeter glass-break detector that protects against intruders accessing a room through a window as well as an ultrasonic interior sensor that detects intruders that have gotten into the room without using the window. Table 14.7 lists and describes types of perimeter and interior sensors.

Fire detection/fire alarm systems consist of different types of fire detection devices and fire alarm systems. These systems may detect fire, heat, smoke, or a combination of any of these. For example, a typical fire alarm system might consist of heat sensors, which are located throughout a facility and which detect high temperatures or a certain change in temperature over a fixed time period. A different system might be outfitted with both smoke and heat detection devices. A summary of several different types of fire/smoke/heat detection sensors is provided in table 14.8.

Once a sensor in an alarm system detects an event, it must communicate an alarm signal. The two basic types of alarm communication systems are hardwired and wireless. Hardwired systems rely on wire that is run from the control panel to each of the detection devices and annunciators. Wireless systems transmit signals from a transmitter to a receiver through the air—primarily using radio or other waves. Hardwired systems are usually lower cost, more reliable (they are not affected by terrain or environmental factors), and significantly easier to troubleshoot than are wireless systems. However, a major disadvantage of hardwired systems is that it may not be possible to hardwire all locations (for example, it may be difficult to hardwire remote locations). In addition, running wires to their required locations can be both time-consuming and costly. The major advantage to using wireless systems is that they can often be installed in areas where hardwired systems are not

Table 14.8   Fire/Smoke/Heat Detection Sensors

Detector Type	Description
Thermal detector	Sense when temperatures exceed a set threshold (fixed temperature detectors) or when the rate of change of temperature increases over a fixed time period (rate-of-rise detectors).
Duct detector	Is located within the heating and ventilation ducts of the facility. This sensor detects the presence of smoke within the system's return or supply ducts. A sampling tube can be added to the detector to help span the width of the duct.
Smoke detectors	Sense invisible and/or visible products of combustion. The two principle types of smoke detectors are photoelectric and ionization detectors. The major differences between these devices are described below: • Photoelectric smoke detectors react to visible particles of smoke. These detectors are more sensitive to the cooler smoke with large smoke particles that are typical of smoldering fires. • Ionization smoke detectors are sensitive to the presence of ions produced by the chemical reactions that take place with few smoke particles, such as those typically produced by fast burning/flaming fires.
Multisensor detectors	Are a combination of photoelectric and thermal detectors. The photoelectric sensor serves to detect smoldering fires, while the thermal detector senses the heat given off from fast burning/flaming fires.
Carbon monoxide (CO) detectors	Are used to indicate the outbreak of fire by sensing the level of carbon monoxide in the air. The detector has an electrochemical cell that senses carbon monoxide but not some other products of combustion.
Beam detectors	Are designed to protect large, open spaces such as industrial warehouses. These detectors consist of three parts: the transmitter, which projects a beam of infrared light; the receiver, which registers the light and produces an electrical signal; and the interface, which processes the signal and generates alarm of fault signals. In the event of a fire, smoke particles obstruct the beam of light. Once a preset threshold is exceeded, the detector will go into alarm.
Flame detectors	Sense either ultraviolet (UV) or infrared (IR) radiation emitted by a fire.
Air-sampling detectors	Actively and continuously samples the air from a protected space and is able to sense the precombustion stages of incipient fire.

Source: USEPA, 2005.

Table 14.9 Types of Buried Sensors

Type	Description
Pressure or Seismic	Responds to disturbances in the soil.
Magnetic Field	Responds to a change in the local magnetic field caused by the movement of nearby metallic material.
Ported Coaxial Cables	Responds to motion of a material with a high dielectric constant or high conductivity near the cables.
Fiber-Optic Cables	Responds to a change in the shape of the fiber that can be sensed using sophisticated sensors and computer signal processing.

Source: Adapted from M. L. Garcia, 2001.

feasible. However, wireless components can be much more expensive when compared to hardwired systems. In addition, in the past, it has been difficult to perform self-diagnostics on wireless systems to confirm that they are communicating properly with the controller. Presently, the majority of wireless systems incorporate supervising circuitry, which allows the subscriber to know immediately if there is a problem with the system (such as a broken detection device or a low battery), or if a protected door or window has been left open.

Exterior Intrusion Buried Sensors

Buried sensors are electronic devices that are designed to detect potential intruders. The sensors are buried along the perimeters of sensitive assets and are able to detect intruder activity both aboveground and belowground. Some of these systems are composed of individual, stand-alone sensor units, while other sensors consist of buried cables.

There are four types of buried sensors that rely on different types of triggers. These are pressure or seismic; magnetic field; ported coaxial cables; and fiber-optic cables. These four sensors are all covert and terrain-following, meaning they are hidden from view and follow the contour of the terrain. The four types of sensors are described in more detail below. Table 14.9 presents the distinctions among the four types of buried sensors.

Exterior Intrusion Sensors

An exterior intrusion sensor is a detection device that is used in an outdoor environment to detect intrusions into a protected area. These devices are designed to detect an intruder and then communicate an alarm signal to an alarm system. The alarm system can respond to the intrusion in many different ways, such as by triggering an audible or visual alarm signal or by sending an electronic signal to a central monitoring location that notifies security personnel of the intrusion.

An intrusion sensor can be used to protect many kinds of assets. Intrusion sensors that protect physical space are classified according to whether they protect indoor or "interior" space (i.e., an entire building or room within a building), or outdoor or "exterior" space (i.e., a fence line or perimeter). Interior intrusion sensors are designed to protect the interior space of a facility by detecting an intruder who is attempting to enter or who has already entered a room or building. In contrast, exterior intrusion sensors are designed to detect an intrusion into a protected outdoor/exterior area. Exterior protected areas are typically arranged as zones or exclusion areas placed so that the intruder is detected early in the intrusion attempt before the intruder can gain access to more valuable assets (e.g., into a building located within the protected area). Early detection creates additional time for security forces to respond to the alarm.

Exterior intrusion sensors are classified according to how the sensor detects the intrusion within the protected area. The three classes of exterior sensor technology include:

- Buried-line sensors
- Fence-associated sensors
- Freestanding sensors

Buried-line Sensors: As the name suggests, buried-line sensors are sensors that are buried underground and are designed to detect disturbances within the ground, such as disturbances caused by an intruder digging, crawling, walking, or running on the monitored ground. Because they sense ground disturbances, these types of sensors are able to detect intruder activity both on the surface and below ground. Individual types of exterior buried-line sensors function in different ways, including by detecting motion, pressure, or vibrations within the protected ground or by detecting changes in some type of field (e.g., magnetic field) that the sensors generate within the protected ground. Specific types of buried-line sensors include pressure or seismic sensors, magnetic field sensors, ported coaxial cables, and fiber-optic cables. Details on each of these sensor types are provided below.

- *Buried-line pressure* or *seismic sensors* detect physical disturbances to the ground—such as vibrations or soil compression—caused by intruders walking, driving, digging, or otherwise physically contacting the protected ground. These sensors detect disturbances from all directions and, therefore, can protect an area radially outward from their location; however, because detection may weaken as a function of distance from the disturbance, choosing the correct burial depth from the design area will be crucial. In general, sensors buried at a shallow depth protect a relatively small area but have a high probability of detecting intrusion within that area, while sensors buried at a deeper depth protect a wider area but have a lower probability of detecting intrusion into that area.
- *Buried-line magnetic field sensors* detect changes in a local magnetic field that are caused by the movement of metallic objects within that field. This type of sensor can detect ferric metal objects worn or carried by an intruder entering a protected area on foot as well as vehicles being driven into the protected area.
- *Buried-line ported coaxial cable sensors* detect the motion of any object (i.e., human body, metal, etc.) possessing high conductivity and located within close proximity to the cables. An intruder entering into the protected space creates an active disturbance in the electric field, thereby triggering an alarm condition.
- *Buried-line fiber-optic cable sensors* detect changes in the attenuation of light signals transmitted within the cable. When the soil around the cable is compressed, the cable is distorted, and the light signal transmitted through the cable changes, initiating an alarm. This type of sensor is easy to install because it can be buried at a shallow burial depth (only a few centimeters) and still be effective.

Fence-associated Sensors: Fence-associated sensors are either attached to an existing fence or are installed in such a way as to create a fence. These sensors detect disturbances to the fence—such as those caused by an intruder attempting to climb the fence or by an intruder attempting to cut or lift the fence fabric. Exterior fence-associated sensors include fence-disturbance sensors, taut-wire sensor fences, and electric field or capacitance sensors. Details on each of these sensor types are provided below.

- *Fence-disturbance sensors* detect the motion or vibration of a fence, such as that can be caused by an intruder attempting to climb or cut through the fence. In general, fence disturbance sensors are used on chain-link fences or on other fence types where a moveable fence fabric is hung between fence posts.
- *Taut-wire sensor fences* are similar to fence-disturbance sensors except that instead of attaching the sensors to a loose fence fabric, the sensors are attached to a wire that is stretched tightly across the fence. These types of systems are designed to detect changes in the tension of the wire rather than vibrations in the fence fabric. Taut-wire sensor fences can be installed over existing fences or as stand-alone fence systems.
- *Electric field or capacitance sensors* detect changes in capacitive coupling between wires that are attached to, but are electrically isolated from, the fence. As opposed to other fence-associated intrusion sensors, both electric field and capacitance sensors generate an electric field that radiates out from the fence line, resulting in an expanded zone of protection relative to other fence-associated sensors and allowing the sensor to detect an intruders' presence before they arrive at the fence line. Note: proper spacing is necessary during installation of the electric field sensor to detect a would-be intruder from slipping between largely spaced wires.

Freestanding Sensors: These sensors, which include active infrared, passive infrared, bistatic microwave, monostatic microwave, dual-technology, and video motion detection (VMD) sensors, consist of individual sensor units or components that can be set up in a variety of configurations to meet a user's needs. They are installed aboveground, and depending on how they are oriented relative to each other, they can be used to establish a protected perimeter or a protected space. More details on each of these sensor types are provided below.

- *Active infrared sensors* transmit infrared energy into the protected space and monitor for changes in this energy caused by intruders entering that space. In a typical application, an infrared light beam is transmitted from a transmitter unit to a receiver unit. If an intruder crosses the beam, the beam is blocked, and the receiver unit detects a change in the amount of light received, triggering an alarm. Different sensors can see single- and multiple-beam arrays. Single-beam infrared sensors transmit a single infrared beam. In contrast, multiple-beam infrared sensors transmit two or more beams parallel to each other. This multiple-beam sensor arrangement creates an infrared "fence."
- *Passive infrared (PIR) sensors* monitor the ambient infrared energy in a protected area and evaluate changes in that ambient energy that may be caused by intruders moving through the protected area. Detection ranges can exceed 100 yards on cold days with size and distance limitations dependent upon the background temperature. PIR sensors generate a nonuniform detection pattern (or "curtain") that has areas (or "zones") of more sensitivity and areas of less sensitivity. The specific shape of the protected area is determined by the detector's lenses. The general shape common to many detection patterns is a series of long "fingers" emanating from the PIR and spreading in various directions. When intruders enter the detection area, the PIR sensor detects differences in temperature due to the intruder's body heat and triggers an alarm. While the PIR leaves unprotected areas between its fingers, an intruder would be detected if he passed from a nonprotected area to a protected area.
- *Microwave sensors* detect changes in received energy generated by the motion of an intruder entering into a protected area. Monostatic microwave sensors incorporate a transmitter and a receiver in one unit, while bistatic sensors separate the transmitter and the receiver into different units. Monostatic sensors are limited to a coverage area of 400 feet, while bistatic sensors can cover an area up to 1,500 feet. For bistatic sensors, a zone of no detection exists in the first few feet in front of the antennas. This distance from the antennas to the point at which the intruder is first detected is known as the offset distance. Due to this offset distance, antennas must be configured so that they overlap one another (as opposed to being adjacent to each other), thereby creating long perimeters with a continuous line of detection.
- *Dual-technology sensors* consist of two different sensor technologies incorporated together into one sensor unit. For example, a dual-technology sensor could consist of a passive infrared detector and a monostatic microwave sensor integrated into the same sensor unit.
- *Video motion detection* (VMD) *sensors* monitor video images from a protected area for changes in the images. Video cameras are used to detect unauthorized intrusion into the protected area by comparing the most recent image against a previously established one. Cameras can be installed on towers or other tall structures so that they can monitor a large area.

Fences

A fence is a physical barrier that can be set up around the perimeter of an asset. Fences often consist of individual pieces (such as individual pickets in a wooden fence, or individual sections of a wrought-iron fence) that are fastened together. Individual sections of the fence are fastened together using posts, which are sunk into the ground to provide stability and strength for the sections of the fence hung between them. Gates are installed between individual sections of the fence to allow access inside the fenced area.

Many fences are used as decorative architectural features to separate physical spaces for each other. They may also be used to physically mark the location of a boundary (such as a fence installed along a properly line). However, a fence can also serve as an effective means for physically delaying intruders from gaining access to a water or wastewater asset. For example, many utilities install fences around their primary facilities, around remote pump stations, or around hazardous materials storage areas or sensitive areas within a facility. Access to the area can be controlled through security at gates or doors through the fence (for example, by posting a guard at the gate or by locking it). In order to gain access to the asset, unauthorized persons would either have to go around or through the fence.

Fences are often compared with walls when determining the appropriate system for perimeter security. While both fences and walls can provide adequate perimeter security, fences are often easier and less expensive to install than walls. However, they do not usually provide the same physical strength that walls do.

Table 14.10 Comparison of Different Fence Types

Specifications	Chain Link	Iron	Wire (Wirewall)	Wood
Height limitations	12'	12'	12'	8'
Strength	Medium	High	High	Low
Installation requirements	Low	High	High	Low
Ability to remove/reuse	Low	High	Low	High
Ability to replace/repair	Medium	High	Low	High

Source: USEPA, 2005.

In addition, many types of fences have gaps between the individual pieces that make up the fence (i.e., the spaces between chain links in a chain-link fence or the space between pickets in a picket fence). Thus, many types of fences allow the interior of the fenced area to be seen. This may allow intruders to gather important information about the locations or defenses of vulnerable areas within the facility.

There are numerous types of materials used to construct fences, including chain-link iron, aluminum, wood, or wire. Some types of fences, such as split rails or pickets, may not be appropriate for security purposes because they are traditionally low fences and they are not physically strong. Potential intruders may be able to easily defeat these fences either by jumping or climbing over them or by breaking through them. For example, the rails in a split fence may be able to be broken easily.

Important security attributes of a fence include the height to which it can be constructed, the strength of the material comprising the fence, the method and strength of attaching the individual sections of the fence together at the posts, and the fence's ability to restrict the view of the assets inside the fence. Additional considerations should include the ease of installing the fence and the ease of removing and reusing sections of the fence. Table 14.10 provides a comparison of the important security and usability features of various fence types.

Some fences can include additional measures to delay, or even detect, potential intruders. Such measures may include the addition of barbed wire, razor wire, or other deterrents at the top of the fence. Barbed wire is sometimes used at the base of fences as well. This can impede a would-be intruder's progress in even reaching the fence. Fences may also be fitted with security cameras to provide visual surveillance of the perimeter. Finally, some facilities have installed motion sensors along their fences to detect movement on the fence. Several manufacturers have combined these multiple-perimeter security features into one product and offer alarms and other security features.

The correct implementation of a fence can make it a much more effective security measure. Security experts recommend the following when a facility constructs a fence:

- The fence should be at least seven to nine feet high.
- Any outriggers, such as bared wire, that are affixed on top of the fence should be angled out and away from the facility and not in toward the facility. This will make climbing the fence more difficult and will prevent ladders from being placed against the fence.
- Other types of hardware can increase the security of the fence. This can include installing concertina wire along the fence (this can be done in front of the fence or at the top of the fence), or adding intrusion sensors, camera, or other hardware to the fence.
- All undergrowth should be cleared for several feet (typically six feet) on both sides of the fence. This will allow for a clearer view of the fence by any patrols in the area.
- Any trees with limbs or branches hanging over the fence should be trimmed so that intruders cannot use them to go over the fence. Also, it should be noted that fallen trees can damage fences, and so management of trees around the fence can be important. This can be especially important in areas where the fence goes through a remote area.
- Fences that do not block the view from outside the fence to inside the fence allow patrols to see inside the fence without having to enter the facility.
- "No Trespassing" signs posted along the fence can be a valuable tool in prosecuting any intruders who claim that the fence was broken and that they did not enter through the fence illegally. Adding signs that highlight the local ordinances against trespassing can further dissuade simple troublemakers from illegally jumping/climbing the fence.

Locks

A lock is a type of physical security device that can be used to delay or prevent a door, a window, a manhole, a filing cabinet drawer, or some other physical feature from being opened, moved, or operated. Locks typically operate by connecting two pieces together—such as by connecting a door to a door jamb or a manhole to its casement. Every lock has two modes—engaged (or "locked"), and disengaged (or "opened"). When a lock is disengaged, the asset on which the lock is installed can be accessed by anyone, but when the lock is engaged, only authorized persons have access to the locked asset.

Locks are excellent security features because they have been designed to function in many ways and to work on many different types of assets. Locks can also provide different levels of security depending on how they are designed and implemented. The security provided by a lock is dependent on several factors, including its ability to withstand physical damage (i.e., can it be cut off, broken, or otherwise physically disabled) as well as its requirements for supervision or operation (i.e., combinations may need to be changed frequently so that they are not compromised and the locks remain secure). While there is no single definition of the "security" of a lock, locks are often described as minimum, medium, or maximum security. Minimum security locks are those that can be easily disengaged (or "picked") without the correct key or code, or those that can be disabled easily (such as small padlocks that can be cut with bolt cutters). Higher security locks are more complex and thus are more difficult to pick or are sturdier and more resistant to physical damage.

Many locks, such as many door locks, only need to be unlocked from one side. For example, most door locks need a key to be unlocked only from the outside. A person opens such devices, called single-cylinder locks, from the inside by pushing a button or by turning a knob or handle. Double-cylinder locks require a key to be locked or unlocked from both sides.

COMPUTER AND DATA SECURITY

The safety and security of a facility's computer system has elements that demand the safety engineer's attention. Familiarity with important issues in computer system and data security is critical. While safety engineers shouldn't be expected to know as much about computers as a systems engineer, ensuring that the IT (Information Technology) staff has the tools they need to provide reasonable levels of system security, ensuring that security measures are in place to limit access to data and equipment, and keeping up with current issues in computer and data security is a growing part of the safety professional's job. Facility administration, IT, and the safety engineer should work together on formulating company policy on computer access, security, auditing, and back-up systems. Current best practice for information security is a layered, defense-in-depth program that involves policy, procedures, and tools.

Best Practices: Layered Security for Defense-in-Depth

According to Larry McCreary (2003) of IT security, as networks become increasingly more complex, security has become much more than off-the-shelf software. Today firewalls, authentication, intrusion detection/prevention, and web filtering are all part of an effective data security program.

Best practices include:

- A regularly updated and reviewed security policy as a formal condition of employment.
- A general security approach that allows access only to authenticated users.
- Encryption of sensitive data accessible by public connections.
- Well-configured and enforced access rules across physical, logical, and social boundaries.
- Traffic control across physical, logical, and social boundaries.
- Role-based authentication/authorization (as needed) including dynamic passwords (generated by hardware or software tokens or sent to mobile devices, etc.) for network administrators, remote users, and employees accessing mission-critical information.
- Filtering content that could cause:
  1. network vulnerabilities (email attachments, hacker tools, viruses, etc.);
  2. legal liability (pornography, criminal skills, anonymizers, cults, etc.);
  3. poor network availability (MP3, streaming content, etc.);
  4. regular third-party audits and assessments.

Computer Security Terminology

In the real world of data systems security, a gap exists between theoretical information security best practices and the reality of implementation. In part, under many circumstances, this gap is widened by IT staff's inability to communicate the necessary information in a way that nontechnical users—from worker level to top administration—can understand. Of course, like any other profession, the computer industry has a huge and ever-growing body of jargon, and the ability to discuss computer security issues with IT staff relies on understanding these terms. For more computer-oriented terminology, visit www.sans.org/resources/glossary.php.

access: The process of interacting with a system (verb or a noun). Users access a system (v). A user gains computer access (n).

access control: Ensures that resources are only granted to those users who are entitled to them.

access management access: Maintenance of access information that consists of four tasks: account administration, maintenance, monitoring, and revocation.

account harvesting: The process of collecting all the legitimate account names on a system.

active content: Program code embedded in the contents of a web page that, when accessed by a web browser, automatically downloads and executes on the user's workstation.

activity monitors: System monitor designed to prevent virus infection by monitoring for malicious activity on a system and blocking that activity when possible.

applet: Java programs; an application program that uses the client's web browser to provide a user interface.

audit: The independent collection of records to assess their veracity and completeness.

audit trail: in computer security systems, a chronological record of when users log in, how long they are engaged in various activities, what they were doing, and whether any actual or attempted security violations occurred. An audit trail may be on paper or on disk.

authentication: The process of confirming the correctness of the claimed identity.

back door: An entry point to a program or a system that is hidden or disguised, often created by the software's author for maintenance. A certain sequence of control characters permits access to the system manager account. If the back door becomes known, unauthorized users (or malicious software) can gain entry and cause damage. Also a tool installed after a compromise to give an attacker easier access to the compromised system around any security mechanisms that are in place.

bandwidth: Commonly used to mean the capacity of a communication channel to pass data through the channel in a given amount of time. Usually expressed in bits per second.

biometrics: Access system that uses physical characteristics of the users to determine access.

bit: The smallest unit of information storage; a contraction of the term *binary digit*; one of two symbols—"0" (zero) and "1" (one)—that are used to represent binary numbers.

broadcast: To simultaneously send the same message to multiple recipients. One host to all hosts on a network.

browser: A client computer program that can retrieve and display information from servers on the World Wide Web.

brute force: A cryptanalysis technique or other kind of attack method involving an exhaustive procedure that tries all possibilities, one-by-one.

buffer overflow: Occurs when a program or process tries to store more data in a buffer (temporary data storage area) than it was intended to hold. Since buffers are created to contain a finite amount of data, the extra information—which has to go somewhere—may overflow into adjacent buffers, corrupting or overwriting the valid data held in them.

byte: A fundamental unit of computer storage; the smallest addressable unit in a computer's architecture. Usually holds one character of information and usually means eight bits.

cache: Pronounced *cash*; a special high-speed storage mechanism. It can be either a reserved section of main memory or an independent high-speed storage device. Two types of caching are commonly used in personal computers: memory caching and disk caching.

ciphertext: The encrypted form of the message being sent.

competitive intelligence: Competitive intelligence is espionage using legal, or at least not obviously illegal, means.

Computer Emergency Response Team (CERT): An organization that studies computer and network IN-FOSEC to provide incident-response services to victims of attacks, publish alerts concerning vulnerabilities and threats, and offer other information to help improve computer and network security.

computer network: A collection of host computers together with the subnetwork or internetwork through which they can exchange data.

computer security: Technological and managerial procedures applied to computer systems to ensure the availability, integrity, and confidentiality of information managed by the computer system.

computer security audit: An independent evaluation of the controls employed to ensure appropriate protection of an organization's information assets.

cookie: Data exchanged between an HTTP server and a browser (a client of the server) to store information on the client side and retrieve it later for server use. An HTTP server, when sending data to a client, may send along a cookie, which the client retains after the HTTP connection closes. A server can use this mechanism to maintain persistent client-side state information for HTTP-based applications, retrieving the state information in later connections.

corruption: A threat action that undesirably alters system operation by adversely modifying system functions or data.

cost-benefit analysis: A cost-benefit analysis compares the cost of implementing countermeasures with the value of the reduced risk.

cryptanalysis: The mathematical science that deals with analysis of a cryptographic system to gain knowledge needed to break or circumvent the cipher text to plaintext without knowing the key.

cryptography: Cryptography garbles a message in such a way that anyone who intercepts the message cannot understand it.

data aggregation: The ability to get a more complete picture of the information by analyzing several different types of records at once.

data custodian: The entity currently using or manipulating the data, and therefore, temporarily taking responsibility for the data.

data-driven attack: A form of attack in which the attack is encoded in innocuous-seeming data that is executed by a user or other software to implement an attack. In the case of firewalls, a data-driven attack is a concern since it may get through the firewall in data form and launch an attack against a system behind the firewall.

data encryption standard (DES): A widely used method of data encryption using a private (secret) key. Some 72,000,000,000,000,000 (72 quadrillion) or more possible encryption keys can be used. For each given message, the key is chosen at random from among this enormous number of keys. Like other private key cryptographic methods, both the sender and the receiver must know and use the same private key.

decryption: The process of transforming an encrypted message into its original plaintext.

defense in-depth: The approach of using multiple layers of security to guard against failure of a single security component to ensure each system on the network is secured to the greatest possible degree. May be used in conjunction with firewalls.

denial of service: The prevention of authorized access to a system resource or the delaying of system operations and functions.

dictionary attack: An attack that tries all of the phrases or words in a dictionary, trying to crack a password or key. A dictionary attack uses a predefined list of words compared to a brute-force attack that tries all possible combinations.

disassembly: The process of taking a binary program and deriving the source code from it.

disaster recovery plan (DRP): A program and system for the recovery of IT systems in the event of a disruption or disaster.

domain: A sphere of knowledge or a collection of facts about some program entities or a number of network points or addresses, identified by a name. On the Internet, a domain consists of a set of network addresses. In the Internet's domain-name system, a domain is a name with which name server records are associated that describe subdomains or hosts. In Windows NT and Windows 2000, a domain is a set of network resources (applications, printers, and so forth) for a group of users. The user need only to log in to the domain to gain access to the resources, which may be located on a number of different servers in the network.

domain hijacking: An attack by which an attacker takes over a domain by first blocking access to the domain's DNS server and then putting his own server up in its place.

domain name: A name that locates an organization or other entity on the Internet.

due care: A minimal level of protection in place in accordance with industry best practice.

due diligence: The requirement that organizations must develop and deploy a protection plan to prevent fraud and abuse and deploy a means to detect them if they occur.

dumpster diving: Obtaining passwords and corporate directories by searching through discarded media.

egress filtering: Filtering outbound traffic.

email bombs: Code that when executed sends many messages to the same address(es) for the purpose of using up disk space and/or overloading the email or web server.

encryption: The process of scrambling files or programs, changing one character string to another through an algorithm (such as the DES algorithm).

escrow passwords: Passwords recorded and stored in a secure location (like a safe) and used by emergency personnel when privileged personnel are unavailable.

event: An observable occurrence in a system or network.

exposure: A threat action whereby sensitive data is directly released to an unauthorized entity.

false rejects: When an authentication system fails to recognize a valid user.

fault line attacks: Attacks that use weaknesses between interfaces of systems to exploit gaps in coverage.

fault tolerance: A design method that ensures continued systems operation in the event of individual failures by providing redundant system elements.

file transfer protocol (FTP): A TCP/IP protocol specifying the transfer of text or binary files across the network.

filtering router: An internetwork router that selectively prevents the passage of data packets according to a security policy, often used as a firewall or part of a firewall. A router usually receives a packet from a network and decides where to forward it on a second network. A filtering router does the same, but it first decides whether the packet should be forwarded at all, according to some security policy. The policy is implemented by rules (packet filters) loaded into the router.

fingerprinting: Sending strange packets to a system to gauge how it responds to determine the operating system.

firewall: A system or combination of systems that enforces a boundary between two or more networks.

flooding: An attack that attempts to cause a failure in (especially, in the security of) a computer system or other data-processing entity by providing more input than the entity can process properly.

fork bomb: Attack that uses the fork() call to create a new process that is a copy of the original. By doing this repeatedly, all available processes on the machine can be taken up.

gateway: A network point that acts as an entrance to another network.

generic utilities: General-purpose code and devices, including screen grabbers and sniffers, that look at data and capture information like passwords, keys, and secrets.

global security: The ability of an access control package to permit protection across a variety of mainframe environments, providing users with a common security interface to all.

hacker: Those intent upon entering an environment to which they are not entitled entry for whatever purpose (entertainment, profit, theft, prank, etc.). Usually it is iterative techniques escalating to more advanced methodologies and use of devices to intercept the communications property of another.

handheld: A small electronic device used to collect data.

hardening: The process of identifying and fixing vulnerabilities on a system.

hijack attack: A form of active wiretapping in which the attacker seizes control of a previously established communication association.

honey pot: Programs that simulate one or more network services that you designate on your computer's ports. An attacker assumes you're running vulnerable services that can be used to break into the machine. A honey pot can be used to log access attempts to those ports, including the attacker's keystrokes. This could give you advanced warning of a more concerted attack.

host: Any computer that has full, two-way access to other computers on the Internet, or a computer with a web server that serves the pages for one or more websites.

host-based security: The technique of securing an individual system from attack. Host-based security is operating-system and version dependent.

hot standby: A backup system configured in such a way that it may be used if the system goes down.

hybrid attack: Attack that builds on the dictionary-attack method by adding numerals and symbols to dictionary words.

hyperlink: In hypertext or hypermedia, an information object (such as a word, a phrase, or an image; usually highlighted by color or underscoring) that points (indicates how to connect) to related information located elsewhere and retrieves it by activating the link.

incident: An adverse network event in an information system or network or the threat of the occurrence of such an event.

incident handling: An action plan for dealing with intrusions, cyber-theft, denial of service, fire, floods, and other security-related events, comprised of a six-step process: preparation, identification, containment, eradication, recovery, and lessons learned.

incremental backups: Backup system that only duplicates files modified since the last backup.

ingress filtering: Filtering inbound traffic.

input validation attacks: Intentionally sending unusual input in the hope of confusing an application.

insider attack: An attack originating from inside a protected network.

integrity: The need to ensure that information has not been changed accidentally or deliberately and that it is accurate and complete.

Internet protocol (IP): The method or protocol by which data is sent from one computer to another on the Internet.

Internet standard: A specification, approved by the IESG and published as an RFC, that is stable and well-understood; is technically competent; has multiple, independent, and interoperable implementations with substantial operational experience; enjoys significant public support; and is recognizably useful in some or all parts of the Internet.

interrupt: A signal that informs the operating system that something has occurred.

intranet: A computer network, especially one based on Internet technology, that an organization uses for its own internal, and usually private, purposes and that is closed to outsiders.

intrusion detection: A security management system that gathers and analyzes information from various areas within a computer or a network to identify possible security breaches, including both intrusions (attacks from outside the organization) and misuse (attacks from within the organization).

IP flood: A denial of service attack that sends a host more echo request ("ping") packets than the protocol implementation can handle.

IP sniffing: Stealing network addresses by reading the packets. Harmful data is then sent stamped with internal trusted addresses.

IP splicing: An attack whereby an active, established session is intercepted and co-opted by the attacker. IP splicing attacks may occur after an authentication has been made, permitting the attacker to assume the role of an already authorized user.

IP spoofing: The technique of supplying a false IP address to illicitly impersonate another system by using its IP network address.

ISSA: Information Systems Security Association.

issue-specific policy: Policy to address specific needs within an organization, such as a password policy.

list-based access control: Associates a list of users and their privileges with each object.

log clipping: The selective removal of log entries from a system log to hide a compromise.

logging: The process of storing information about events that occurred on the firewall or network.

log retention: How long audit logs are retained and maintained.

loopback address: Pseudo IP addresses that always refer back to the local host and are never sent out onto a network.

malicious code: Software (e.g., Trojan horse) that appears to perform a useful or desirable function but actually gains unauthorized access to system resources or tricks a user into executing other malicious logic.

malware: A generic term for a number of different types of malicious code.

masquerade attack: A type of attack in which one system entity illegitimately poses as (assumes the identity of) another entity.

monoculture: A large number of users who run the same software and are thus vulnerable to the same attacks.

National Institute of Standards and Technology (NIST): A unit of the U.S. Commerce Department that promotes and maintains measurement standards. It has active programs for encouraging and assisting industry and science to develop and use these standards.

natural disaster: Any "act of God" (e.g., fire, flood, earthquake, lightning, or wind) that disables a system component.

network-level firewall: A firewall in which traffic is examined at the network protocol packet level.

network mapping: To compile an electronic inventory of the systems and the services on a network.

network taps: Hardware devices that hook directly onto the network cable and send a copy of the traffic that passes through it to one or more other networked devices.

network worm: A program or command file that uses a computer network as a means for adversely affecting a system's integrity, reliability, or availability. A network worm may attack from one system to another by establishing a network connection. It is usually a self-contained program that does not need to attach itself to a host file to infiltrate network after network.

null session: Known as *anonymous logon*, it is a way of letting an anonymous user retrieve information such as user names and shares them over the network connection without authentication.

octet: A sequence of eight bits. An octet is an eight-bit byte.

operating system: System software that controls a computer and its peripherals. Modern operating systems such as Windows 95 and NT handle many of a computer's basic functions.

orange book: The Department of Defense Trusted Computer System Evaluation Criteria. It provides information to classify computer systems, defining the degree of trust that may be placed in them.

overload: Hindrance of system operation by placing excess burden on the performance capabilities of a system component.

partitions: Major divisions of the total physical hard disk space.

password: A secret code assigned to a user and known by the computer system. Knowledge of the password associated with the user ID is considered proof of authorization.

password authentication protocol (PAP): A simple, weak authentication mechanism in which a user enters the password and it is then sent across the network, usually in the clear.

password cracking: The process of attempting to guess passwords, given the password file information.

password sniffing: Passive wiretapping, usually on a local area network, to gain knowledge of passwords.

patch: A small update released by a software manufacturer to fix bugs in existing programs.

penetration: Gaining unauthorized logical access to sensitive data by circumventing a system's protections.

penetration testing: Used to test the external perimeter security of a network or facility.

perimeter-based security: The technique of securing a network by controlling access to all entry and exit points of the network.

personal firewalls: Firewalls installed and run on individual PCs.

PIN: In computer security, a personal identification number used during the authentication process. Known only to the user.

plaintext: Ordinary readable text before being encrypted into ciphertext or after being decrypted.

policy: Organizational-level rules governing acceptable use of computing resources, security practices, and operational procedures.

polymorphism: The process by which malicious software changes its underlying code to avoid detection.

port scan: A series of messages sent to learn which computer network services, each associated with a "well-known" port number, the computer provides. This computer cracker favorite gives the assailant an idea where to probe for weaknesses. By sending a message to each port, one at a time, the response received indicates whether the port is used and can therefore be probed for weakness.

program policy: A high-level policy that sets the overall tone of an organization's security approach.

promiscuous mode: Occurs a machine reads all packets off the network, regardless of who they are addressed to. This is used by network administrators to diagnose network problems, but also by unsavory characters who are trying to eavesdrop on network traffic (which might contain passwords or other information).

proprietary information: Information unique to a company and its ability to compete, such as customer lists, technical data, product costs, and trade secrets.

protocol: A formal specification for communicating. Protocols exist at several levels in a telecommunication connection.

proxy server: A server that acts as an intermediary between a workstation user and the Internet so that the enterprise can ensure security, administrative control, and caching service. A proxy server is associated with or part of a gateway server that separates the enterprise network from the outside network and a firewall server that protects the enterprise network from outside intrusion.

race condition: Exploits the small window of time between a security control being applied and when the service is used.

reconnaissance: The phase of an attack where attackers find new systems, map out networks, and probe for specific, exploitable vulnerabilities.

registry: In Windows operating systems, the central set of settings and information required to run the Windows computer.

remote access: The hookup of a remote computing device via communications lines such as ordinary phone lines or wide area networks to access network applications and information.

resource exhaustion: Attacks that tie up finite resources on a system, making them unavailable to others.

reverse engineering: Acquiring sensitive data by disassembling and analyzing the design of a system component.

reverse lookup: Finding out the hostname that corresponds to a particular IP address. Reverse lookup uses an IP (Internet Protocol) address to find a domain name.

risk: The product of the level of threat with the level of vulnerability. It establishes the likelihood of a successful attack.

risk analysis or assessment: The analysis of an organization's information resources, existing controls, and computer system vulnerabilities. It establishes a potential level of damage in dollars and/or other assets.

rogue program: Any program intended to damage programs or data.

rootkit: A collection of tools (programs) that a hacker uses to mask intrusion and obtain administrator-level access to a computer or computer network.

router: Device to interconnect logical networks by forwarding information to other networks based upon IP addresses.

scavenging: Searching through data residue in a system to gain unauthorized knowledge of sensitive data.

secure shell (SSH): A program to log into another computer over a network, to execute commands in a remote machine, and to move files from one machine to another.

security policy: A set of rules and practices that specify or regulate how a system or organization provides security services to protect sensitive and critical system resources.

sensitive information: As defined by the federal government, any unclassified information that, if compromised, could adversely affect the national interest or conduct of federal initiatives.

server: The control computer on a local area network that controls software access to workstations, printers, and other parts of the network.

shadow password files: A system file in which encryption user passwords are stored so that they aren't available to people who try to break into the system.

share: A resource made public on a machine, such as a directory (file share) or printer (printer share).

smurf: An attack that works by spoofing the target address and sending a ping to the broadcast address for a remote network, which results in a large amount of ping replies being sent to the target.

sniffer: A tool that monitors network traffic as it is received in a network interface.

sniffing: A synonym for "passive wiretapping."

social engineering: A euphemism for nontechnical or low-technology means, including lies, impersonation, tricks, bribes, blackmail, and threats, used to attack information systems. Typically carried out by telephoning users or operators pretending to be an authorized user to attempt to gain illicit access to systems.

spam: Electronic junk mail or junk newsgroup postings.

spoof: Attempt by an unauthorized entity to gain access to a system by posing as an authorized user.

stack mashing: Using a buffer overflow to trick a computer into executing arbitrary code.

stealthing: Approaches used by malicious code to conceal its presence on the infected system.

system security officer (SSO): A person responsible for enforcement or administration of the security policy that applies to the system.

system-specific policy: A policy written for a specific system or device.

tamper: To deliberately alter a system's logic, data, or control information to cause the system to perform unauthorized functions or services.

TCO: Total cost of ownership, a model to help IT professionals understand and manage the direct and indirect costs of acquiring, maintaining, and using an application or a computing system, including training, upgrades, and administration as well as the purchase price.

TCP/IP: A synonym for "Internet Protocol Suite," of which the transmission control protocol and the Internet protocol are important parts. The basic communication language or protocol of the Internet, TCP/IP is often used as a communications protocol in a private network (either an intranet or an extranet).

threat: A potential for violation of security that exists when a circumstance, capability, action, or event could breach security and cause harm.

threat assessment: The identification of types of threats that an organization might be exposed to.

threat model: Used to describe a given threat and the harm it could to do a system if it has a vulnerability.

threat vector: The method a threat uses to get to the target.

transmission control protocol (TCP): A set of rules (protocol) used along with the Internet protocol to send data in the form of message units between computers over the Internet. While IP takes care of handling the actual delivery of the data, TCP takes care of keeping track of the individual units of data (called packets) that a message is divided into for efficient routing through the Internet. While the IP protocol deals only with packets, TCP enables two hosts to establish a connection and exchange streams of data. TCP guarantees delivery of data and also guarantees that packets will be delivered in the same order in which they were sent.

Trojan horse: A computer program that appears to have a useful function but that also carries a hidden and potentially malicious function for invading security mechanisms, sometimes by exploiting legitimate authorizations of a system entity that invokes the program.

turn commands: Commands inserted to forward mail to another address for interception.

uniform resource locator (URL): The global address of documents and other resources on the World Wide Web.

unprotected share: In Windows terminology, a "share" is a mechanism that allows a user to connect to file systems and printers on other systems. An "unprotected share" is one that allows anyone to connect to it.

user: A person, organization, entity, or automated process that accesses a system, whether authorized to do so or not.

user contingency plan: The alternative methods of continuing business operations if IT systems are unavailable.

virus: A hidden, self-replicating section of computer software, usually malicious logic, that propagates by infecting—i.e., inserting a copy of it into and becoming part of—another program. A virus cannot run by itself; it requires that its host program be run to make the virus active.

vulnerability: A flaw or weakness in a system's design, implementation, or operation and management that could be exploited to violate the system's security policy.

war dialing: A simple means of trying to identify modems susceptible to compromise in attempting to circumvent perimeter security.

web server: A software process that runs on a host computer connected to the Internet, intended to respond to HTTP requests for documents from client web browsers.

wiretapping: Monitoring and recording data flowing between two points in a communication system.

worm: A computer program that can run independently, propagating a complete working version of itself onto other hosts on a network and consuming computer resources destructively.

wrap: Using cryptography to provide data confidentiality service for a data object.

THREATS TO COMPUTER SYSTEM SECURITY

Risks to computer systems data include physical theft, logical usurpation, application damage through viral attack, and "social engineering" attacks. For example, the possibility is very high that any organization with more than twenty workers will have lost laptops or handhelds in the last year. The associated costs for such losses are usually determined only for replacement costs—which doesn't include the value of the lost data. Count up all of the replacement costs, including equipment, software, and restoring or replacing data, and the reasoning behind why staff should be taught clear policies, procedures, and the importance of computer security becomes clear.

Evaluating System Needs for Security

Computer system risks mean that any complete security program pays attention to all risk areas. That means developing ways to protect:

- The physical equipment itself, including desktop computers (whether office-based or home computers supplied by the company) and the associated peripherals (output devices like printers, input devices like scanners, and data transfer, back-up, and storage devices like Zip drives, CD or DVD drives, whether external or internal); laptop computers and their associated gear; and an ever-widening body of handheld devices, including PDAs, bar code readers, and many different types of data readers.
- Logical assets, including IP and other network configuration information.
- Application and data integrity.
- Password integrity.

Different types of equipment present different security needs. Laptops are easier to physically steal than desktop units but are more apt to be the targets of incidental theft, not a particular unit being stolen deliberately because of the data it contains. More important is the protection of the computer system and the information held within the system, which may have value, or may simply be vulnerable to hackers.

Developing the Security Process

When developing a computer and data protection program, vulnerability management is a primary concern. A vulnerability management process should be put in place to ensure that all desktops remain secure from potential vulnerabilities. According to Trinity Security Services (2003), the various stages of the process include taking inventory, gathering intelligence, auditing, fixing vulnerabilities quickly, reviewing fixes, and writing company policy.

Inventory: You can't determine vulnerability of your equipment and systems without knowing what you have. Begin, regularly update, and maintain lists of all equipment, types, manufacturers, capacity, special features, operating systems, important applications, and company purpose and operators and other users, with serial numbers. This data can be critically important in tracking missing or stolen equipment, in helping to pinpoint certain problems, and for insurance and replacement purposes.

Gather intelligence: You can't protect something you don't know is at risk. Being aware of the latest vulnerabilities is key to maintaining a high-security posture. Since new dangers can occur at any time, track potential problems specific to your inventory.

Audit: Use automated vulnerability-scanning tools to regularly check system status for known vulnerabilities. (This can give you 90 percent protection.) Develop internal auditing practices to increase your protection percentage.

Fix: Once vulnerabilities have been identified, fix them quickly (before they allow damage), through patches, upgrading, or removal of software. Test any changes before general release to personnel.

Review: Review the findings and fixes for the process to identify where risk reduction changes are possible. Develop a plan for regular implementation of changes.

Company policy: With computer and data protection programs, just as with any other company safety or security program, a clearly written policy and management buy-in are essential to the program's success. The written policy should cover equipment and data, of course, and should also address issues like individual user's personal use of facility equipment, including email and Internet use as well as data copying and transfer.

NETWORKS AND STAND-ALONES

Networked computer systems offer so many benefits that most places are heading in that direction, if they aren't there already. These days, servers and an IT staff handle computer operations at many facilities, and while many facilities might have stand-alone computers dedicated to specific purposes, facilities with only stand-alone computers are rapidly becoming a thing of the past. The differences between stand-alone computers and a networked system make some aspects of data protection easier, and others harder.

Server systems allow IT staff to automatically and regularly back up all data, keep track of employee Internet and email usage (in part to protect from data theft or loss) and to help diagnose problems. Servers also allow IT staff to monitor many types of employee computer activity. How far management goes in keeping tabs on worker computer use is up to individual facility policy, but networked systems provide the possibility.

Setting up the capability for and gathering that information is possible on stand-alone computers, too, but it is much more labor-intensive. Duplication of equipment is also an expensive factor for stand-alones.

One real advantage of stand-alones is that they are isolated if a computer virus strikes. Even if a viral infection occurs (from a modem or data transfer device), it is isolated and can be confined. Data exchange and backup are much more time-consuming, however. With a stand-alone system, choosing each workstation's peripherals in light of the other stand-alone workstations keeps data transfer and backup both possible and simple.

Portable Computing

Laptops and handheld computing devices offer convenience, portability, flexibility, and many other benefits. They also take both equipment and data out into the cold, cruel world, where they are vulnerable to loss and theft. Personnel who have had equipment lifted from under their noses or who have absentmindedly walked off and left a laptop in the taxi should be more than embarrassed at the loss—because company policy and training should alert them to the possibility of such losses. The bottom line is that whatever the security issue, the biggest threat comes from the user. The best protection is to implement effective and well-managed safe work practices for all employees—education and training is key to protecting the network elements that go into the public domain.

Laptops and handhelds, including PDAs, hold information and provide access to email and other network resources. Even a casual thief (not only one who has deliberately targeted a particular company as a source for valuable information) could benefit from information stored in a poorly secured device. Enforcing standardization and policy among PDA users is especially difficult since the users are generally the owners of the device. Since they don't see the PDA as computers and since the PDA is personal, not company property, workers are sometimes much more careless with them than they would be with a corporate laptop.

In some ways (especially physically) laptops are more difficult than full-sized desktop computers or workstations to protect. *A Guide to Rolling Out Handhelds in the Corporate Environment* lists the principle laptop security issues as:

- Physical theft
- Loss of information
- Hacking
- Installation of noncorporate software

Physical theft or loss is the most likely security risk for laptops, and the key to preventing such losses is user education. When workers take laptops off-site, either occasionally or as their usual routine, they must be aware that cars, airport lounges, taxies, trains, and hotel rooms are high-risk areas. An unattended laptop is a theft waiting to happen.

Cable locks (similar in many ways to bicycle locks) are used to attach laptops to stationary objects to prevent theft. Newer machines have the capability of logical security measures to protect data from casual thievery—company and user passwords, for example. The data the laptop contains could easily be more valuable (to the facility, if to no one else) than the cost of the equipment, so regular backup before taking any laptop or other portable device off-site should be standard practice. Data encryption provides a deeper level of protection for sensitive data.

Common practice is that laptops are equipped with Internet connection capability as a transport mechanism for connecting back to the office. However, using the Internet itself can be risky. A desktop-level firewall protects the laptop during connection to any "foreign" network (i.e., the Internet).

Handheld Devices

Handheld devices offer low cost, ease of use, and portable convenience. While contemporary laptops can have all the power and hard disk capacity of an office PC, a handheld device is a palmtop whose weight is measured in ounces, not pounds. Limited to one or a few dedicated purposes, the rechargeable batteries last three to four times longer than a PC's. Dedicated handhelds (the UPS guy's delivery pad, for example, or a bar-code reader) are less attractive to casual theft than laptops, too. Handheld devices are a highly effective addition to mobile working practices, especially for data collection. Administrators and sales staff (and others) love their PDAs for the huge amount of useful data such devices can make portable.

The combination of low cost, portability, and high functionality makes handheld computing equipment more and more attractive for companies with the need for a mobile workforce. However, many IT departments haven't fully incorporated handhelds into computer and data security programs. A handheld computing component that addresses personnel, technical, and security issues should be included for facilities that issue such equipment.

Handheld equipment's small size doesn't necessarily mean low capacity. IPods (now used mostly for music but capable of a startling array of functions for other purposes) can hold thirty gigabytes of memory in a case about the size of, but much thinner than, a pack of cigarettes. The large data storage capacity of both laptops and handhelds means that staff members have the potential of carrying large amounts of company information around with them. Best practice is to upload data to the server on a regular basis (determined by value of the data and opportunity), not storing it for long periods of time, so that any loss is mostly equipment loss, not data loss. Collecting the data again can be impossible (for time-sensitive measurements, for example) or difficult (would that UPS driver have to go back and collect all those signatures again?). The involved costs are hard to accurately calculate.

Other than loss or theft, the biggest problem areas for handhelds are in file transfer and synchronization. Tools that enable all devices to synchronize to a single server means that information available to mobile users can be easily monitored. Password protection on many devices is two-level—a password for the device and one for individual applications or files is common.

So far, handheld devices are not heavily affected by computer viruses. Currently, only a few Trojans affect them, most of them irritating but harmless—like randomly switching backlights on and off or posting nasty and hard-to-irradicate screen messages. However, sooner or later, creating more troublesome handheld-directed trouble will occur to hackers, so keeping an eye out for information on new vulnerabilities is important. Many hardware vendors have established partnerships with leading antivirus organizations already.

Equipment can be infected by malicious code through email attachments or during synchronization with infected systems. Antivirus software at all synchronization points on the network cuts down the possibility of mobile workers infecting the corporate system and vice versa. In fact, some operating systems can be made resistant to virus attacks, though this can restrict the functionality of the device—in general, the more you increase the security, the more you restrict what the user can do. (Some of the previous information came from *A Guide to Rolling Out Handhelds in the Corporate Environment*.).

THE HUMAN ERROR FACTOR

People cause the biggest risks to information systems, whether deliberate or accidental. In fact, people—the users themselves, are a commonly overlooked possible breach in any security infrastructure. They'll write down passwords and leave them in obvious places. They'll set their software to remember passwords. They'll open email with strange attachment file endings and cute taglines, and the resulting avalanche of virally induced spamming will bring down the whole system and help to spread the infection at top speed throughout the world. Educating them on the implications of their actions is ongoing, an integral part of employee training. IT staff and safety personnel should work together to educate administrations and the workforce in computer and data protection. Training and worker education helps to control a wide range of problems. For example, as safety engineer, you should know that the workers in your organization have been trained to follow IT instructions on updating viral protection regularly, that they know better than to open suspicious email, that they are aware that a random stranger on-site could be seeking vulnerable equipment or data, and that passwords must never be shared with anyone outside of a verified IT staff member. In short, one of the best protections for computer security are users who are thinking.

Security Measures for IT Staff

While the IT staff handles the background of computer security, they are often the ones responsible for communicating security needs to workers. According to Dancho Danchev, IT security issues handled on a worker level may include:

Table 14.11   Tips for Information Security

Four fundamentals of security...

1	Identification and authentication	Who are you and can you prove it
2	Access control	Who can access, edit, create, delete, or copy data
3	Confidentiality and integrity	Who sent this, who can read it, has the data been changed
4	Administration and audit	Who did what, when, and where

Know your "enemies."

1   The malicious insider
2   The thief
3   The hacker
4   IGNORANCE and COMPLACENCY

Know what your enemies use ...

1   Social engineering
2   Password cracking
3   Network monitoring (sniffing)
4   War dialing (brute force modem attacks)

Top Ten Elements of Security...

1	POLICY	Define, implement, and measure systems, networks, and applications for compliance
2	EDUCATION	Part of (1) above is to ensure your users understand the consequences of their actions and the importance of information security to the organization
3	WARNINGS	Implement online warnings informing users (internal and external) of the rules of access to the systems
4	ASSESS	Determine what you are trying to protect, why, and what value it has
5	PROTECT	Based on (4) above, implement firewalls, antivirus, and intrusion detection (IDS) tools
6	AUDIT	Regularly audit systems for access violations, latest applicable patches, and integrity of operating systems and applications
7	PASSWORD	Enforce strong password policy and run password guessing tests to determine those with easy-to-guess passwords
8	SCAN	Periodically check for known vulnerabilities and the latest exploits
9	BACKUP	Ensure regular, clean backups are made to enable speedy recovery
10	RESPONSE	Set up an incident response team with complimentary measures

Adapted from *Tips for Information Security* by Malcolm Skinner of AXENT Technologies (2002). www.axent.com.

- Physically secure terminal and working places. Make staff aware of the dangers of malicious "snoopers" walking around workplaces or having access to terminals.
- Set terminals for automatic logout, or set up a time out so that the system is protected once it detects the operator is not in front of the keyboard.
- Limit the use of notes and papers for any sensitive information such as passwords, IPs, and anything that might help a potential intruder gain access to systems. Advise staff to shred and destroy password notes each time they leave their workplace. This is a well-known workplace weakness, so limit or completely eliminate the use of these notes.
- Remember that the general staff may or may not have real experience with computer use. What seems simple to an IT staff member may be only halfway comprehensible to the person at the workstation—or that person may be totally lost. Effective communication both ways is key to performance.

Internal IT Security Measures

- Conduct a complete security audit of the system before physically connecting it to the network.
- Make sure the system has the latest versions of the software installed and securely configured.
- Consider blocking access to the test system from the Internet during network auditing and testing.
- Verify that the system to connect doesn't contain any sensitive data.
- Install an intrusion detection system to see how often the new system is probed for various vulnerabilities.

- Subscribe to appropriate security-related newsletters and mailing lists for updates on the latest vulnerabilities.
- Visit the appropriate websites as an early warning system for potential intrusions from outdated or unpatched software.
- Read the latest security-related white papers as an essential step in self-education. (Danchev 2003)

System Administrator Mistakes

System administrators manage the continued operation of system computers and network functionality. In many organizations they also handle equipment security, potential and real intrusion detection, and network security. Holding many responsibilities increases workloads as well as levels of stress and distraction from multiple projects, and it can increase the probability of accidental human error. Common system administration problems include:

- The lack of a well-established personnel security policy
- Connecting misconfigured systems to the Internet
- Relying on tools alone, without auditing and analysis
- Failing to monitor the logs
- Running extra and unnecessary services/scripts (Danchev 2003).

Company Executive Mistakes

As top management sees what the Internet offers, they realize that many possibilities are at hand—the potential for new markets, higher profits, and so forth. However, often they don't see that the global connectivity isn't all golden. Connectivity represents a threat to sensitive information without solid system security. Ensuring that executives know the cost of doing business on the Internet includes effective security measures is critical for data security program success. Common mistakes include:

- Employing untrained and inexperienced experts
- Failing to realize the consequences of a potential security breach
- Trying to handle the information security issue on the cheap
- Relying too heavily on commercial tools and products
- Thinking that security is a one-time investment (Danchev 2003)

End User Mistakes

While end users can't be blamed for IT or management problems, on both network systems and stand-alone computers, trouble often arrives via a run-of-the-mill user's workstation. Training and communication from both safety personnel and IT staff is essential to keeping systems secure. Workers cause problems by:

- Violating the company's security policy
- Forwarding sensitive data to home computers
- Writing down account data
- Downloading contaminated email or from infected websites
- Failing to pay serious attention to physical security (Danchev 2003)

COMPUTER SECURITY AUDITING

Issues and actions for computer and data systems security include many different elements to consider. Using a computer security audit checklist provides steps to securing computer and information systems. Just like any other safety area, computer and data security is not static—no once-and-done solution is possible. The equipment to protect, the risks, and the threats change rapidly. This section addresses common audit items and the audit questions to help determine system needs.

- Who within the organization should have access to what information? Ensure that people do not unnecessarily have full access to the systems.
- What information is most important to protect in terms of confidentiality?
- Does the acceptable use policy explain what everyone in the organization should and should not do with the information systems?
- Is a current equipment and configuration list maintained for the organization?
- Is a current software installation list available for all computers and equipment in the organization?
- Are all passwords used in the IT systems changed from the default?
- Are log-on passwords effectively maintained as confidential, changed regularly, and in a format that cannot easily be guessed?
- Do any workers take laptops and other portable devices off the network and outside of standard security procedures?
- Will any remote workers need to connect to the network?
- What is the organization plan for security requirement evaluation?
- Is someone responsible for network and information security? Does that person have the authority to take action to respond to security breaches and vulnerabilities?
- Can abnormal network activity on your network be identified?
- How is company physical waste disposed? Secure computer data revealed through printout paper waste instead of system intrusion is just as disruptive (Rose 2002).

DATA PROTECTION

Once your system is in place, daily maintenance must occur to prevent trouble. Individual tasks must be scheduled to fit the individual needs of the organization and should include a number of different measures and concerns. These include backing up data, ensuring continued operation during high-level problems, and using viral protection in a number of different ways.

Backup Protection

Both equipment and data are important issues in computer security. While computer equipment is valuable, attractive to thieves, expensive to purchase or replace, and certainly vulnerable, if your system is properly protected by backup, the equipment itself is generally fairly easily replaceable. Often equipment theft is covered by insurance and can be replaced. More important than the equipment itself is the data the computer equipment is used to generate, control, store, format, and distribute. The computer equipment is there for a concrete purpose—computers are elaborate tools for data management—and the data is not so easy to replace if irretrievably lost. The downtime related to replacement of both equipment and data is another important area of concern.

Data protection starts with backup. Set up equipment and systems for easy data backup (either automatically, or easy for individual users to perform), and keep duplicate backups off-site as well as on-site. Servers allow automated backup—but automated backup doesn't matter if the place burns down and you don't have a current data backup off-site. Stand-alone computers should be outfitted with a company-standardized means of backing up data, and the backup media should be collected regularly and stored off-site. Obviously, a scrubbable media is more cost-effective than many one-time use media (writeable CDs, for example), but currently, prices for writeable and even rewriteable CDs are low. Service agencies on the Internet offer automated backup of data onto remotely located servers over broadband Internet connections—these have advantages and disadvantages, but they are certainly off-site and simple. Be sure you secure your backups as part of your data protection policy.

If facility data isn't adequately backed up, restoration may be difficult, time-consuming, and costly. How much of the data in your facility computer system was entered by hand? With a massive loss of data, people may have to go back in and tediously re-enter it, which not only is time-consuming but also introduces errors, which makes the loss that much more expensive. Was the data read from an electronic reader of some sort, a handheld data collector, or an in-the-field laptop? How much data does that electronic reader or handheld hold, anyway? Loss of either equipment or the data on the equipment is tough to handle.

Backup solutions should meet the requirements of any user, from a periodic (weekly or monthly) copying of some information onto a CD or other external media to fully replicated server system with two parallel systems operated in separate parts of the world. The more extensive the system is, the more apt it is to be properly backed up but the bigger the tangle of data can be to sort out. Essential for any backup system are regular tests of the retrievability of the backed-up data. IT should keep updated lists of all the programs on each workstation hard drive. Data files without the application used to generate or read the files are more difficult to use, or useless. Don't rely on individual workers to know the ins and outs of their workstation's applications. IT should maintain records for each workstation and all the related devices of the system. The backup storage system should also ensure that the stored data is traceable back to the workstation it was copied from. Factors to consider in arranging backups include:

- What information or data should be backed up?
- How often should I back up information?
- What is the most efficient and cost-effective backup medium for my system?
- Can I restore information in the event of a loss of the computer(s), the backup machine, and the backup software?
- Can I restore data when I check to make sure my backup processes are working?
- Where will I store my backups? Are the backups themselves secure?

Worst-Case Scenario ... Disaster Recovery

The need for adequate backup measures, both on- and off-site, is an out-of-the-mainstream lesson of 9/11. After the initial losses, many businesses found themselves in a worst-case scenario beyond their worst nightmares in terms of disaster recovery.

The TowerGroup estimates that to replace the technology destroyed for securities firms affected by 9/11 cost approximately $3.2 billion: $1.7 billion spent on hardware—trading stations, sales stations, workstations, PCs, servers, printers, minicomputers, storage devices, cabling, communications hubs, routers, switches, etc.; $1.5 billion for software and services to install and connect the network, operating system, and software applications for these firms. This is only for securities firms—not for all those affected.

Usually, disaster recovery planning provides desk space for only a business's more critical elements—usually about 20 percent of an organization's workforce. That means the other 80 percent of the workforce have no access to desks and computer systems. Getting the 80 percent to work again in the shortest possible time is essential for continuity.

Such planning was a hard sell before 9/11. Answering "what do I get for my money" was hard to justify, though many organizations had some reasonable level of disaster planning. Funding difficulties led to a two-tier approach: separate plans to cope with an immediate (short-term) incident (the disaster recovery plan), and longer-term major incident recovery methods (continuity plan).

After 9/11, when all of the affected organizations invoked their disaster recovery plans (in one way or another), the contracts with their disaster recovery service providers went into effect: emergency desk space previously put into a state of semireadiness were supposed to be available. The sheer scale of the disaster caused problems. These providers operate somewhat like insurance policies—they plan their operations by statistical analysis, and statistically, such massive failures are uncommon. While organizations had contracted for a specific space, the service providers had no way to provide for disaster on that scale—so many different companies invoking their contracts for use at once. Such companies usually provide a shared space with multiple companies contracting for the same space, and using it at different times, at need. After 9/11, everyone needed all of the space, at once, and not for just the usual couple of days or weeks expected after a company has problems. Many clients wanted the full use of their contract for the full terms of their contracts—usually a maximum of six months.

Post-9/11 thinking for continuity planning involves the possibility of alternate premises, sometimes independent facilities (sometimes moving to another company site within the same organization) to allow the full staff a quick return to a normal workload.

Setting up a new installation from scratch takes time, often more than the six months most disaster recovery service contracts offer. While a client organization is heavily involved in restoring business as usual, the disaster recovery service provider must deal with many other clients who may need to share the same limited resources. The disaster recovery service suppliers end up seeking alternate space as well, possibly competing with their own clients for suitable housing (McCreary 2003).

Viral Protection

IT personnel should regularly update and share viral protection software as well as educate employees on the dangers of stranger-generated email. As some folks have to climb a mountain "because it's there," other folks seem to feel obligated to trash computer systems on a widespread basis to prove they can, or for financial gain. Computer server security systems involve solid firewall protection, but various Trojan horses, viruses, parasites, worms, and bugs are being created and spread all the time. A really massive infestation of some of the more wicked bugs can bring an organization to its knees for days and cost thousands of dollars of wasted work hours and ruined data and applications as well as hundreds of hours of IT staff time to fix. Make sure viral protection software is on all workstations and that it is regularly and completely updated. Ensure that antivirus software is configured to identify viruses by all of each computer's means of entry—email, web browsing, floppy disks, CDs, archives, etc.

Email

Email is a wonderful tool for business and industry in many ways, but it is also one of the biggest security threats in many systems. Email systems demand effective management—one of the security policy's key elements. IDC predicted that by 2005, cyberspace and the physical components in people's offices and homes would be running 1.2 billion email boxes (138 percent compound growth) and 36 billion person-to-person emails daily. Instant messaging was a high-growth area, too—an estimated growth rate of 100 percent compounded with 150 million business users by 2004.

Think about the number of email-based or delivered bug scares in the last year. Now consider what such growth may mean in terms of security problems. Virus defenses, email data security, content filtering, and spam protection are key—and keeping them up-to-date cannot be ignored. Software solutions for email protection help define and enforce an email security policy and provide filtering, monitoring, and reporting. Effective software works in tandem with antivirus software, providing additional protection companywide, including protection designed to safeguard intellectual property and confidential information and methods of dealing with spam. Contextual searching picks up emails that might contain offensive materials and can determine whether employees are mailing out sensitive information such as customer lists or research data.

The quick growth of the Internet and email use has made employees' privacy rights vs. corporate security/responsibility a pertinent and controversial issue. As email and Internet monitoring grow, more and more disciplinary actions ensue, leaving individuals unsure about their rights to workplace privacy. The struggle to balance individual right to privacy and security with organizational responsibility to protect their own intellectual assets has become increasingly intense.

As a middle ground between absolute freedom for employees to email and surf at will and the "Big Brother Sees All" effect of high levels of employee monitoring, a sensible and reasonable company policy offers compromise. Written company policy, effectively established, regularly discussed with employees and consistently enforced through the appropriate technology, is critical to maintaining consistency and predictability in the process of monitoring the workplace and helps address the privacy issue.

Establishing a Company Email Policy

Scanning and enforcing: An indication of the measures in place to show determination to enforce; the penalties for/consequences of breaching the policy; and why the policy exists (protection of staff, company reputation, and IT security).

Responsibility: Name the individuals responsible for training and indicate where to go for answers to queries regarding the policy. Employees should be fully aware at all times of what constitutes the policy and know the consequences should it be abused.

Liability: Details of potential legal liability should the policy be breached (to reinforce the seriousness of the problem).

Definitions: A definition of what is appropriate business and personal Internet usage if an Internet connection is provided for business use. This should include details of sites that should never be accessed using business facilities and time; a clear statement on whether the company allows private web-based emails, and if so, under what circumstances (no attachments; video or music files; only for making arrangements; times of day; costs; etc.).

Business materials: Company policy on the circulation of business materials, in particular those likely to be generated within an employee's own field of work; potential pitfalls in the circulation of materials (how to avoid accidental distribution of confidential information and copyright infringement).

Lost productivity: How to avoid it and what the consequences might be.

Forbidden files: What types of files may not be circulated via emails or downloaded or uploaded via the web.

Data theft: How to avoid it (cyberwoozles, cookies, and how to spot them).

Viruses: How to avoid them.

Acknowledgement: A form to sign acknowledging that the employee has understood their obligations, making adherence to the policy part of their terms and conditions of employment (Panesar 2008).

AUTOMATED TOOLS FOR DATA SECURITY SYSTEMS

Computer and data security systems must be suited to the needs of the company. Levels of protection should correspond to the value (both monetary and replacement value) of the equipment and data of the business. Remember, however, that while automated tools do a great job in many areas, they can't do everything. Regular human-initiated checks must also be in place. Common automated tools for system protection include:

- Firewall reporting: Firewall reports give complete accounts of everything that goes through a firewall and highlight attempts to breach security.
- Vulnerability testing: Vulnerability testing tools regularly test systems in different ways to illuminate system strengths and weaknesses.
- Intrusion detection: Intrusion detection identifies and responds rapidly to intrusion attempts.
- Web filtering: Limiting access to unneeded and unrelated websites lowers productivity costs generated by web abuse.
- Honey pot system: Decoy servers or systems set up to attract intrusion attention and to gather information regarding an attacker or intruder are available for internal, external, and remote access systems.

Firewall

With the number of port scans carried out against computers while connected to the Internet, even single-user computers should have suitable firewall protection.

For personal use, simple software personal firewalls offer a reasonable degree of protection when correctly configured to the individual computer and operating system. Even with a firewall in place, open communication channels through the firewall can be exploited by hackers to gain access to information. Firewalls are even more critical in network situations. Sophisticated software/hardware combinations offer a higher level of protections for networks. Always-on connections offer excellent opportunities for hackers to gain access to systems and are also a route by which spam can be propagated, increasing the need for firewall protection.

Access Control

Log-on systems protect data on both networked workstations and stand-alone machines as well as laptops. Larger networks and companies should determine what needs protection and the level of data protection needed.

Internal Threats

While external threats are an ever-present problem in computer security these days, theft of data by a company employee is also a widespread problem, a form of theft made easier by the worldwide ease of information transmission. Any facility that uses or generates sensitive information should have systems in place to prevent internal theft. Industrial espionage is a reality, even if it probably isn't as sexy as it seems in books and movies. Limiting access to sensitive materials, encryption, special authorization processes for users, monitoring computer activity for those who handle sensitive materials, and blocks on electronic transmission of certain types of data are some ways to protect data. Clearly stated and specific company policy provides legal protection for the employer.

COMPUTER SECURITY ... CRADLE TO GRAVE ...

In our throwaway society, we sometimes forget that one man's trash is another man's treasure. Managing data on hard drives at the end of the PC lifecycle is commonly ignored. However, data recovery is one growing and quickly expanding element in the computer industry. Data recovery companies can pull useable data off hard drives that were data cleansed, from laptops pulled from shipwrecks, hard drives that went through fires, or were otherwise severely physically abused. What can businesses do to ensure their old data is gone—and not available to someone who's just bought an old hard drive?

Several years ago, two students at the MIT Laboratory of Computer Science purchased 158 used hard drives with a scrap value of $1,000. On some of those drives, they found 5,000 credit card numbers and many medical reports, as well as corporate and financial information. Fortunately for the companies concerned, the students used this project as data in a report on their findings. In short, they concluded that someone buying any ten random drives on the used market would have a 30 percent chance of finding useful confidential information (Riches 2003).

How can a company destroy the information on redundant equipment—or at least protect it from recovery? Keeping such equipment forever isn't a great option—secure space is costly, and besides, the equipment is subject to theft. Be sure data removal tools are used on redundant equipment and remember that how far IT should go to protect that data should depend on the value of that data to outside sources.

IN-HOUSE OR OUTSOURCED SECURITY?

Just as outsourcing emergency services is more difficult to achieve than it sounds, outsourcing security poses a unique set of problems. Outsourced computer system security is bound to be expensive, and as with other safety needs, has no direct financial benefit to the organization in the eyes of most financial officers. Security is an insurance policy—those financial officers will have very little way to know what any security system has saved them—unless the savings are detailed to them. When determining whether computer and data security should be accomplished in-house or outsourced, estimating the costs of both is essential.

Determining the level of expertise already available in-house is the next consideration. If local talent isn't capable on the job, the cost of new hires must be figured in. While lack of internal recourses might make outsourcing seem more attractive, the element of trust is higher for in-house systems. Security is usually outsourced when company resources can't handle it. Realize that outsourcing cannot be accomplished without in-house involvement at some level. Outsourcing does not remove the ultimate security responsibility from your organization—so the question becomes "Will my company computer system be more secure with in-house or outsourced security?"

Outsourcing concerns include return-on-investment considerations, risk assessment, security spending verses value of the data to be protected, and the possible consequences of a security breach.

One of the most important outsource concerns is training. Who will be responsible for staff training? Will training be in-house, or through outside consultants? How much will training and staff education cost—for both in-house and outsourced? Will you run seminars? Will you produce your own study materials?

What to Ask When Buying Security

- Do you know what you need to protect?
- Do you know what you are protecting it from?
- Have you completed a risk assessment?

- Have you researched the legal obligations of your company?
- Does your security strategy mirror your business strategy?
- Have you considered the training implications for users?
- Do you have appropriate skills and resources in-house to fulfill the project, or do you need outside help?
- Have you planned for the use of mobile technology within the organization?
- Have you thoroughly investigated all the options for both in-house or outsourced security? (Franklin 2003)

REVIEW QUESTIONS

1. What lessons in emergency response can we learn from 9/11 and the power outages and hurricane of summer 2003?
2. What are some of the drawbacks to ethnic profiling? Why is it so popular for enforcement?
3. Discuss the financial considerations at work in putting a security system.
4. What role do hiring practices play in security and vulnerability assessment?
5. How does a threat assessment team accomplish its mission? What tools can they use? How can complete and accurate recordkeeping help?
6. Discuss hazard analysis and workplace security analysis and hazard control and prevention in terms of vulnerability assessment.
7. What building-level considerations should safety engineers examine?
8. Discuss the usefulness and need for five of the work practice controls and procedures.
9. What role does knowing your workforce play in facility security?
10. Discuss the security survey as an assessment tool.
11. What is the purpose of facility security systems? Why are they important? What do they protect against?
12. How can you accurately define "level of need"? Discuss its importance for security assessment.
13. Discuss "three-way protection" and security systems.
14. What roles do fencing, signage, exterior lighting, and facility exterior visibility play in facility safety?
15. Discuss theft of materials and theft of time or resources in light of "Big Brother."
16. Discuss the concept of the implied contract of workplace trust between employer and employee.
17. What advantages do 24/7 operations have in terms of building security?
18. What are the special concerns regarding chemical security?
19. What lessons can we learn from anhydrous ammonia bootleggers for meth labs?
20. Why must we be concerned about chemical security in the face of terrorist activity?
21. Discuss the "protection dilemma."
22. How can you determine if your facility's chemical security is adequate to your needs?
23. What special measures should be taken for remote locations? Why?
24. What role do good customer and community relations have in chemical and facility security?
25. How much should the public know about the details of your security plan?
26. What should you plan for rapid shutdown procedures?
27. What level of knowledge should safety engineers have on computer systems and data security?
28. How can the safety engineer and IT establish good cross-departmental communication?
29. What are the elements of best practice for data security?
30. What problems can be caused by spoofing, spamming, sniffing, worms, bugs, and Trojan horses?
31. What threats to computer and data security demand attention?
32. How can the safety engineer and IT evaluate system vulnerability? What roles should each play?
33. What role does inventory play in system security? What information should the inventory include?
34. What's the place of—and what is—intelligence gathering?
35. Discuss the issue of information and system backup. Consider frequency, method, media, backup storage, backup safety, and restoration.
36. Discuss 9/11 and computer system's "worst-case scenario."
37. What services can a disaster recovery service provide? Why can this be important?

38. What special security problems do connection to email and the Internet create?
39. What can viral protection software do to protect systems and data? What's the downside?
40. What's a firewall, and why would you need one?
41. What controls and considerations should be placed on email? What personal privacy issues are involved?
42. What are some of the elements to consider when establishing a company email use policy?
43. What are the major external threats to computer systems and data?
44. What are the major internal threats to computer systems and data?
45. Name and discuss some common automated computer security tools.
46. What issues should be considered when disposing of outdated computer equipment, especially hard drives? Why?
47. What are the advantages and disadvantages of in-house and outsourced computer security? How can you determine cost-effectiveness of each? What factors make one or the other more likely to be effective?
48. Discuss the best level of protection.
49. What role does training play in computer systems and data security?

REFERENCES AND RECOMMENDED READING

A guide to rolling out handhelds in the corporate environment. 2009. http://www.itsecurity.com/ papers/psion1.htm. Site now expired.

Anderson, R. J. 2008. *Security engineering: A guide to building dependable distributed systems.* 2nd ed. New York: Wiley.

Cox, P. 2009. *Security evaluation: The common criteria certifications.* http://www.itsecurity.com/papers/border.htm. Site has expired.

Danchev, D. 2003. *Reducing "human factor" mistakes.* http://www.itsecurity.com/papers/ dancho1.htm (August 5).

Ellison, R. J., et al. 2009. *An approach to survivable systems.* CERT® Coordination Center, Software Engineering Institute, Carnegie Mellon University, Pittsburgh, PA.

Franklin, I. 2003. *How to buy security.* http://www.itsecurity.com/papers/entercept3.htm (April 27, 2003).

Garcia, M. L. 2001. *The design and evaluation of physical protection systems.* New York: Butterworth-Heinemann.

Henry, K. 2002. New face of security. *Gov. Security* (April): 33.

IBWA. 2004. *Bottled water safety and security.* Alexandria, VA: International Bottled Water Association.

Kilpatrick, I. 2003. *Security patching: Are you guilty of corporate stupidity?* http://www.itsecurity.com/papers/wick5.htm (accessed June 5, 2003).

Kilpatrick, I. 2002. *Security: The challenges of today and tomorrow.* http://www.itsecurity.com/papers/wick3.htm (accessed August 16, 2002).

Mathews, S. 2002. *Security can be simple and secure.* http://www.itsecurity.com/papers/articsoft15.htm (accessed September 10, 2002).

McCreary, L. 2003. *A proven paradigm for best practices in information security.* http://www.itsecurity.com/papers/securecomp2.htm (accessed April 14, 2003).

Millard, N. 2003. *Achieving policy management best practice: A guide to improving corporate governance and realizing more value from policies.* http://www.itsecurity.com/papers/extend1.htm (accessed March 8, 2003).

OSHA. 2009. Hazardous waste operations and emergency response, CFR 1910.120. http://www.osha.gov/pls/oshaweb/ owadisp.show_document?p_table=standards&p_id=9765 (accessed June 24, 2009).

Panesar, R. 2008. Employer responsibility vs. employee privacy. www.itsecurity.com/papers/mine8.htm.

Perman, S. 2001. Best-case scenario: The El Al approach: A look at the Israeli airline's security procedures. *Business 2.0* (November 2001).

Rasiah, S. *Putting a price on leaked information.* http://www.itsecurity.com/papers/pentasafe2.htm (October 27, 2002).

Riches, K. 2003. *Data protection.* http://www.itsecurity.com/papers/tam1.htm (accessed June 22, 2003).

Rose, C. 2002. *Computer security audit checklist.* http://www.itsecurity.com/papers/iomart2.htm (accessed April 8, 2002).

SANS Institute. 2002–2003. *Terminology.* http://www.sans.org/resources/glossary.php.

® CERT and CERT Coordination Center are registered in the U.S. Patent and Trademark Office.

Shore, D. 2002. *Lessons emerging from disaster.* http://www.itsecurity.com/papers/shore1.htm (September 20, 2002).

Skinner, M. 2009. *Tips for information security.* http://www.itsecurity.com/papers/axent.htm.

Tickle, I. 2002. *Data integrity: The unknown threat.* http://www.itsecurity.com/papers/tripwire1.htm (accessed December 6, 2002).

Training and awareness programme. 2002. http://www.itsecurity.com/papers/trinity8.htm (accessed May 19, 2002).

Trinity Security Services. 2003. Desktop security. http://www.itsecurity.com/papers/trinity13.htm (accessed April 27, 2003).

U.S. Department of State. 2002. September 11, 2001: Basic facts. Washington, DC: U.S. Department of State.

USEPA. 2005. *Water and wastewater security product guide.* http://cfpub.epa.gov.safewater/watersecurity/guide (accessed June 6, 2009). Site has expired.

15

Safety Engineering Management Aspects

Leadership, leadership, leadership: that nebulous characteristic that unlocks the mystery of managing people effectively.

You can't make something safe if you don't know how to make it unsafe.

INTRODUCTION

What is the difference between a newly degreed safety engineer and a newly degreed management specialist? The main difference is obvious: the safety engineer is a technical expert, and the management specialist is a management expert. Which of these two newly degreed specialists will make the best manager? This question is not easy to answer. One might think that the management specialist has the advantage over the safety engineer in managing people. But is this really the case?

To begin with, the individual person is a key factor. If the individual is frank, decisive, and assumes leadership readily, he or she might become a good manager. If the individual quickly perceives illogical and inefficient procedures and policies and develops and implements comprehensive systems to solve organizational problems, he or she might become a good manager. If the individual enjoys long-term planning and goal setting, he or she might become a good manager. If the individual is well informed, well read, enjoys expanding his or her knowledge and passing it on to others, he or she might become a good manager. If the individual is forceful in presenting his or her ideas, he or she might become a good manager.

Along with the parameters listed above, each is qualified by the key phrase: "might become" a good manager. Simply, no individual, even a management specialist, comes with a guarantee for becoming an effective manager or having good management ability. One can meet all the parameters listed above (and others) and still lack that special innate skill (or set of skills) that is management ability (laced with a great deal of common sense). We have witnessed many newly degreed managers step into the workplace for the first time, thinking that because they have a college degree they are "instant" managers—when in reality they were managers in title only. Unfortunately, many of these well-intentioned individuals fall on their faces when trying to make and orchestrate even the most basic managerial decisions.

Management skill is a blessing, an asset, a cherished, nebulous, amphorous, intangible trait that we all recognize but find difficult to define. Some say that you either have it or you don't. "It," of course, is the natural leadership ability to manage almost any situation and any worker. Others state that managers are born and not made. The truth lies somewhere in the middle. Someone with the desire to manage might have the innate ability to do so. However, to manage and to manage effectively are opposite edges of a double-edged sword.

We don't have a template for the recent college grad that will assure management success—we don't think there is such a template. However, we are certain of a few things: to manage effectively, education (formal or otherwise) is important; the desire to manage is very important; innate leadership qualities are even more important; and the actual ability to manage effectively is priceless.

THE RIGHT WAY

Go placidly amid the noise and the haste, and remember what peace there may be silence.

Desiderata (Ehrmann 1927)

What is the right way to manage?

This question has generated so many different responses and resulted in so many different theories that keeping up with them is difficult. From the "old" days, when the one flagship way to manage was lead, follow, or get out of the way, to the more sophisticated mantras of participative, collective, and empowered management styles carrying the flag today, determining the "right way to manage" has been an important question.

The right way? Take your choice. The management style probably does not matter anyway if it includes some basic ingredients. The recipe for success is dependent on these ingredients—without them, the would-be manager would be likely to fail—and in any given circumstance, several different effective managers could operate in several different management styles with success—if the ingredients for success are available.

So, what are the basic (or "magic") ingredients?

The ingredients that have worked for us are a combination of the old (proven) and a few (very few) relatively new (behavioral approach). Actually, for us, success lies somewhere along the continuum between these two.

Let's look at some of the old (proven factors) first. These include the standard bearers of planning, organizing, controlling, and directing.

- **Planning** simply advocates the old adages of "a job well planned is a job well done" and "a job well planned is a job half-finished." Planning is essential in safety engineering, especially in a proactive approach to safety, where waiting for accidents to occur is not an option. The planning factor allows for anticipating and determining how to deal with problems before they occur.

  Keep in mind that planning a safety and/or security program or policy is not a one-off project—it is a work in progress, on a continuing basis. Safety managers often must build their programs from scratch; this can be an accident-prone undertaking—many plans have fizzled or crash and burned. We've seen cases where unplanned for (not thought through) change attempts to force through full OSHA compliance models have so disrupted work routines that morale was seriously undermined. In the design of any plan, the manager (of any type) would be wise to follow Herb Simon's (1996) classic model of design, whereby you start off from a goal, desiderata (needs and wants), a utility function and budget constraints (remember that safety is often viewed by the narrow-minded as a cost and not a contributor to the bottom line), then work through a design tree of decisions until you find a design that's "good enough," then iterate (repeat, do again) until you find the best design or run out of time.

- **Organizing** is all about fashioning a safety program that will be accepted and followed by everyone in the organization. This is only possible, of course, if upper management fully supports the safety program and the safety engineer. Generally looked upon as a staffer and not a liner, the safety engineer must not only know his or her safety program front to back but must also have full understanding of the entire organization and what makes it tick. If a safety engineer has no idea where he or she fits into the organization, that person needs to look for a new line of work in a different organization.

- **Controlling** the safety program is critical to success. A good safety program with good support is doomed if not controlled (managed). Some form of metrics must be involved to continually measure for evaluation the performance of the program versus the number, type, and severity of on-the-job accidents, chemical spills, fires, incidents of workplace violence, and so on.

- **Directing** is important if used correctly. Safety engineers walk a thin line of authority. With management support, they carry a great deal of authority, but the safety engineer's best policy is to "speak softly and carry a big stick." Direction must come from the safety engineers to the supervisors, and the supervisors must buy into that direction. If the new safety engineer thinks he or she is going to walk into an organization and start directing workers to do this or that without working through the workers' supervisors, he or she is headed for certain failure. About the only immediate directing the safety engineer should do is of his or her own staff (if he or she has one). Being an advocate for safety is different than being an enforcer—one line the safety engineer never wants to cross.

The big four (planning, organizing, controlling, and directing) are old diehards that served managers well in the past. However, to synergize these four important ingredients, the safety engineer must be able to communicate. Reticence and safety engineering mix like gasoline and sand. We don't mean that the safety engineer needs to be extroverted to the point of setting the standard of demonstrative actions throughout the organization. On the contrary, the safety engineer almost needs the skills of a used car salesperson and diplomat all rolled up into one effective delivery methodology. Because of the staff alignment in which most safety engineers find themselves, the safety engineer needs to bridge the gap between line and staff by communicating effectively. In our experience, when a safety problem is apparent, asking the supervisor for his or her recommendations is best—we call this reverse empowerment. However, no matter what you call it, the ability to communicate is important. Remember, the goal is to protect workers from harm, not for the safety engineer to be right. Whatever communication technique works best should be employed on a consistent and straightforward basis.

The right way of safety engineering management also includes ensuring compliance with federal, state, and local safety and environmental health regulations. This can be a tricky proposition, however. The safety engineer who walks into a manager's office and throws a copy of some new OSHA, EPA, or DOT regulation on his or her desk and states that from now on we must do it (whatever it might be) the regulator's way or else is using the wrong approach.

Few people like rules and regulations, and fewer like demands made upon them (especially us Americans). While making sure the organization is in compliance with the regulators is important, the safety engineer needs to put on his or her used car salesperson's hat because implementing any regulation in the workplace is a selling job. The managers and workers must buy into the proposition before they will comply with it.

Managing under the precepts of participative, collective, and empowered systems is not a pie-in-the-sky management solution. However, like anything else, these concepts have good parts and bad parts. The key is to use the good and trash the bad.

Participative management requires the safety engineer to:

1. have confidence in his or her subordinates
2. establish open two-way communication with subordinates (remember, speak your truth quietly and clearly and listen to others, even to the dull and the ignorant; they too have their story) (Ehrmann 1927)
3. seek different ideas, different views
4. establish communication in all directions
5. ensure accurate upward communication
6. establish decision-making as a team effort
7. ensure goal-setting is performed by group action
8. share control widely and work to make sure that informal organized resistance (internal cliques that try to undermine the organization's goals) does not exist. This management style (approach) is do-able and can be effective but requires high maintenance; no element can be ignored.

Based on our personal professional experience, we do not recommend the collective style of leadership. Why? We can sum this one up quite simply. In any organization, when everybody is accountable, nobody is accountable, and when getting things done matters, the collective approach does not add to but detracts from individual decision-making—the antithesis of participative management. In fact, in many respects, collective leadership reminds us of the old saying "The camel is a horse that was designed by a committee."

We discuss empowerment in detail in our discussion of total quality management (TQM). However, for now let us point out that empowerment is one of the latest buzzwords in a long litany of buzzwords currently being broadcast about various schools of management. Simply, empowerment of workers allows workers to do their jobs without over- or undermanagement (if you get off my back, you'll be amazed how productive I can be). This buzzword, as with buzzwords from the past and probably in the future, in practice works better for inspiring some workers than for others.

Behavior-based Models

Few topics invoke more dissent among safety professionals than the benefits to be derived from behavioral safety versus the traditional three Es of safety (education, engineering controls, and enforcement) strategy employed since the 1930s. Compelling arguments on both sides keep the dissent and discussion lively. For example, the gurus of behavioral safety, namely E. Scott Geller and associates, argue that we should not fix the worker but instead fix the workplace. The other side counters with "the only thing that makes safety work and work safe is fear" (i.e., we are safe only because of the fear of losing a finger, arm, eye, leg, life, reputation, respect, or job, and so forth). Which side is right? The answer is up for grabs. While both viewpoints have positive and negative points, in our experience, people aren't necessarily always and only motivated by just one factor over another. Fear isn't the only factor that influences worker behavior, for example, nor are rewards what drives all positive worker behavior. We won't argue for or against any particular safety management model. Instead, we simply discuss the tenets of the various models and leave final judgment on which model is best for a particular application to the reader.

What exactly is behavior-based safety? According to Mathis and McSween (1996), behavior-based safety is the infamous quick fix. They argue that "common wisdom in safety has it that the quick fix doesn't work." But then they ask the question: "Why?" We leave the answer to the experts. In the meantime, consider what behavior safety is all about—what exactly is behavior-based safety?

Simply, behavior-based safety is all about using positive reinforcement to change unsafe individual behaviors. Opponents of this model argue that this is nothing more than the "Hawthorne effect"—that environmental conditions impact work input and worker happiness.

The behavior-based safety model begins with a behavioral analysis to identify "at-risk" behaviors. These can be determined by using:

1. Accident/incident/injury/near-miss reviews
2. Use of incident/antecedent reports
3. Employee interviews
4. Job Hazard Analysis (JHA)
5. Brainstorming

The tenets of the behavior-based model are loosely founded on B. F. Skinner's basics of behavioral modification. Skinner, a famous psychological theorist, examined the way people act—why they react to things in the way they do. He, and others, did not think bothering with the inner workings of the mind was important. Instead, they elected to focus on the observable behaviors of their subjects. From their work emerged a body of knowledge that explains behavior in terms of stimulus, response, and consequences. Behavior modification theory seeks to break down all human behavior into these parts. The actual behavior modification methodology involves four techniques (Skinner 1965). Two promote behaviors and two discourage them.

Promote Behavior

1. Positive Reinforcement—this technique rewards correct behavior with a positive consequence. When the desirable behavior occurs, something follows that is pleasurable. For teaching the correct behavior, this technique is unsurpassed. The pleasurable consequence follows the correct behavior and whoever receives the consequence knows exactly what behavior to repeat next time to receive the pleasure. One critically important condition is that the closer the reinforcement is to the behavior, the more power it has on future behaviors.
2. Negative Reinforcement—from the pain/pleasure principle, we know that behaviors that lead to pleasure will be repeated and behaviors that lead to pain will be avoided. Negative reinforcement relies on the second half of this principle, the tendency to avoid pain. As with positive reinforcement, behaviors that are negatively reinforced are more likely to be repeated when a similar circumstance is encountered. However, in the case of negative reinforcement, the consequence is not pleasurable. Instead, negative reinforcement relies on the removal or avoidance of negative consequences as the result of a behavior to reinforce the behavior. In other words, when some behavior leads to the removal of a negative consequence, that behavior is likely to be repeated when similar situations arise in the future. For example, if a person is cold (negative consequence), and turns on the heat (behavior), they are not cold any more. Next time the circumstances are similar, they will likely behave in the same way.

Discourage Behavior

1. Removal of Positive Consequences as a result of undesirable behavior (Extinction)—the first technique used to reduce the likelihood that a behavior will be repeated is called extinction. Responses that are not reinforced are not likely to be repeated. (Ignoring worker misbehavior should extinguish that behavior.) By taking away the pleasurable consequence, the behavior eventually goes away and is replaced by new behaviors that do lead to pleasurable consequences.
2. Negative Consequences/Punishment—responses that bring painful or undesirable consequence will be suppressed, but many reappear if reinforcement contingencies change. (Penalizing late workers by withdrawing privileges should stop their lateness.) The behavioral modification technique of punishment is neither the end all, be all, nor the enigma of discipline. When used correctly it is no more out of place than rewards (which taken to the extreme can be misused as well). With this justification aside, consider exactly how punishment fits into the model of behavioral modification.

The behavior-based safety system is considered the only proactive safety system. It involves direct intervention to change unsafe behavior, using scientific methodology to reinforce desired actions. Behavior-based safety involves four linked steps:

1. Identify safety-related behaviors critical to performance excellence.
2. Gather data on work group conformance to safety excellence.
3. Provide ongoing, two-way performance evaluation.
4. Use accumulating behavior-based data to remove barriers to continuous improvement.

Behavior-based safety focuses on human factors: behavior = observable effect (no connotation of good or bad). It involves no mind reading or assumptions. The employee is simply a Black Box: This is the situation (input); this is the action (output). One of the critical factors in behavior-based safety is observation. Observers apply the "ABC" principles of behavior-based safety observation (Krause, Hidley, and Hodson 1990):

- Antecedent: the issue, which precedes actions, but is the root cause
- Behavior: the action, which is the result of an antecedent
- Consequence: what happens, if anything, as a result of a behavior (positive or negative)

Observers learn to look at how the job is done and identify all task components on the checklist that are completed safely. Any step in the process completed unsafely—or any near miss—is marked as "at risk." The observer also provides feedback, emphasizing positives to reinforce safe performance and discussing at-risk performance. The observer looks for "footprints," or antecedents, to use in reaching agreement with the work team member on the need to change at-risk components.

So, is the behavior-based model the "right way" to manage safety in the workplace? Judge for yourself; for many, the jury is still out. There seems to be little doubt that the behavior-based model will work in some applications—the documented success stories seem unlimited. However, there are also some disadvantages of behavior-based safety. Specifically, this model requires "change" in the way things are done. We all know about change and its ramifications. We are also wary of any theory that treats human subjects and human intelligence strictly on a behavioral basis. In our experience, people don't always follow behaviorist principles in how they act. On a more practical note, this model takes a lot of initial effort to train observers and make observations, and that initial effort will cost—consultants are expensive. Moreover, we have heard it repeatedly said by safety professionals in the field that the behavior-based model is a gimmick based on the Heinrich model (i.e., approximately 90 percent of accidents are the workers' fault). Based on our real-world experience in the workplace, we realize that injuries and illnesses are caused by exposure to hazards … and that it is the goal of the safety engineer to eliminate or reduce exposure to hazards. One fact is clear; if no hazard is present, no injury is likely to occur.

Of course, as with any safety model, to work in the reality of the workplace, management must support the model wholeheartedly and employees must buy in to it. In regard to management, many managers view behavior-based models as their way out—a manager's delegation technique to place the blame elsewhere for accidents occurring in the workplace. We advise against this view simply because experience has shown us that the responsibility for the safety of workers can't be delegated.

On the other side of the coin, McSween sings the praises of behavior-based models: "The behavior-based safety process is the only empirical approach to improving safety that has proven to be effective" (McSween 1998).

The bottom line: No matter what model one implements to manage an organizational safety program, the question remains: How does one achieve lasting results? The unfortunate reality is fairly simple—lasting change takes lasting effort. In our experience, lasting effort is not easy.

Benchmarking

Benchmarking, another relatively new buzzword, is a valuable tool for use under any management system. For safety management, benchmarking is defined as a process for rigorously measuring your safety program vs. "best-in-class" programs and for using the analysis to meet and exceed the best in class. Benchmarking vs. best practices gives organizations a way to evaluate their safety programs—how effective and how cost-effective they are. Benchmarking also shows companies both how well their programs stack up and how well those programs are implemented. Simply, 1. benchmarking is a new way of doing business; 2. it is an objective-setting process; 3. it forces an external view to ensure correctness of objective-setting; 4. it forces internal alignment to achieve company safety goals; and 5. it promotes teamwork by directing attention to those practices necessary to remain competitive. The benchmarking process is shown in figure 15.1.

START ➡ PLAN ➡ RESEARCH ➡ OBSERVE ➡ ANALYZE ➡ ADAPT

Figure 15.1   Benchmarking Process Diagram

Our focus in this text is on the use of the benchmarking tool to improve safety management. Benchmarking vs. best practices gives the safety engineer a way to evaluate his or her operations overall.

What Benchmarking Can Reveal

- how effective the organization or process is
- how cost-effective the organization or process is
- how well the safety engineer's operations stack up and how well those operations are implemented

Potential Results of Benchmarking

- Benchmarking is an objective-setting process.
- Benchmarking is a new way of doing business.
- Benchmarking forces an external view to ensure correctness of objective setting.
- Benchmarking forces internal alignment to achieve plant goals.
- Benchmarking promotes teamwork by directing attention to those practices necessary to remain competitive.
- Benchmarking may indicate a direction of required change rather than specific metrics.
  1. costs must be reduced
  2. customer satisfaction must be increased
  3. return on assets must be increased
  4. improved maintenance
  5. improved operational practices
  6. best practices translated into operational units of measure

Targets

- Consideration of available resources converts the benchmark findings to targets.
- A target represents what can realistically be accomplished in a given time frame.
- A target can show progress toward benchmark practices and metrics.
- Quantification of precise targets should be based on achieving the benchmark.

Table 15.1   Benchmarking Steps

Step 1	Planning	Managers must select a process (or processes) to be benchmarked. A benchmarking team should be formed. The process of benchmarking must be thoroughly understood and documented. The performance measure for the process should be established (i.e., cost, time, and quality).
Step 2	Research	Information on the "best-in-class" performer must be determined through research. The information can be derived from the industry's network, industry experts, industry and trade associations, publications, public information, and other award-winning operations.
Step 3	Observation	The observation step is a study of the benchmarking subject's performance level, processes, and practices that have achieved those levels, and other enabling factors.
Step 4	Analysis	In this phase, comparisons in performance levels among facilities are determined. The root causes for the performance gaps are studied. To make accurate and appropriate comparisons, the comparison data must be sorted, controlled for quality, and normalized.
Step 5	Adaptation	This phase is putting what is learned throughout the benchmarking process into action. The findings of the benchmarking study must be communicated to gain acceptance, functional goals must be established, and a plan must be developed. Progress should be monitored and, as required, corrections in the process made.

Note: Benchmarking should be interactive. It should also recalibrate performance measures and improve the process itself.

Note: Benchmarking can be performance-based, process-based, or strategic-based and can compare financial or operational performance measures, methods, or practices, or strategic choices.

Benchmarking: The Process

When forming a benchmarking team, the goal should be to provide a benchmark that allows the safety engineer to evaluate and compare. For example, a benchmarking process could be used to evaluate privatized and reengineered public service entities and compare them to your operation as a way to determine how your facility compares; how it can be more efficient; remain competitive; and make continual improvements. Benchmarking is more than simply setting a performance reference or comparison; it is a way to facilitate learning for continual improvements. The key to the learning process is looking outside one's own plant to other plants that have discovered better ways of achieving improved performance and determining how to apply those more effective methods to your own operation.

Benchmarking Steps

As shown in figure 15.1 and table 15.1, the benchmarking process consists of five steps.

TOTAL QUALITY MANAGEMENT (TQM) PARADIGM

We want to be the best. What else is there?

—Lee Iacocca

What is total quality management (TQM)? TQM is not a traditional part of American culture. In the old days, if an organization had an abundance of resources, waste was affordable. At present, with increasing competition and an economy growing at a slower rate, doing it right the first time is more important. Simply, "TQM is an integrated system of principles, methods, and best practices that provide a framework for organizations to strive for excellence in everything they do" (Total Quality Management 2003).

No two organizations have the same TQM implementation, but certain characteristics or elements are uniform. Consider the following:

- TQM incorporates dynamic people concepts.
- TQM uses strategic planning with objectives and measurements.
- TQM uses benchmarking and program evaluation techniques.
- TQM focuses on continuous improvement.
- TQM evolves around four key concepts: customers, waste, time, and excellence.

To determine and define "customers," answer the following questions:

- Who are they?
- What do they need and want?
- What do they expect?
- Are they internal or external customers?

To determine and define "waste," consider the following:

- Waste is not only what is thrown away.
- Waste is anything that is inefficient, that wastes resources, or that generates waste. Waste includes product rework, excess inventory, disposal costs, excess labor, and injuries.
- Waste drives up cost.

To determine and define "time," consider the following:

- Cycle time: from start of an activity to completion.
- Time is money.
- Wasted time affects both product and operation cost.

To determine and define "excellence," consider the following:

- Excellence determined by customers (in TQM process) in comparison to competition.
- Continual pursuit.
- Outcome of total quality process.

TQM incorporates dynamic people concepts, such as participation, empowerment, and ownership.

What does all this have to do with safety? Accidents and incidents can be viewed as defective safety products. TQM is a significant management paradigm for nonline, special functions like safety, in which high participation and holistic management can be used to advantage. Remember that safety does not generate income (however, it can certainly save income), so it must be cost-effective.

THE SAFETY ENGINEER'S TOOLBOX: SYSTEM SAFETY

Along with a strong backbone, accountability, trustworthiness, decision-making ability, knowledge, creativity, integrity, sense of humor, great communication skills, and an over-abundance of common sense, safety engineers must reach into their toolboxes and also draw on various analytical tools to help them succeed in professional practice.

Many practicing safety professionals draw upon the tenets of system safety to accomplish their goals. *System safety* is a doctrine of a management practice that mandates that hazards be found and risks controlled; it is a collection of analytical approaches with which to practice the doctrine (NIOSH 2009). This process is methodical, careful, and purposeful. The purpose is to provide information for informed management decisions. System safety analysis techniques can be categorized as either inductive or deductive. Note: the information provided in the following is from NIOSH (2009).

Inductive Methods of System Safety

Inductive methods use observable data to predict what *can* happen. These techniques consider systems from the standpoint of the component parts and determine how a particular mode of failure of component parts will affect the performance of the system. Major inductive methods are preliminary hazard analysis (PHA), job safety analysis (JSA), failure modes and effects analysis (FMEA), and systems hazard analysis (SHA).

Preliminary hazard analysis (PHA) is a qualitative study conducted during the conceptual or early developmental phases of a system's life. Its objectives are to:

1. identify known hazardous conditions and potential failures
2. determine the causes(s) of these conditions and potential failures
3. determine the potential effect of these conditions and potential failures on personnel, equipment, facilities, and operations
4. determine the probability of each failure
5. establish initial design and procedural requirements to eliminate or control these hazardous conditions and potential failures.

In some cases an additional step, the estimation of the probability of an adverse event due to the hazard, is performed between steps 3 and 4 above.

The PHA is often based on a limited number of hazards determined as soon as initial facts about the system are known. These basic hazards must be dealt with even though many different circumstances might lead to the different hazards. The design process may be monitored to determine whether these hazards have been reduced or eliminated and, if not, whether the effects can be controlled.

Job safety analysis (JSA)— is a written procedure designed to review job methods, uncover hazards, and recommend safe job procedures. The following four basic steps are included in making a JSA:

1. Select the job, usually basing selection on potential hazards or high incidence rates.
2. Break the job down into a sequence of steps. Job steps are recorded in their normal order of occurrence. Steps are described in terms of *what* is done ("lift," "attach," "remove") not *how* it is done.
3. Identify the potential hazards. To determine what events can happen or should: observe the job, discuss the job with the operator, and check accident records.
4. Recommend safe job procedures to avoid the potential incident or injury.

A basic JSA form should include the job steps, the hazards associated with these steps, and recommended safe procedures. The form may be altered if needed to meet specific organizational needs by including such information as the name of the person performing the analysis, name of the operator and supervisor, or the name of reviewers or approvers of the analysis.

Failure mode and effect analysis (FMEA) is both a system safety and reliability analysis used to identify critical failure modes that seriously affect the safe and successful life of the system or failure modes that could prevent a system from accomplishing its intended mission. This technique permits system change to reduce the severity of failure effects. FMEA is organized around the basic question "What if?" The areas covered and the questions asked move logically from cause to effect.

1. Component—What individual components make up the system?
2. Failure mode—What could go wrong with each component in the system?
3. System causes—What would cause component failure or malfunction?
4. System effects—What would be the effect of such a failure on the system and how would this failure affect other components in the system?

5. Severity index—Consequences are often placed into one of four reliability categories:

   1—Catastrophic: may cause multiple injuries, fatalities, or loss of a facility.
   2—Critical: may cause severe injury, severe occupational illness, or major property damage.
   3—Marginal: may cause minor injury or minor occupational illness resulting in lost workday(s) or minor property damage.
   4—Negligible: probably would not affect the safety and health of personnel but is still in violation of a safety or health standard.

6. Probability index: How likely is the event to occur under the circumstances described and given the required precursor events? These probabilities are based on such factors as accident experience, test results from component manufacturers, comparison with similar equipment, or engineering data. Probability categories, which may be developed by individual companies or analysts, are sometimes classified as:

1—probable (likely to occur immediately or within a short period of time)
2—reasonably probable (probably will occur in time)
3—remote (possible to occur in time)
4—extremely remote (unlikely to occur)

7. Action or modification: After the failure modes, causes and effects, severity, and probability have been established, the system must be modified to prevent or control the failure.

The severity index, probability index, and a third index relating to personnel exposure may be used to determine the overall risk. A review of the above steps makes the objectives of FMEA clear. FMEA is intended to rank failures by risk (severity and probability) so that potentially serious hazards can be corrected.

Systems hazards analysis (SHA) includes the human component (a strength of job safety analysis) and the hardware component (a strength of failure mode and effects analysis).

SHA concentrates on the worker/machine interface. What process is being performed on what equipment? What major operations are required to complete the process? What tasks or activities are required to complete an operation? The thesis of SHA is that failures (undesired events) may be eliminated by systematically tracking through the system, looking for hazards that may result in a failure situation.

In the language of SHA, the terms *process*, *operation*, and *task* have specific meaning. *Process* means the combination of operations and tasks that unite physical effort and physical and human resources to accomplish a specific purpose. An *operation* is a major step in the overall process (e.g., drilling and countersinking stock on a drill press). A *task* is a particular action required to complete the operation (e.g., placing a cutting tool in a holder before sharpening the tool on the grinder).

Once the process to be analyzed has been identified, it is subdivided into its operations and tasks. To do this, the analyst must be familiar with the tasks involved in the operation and the interactions between and within the system being analyzed and associated systems and subsystems. Often a flow diagram is constructed to record what is taking place throughout the flow of operations and tasks that fulfill process demands. This enables the analyst to see the pertinent subsystems, methods, transfer operations, inspections techniques, and human/machine operations.

One type of hazard analysis process (adapted from Firenze 1991) includes identifying and recording information relating to:

1. process (turning between centers on a machine lathe)
2. major unit operations required to complete the process (rough-turning steel stock)
3. tasks to complete an operation (select cutting tool and place in holder)
4. variance from safe practices with the potential to cause hazards (incorrect cutting tool used)
5. hazard that has the potential to cause an injury (worker in proximity to lathe when incorrect cutting tool used)
6. triggering event causing hazards to result in incidents, brought about by human error or situation or environmental factors (starting the lathe)
7. incident resulting from effect of triggering event on hazard (stock comes off centers when lathe is running)
8. effect indicating type of injury or damage resulting from the incident (eye injury)
9. hazard consequence classification (called the severity index in FMEA)
10. hazard probability (same as in FMEA)
11. procedural requirements to eliminate or reduce hazards in the workplace
12. safety and person protective equipment requirements to reduce the possibility of injuries and illnesses while performing operations and tasks
13. instructions/recommendations to ensure safety and health in the workplace

Deductive Methods of Systems Safety Analysis

As mentioned, inductive methods of analysis analyze the components of the system and postulate the effects of their failure on total system performance. Deductive methods of analysis move from the end event to try to determine the possible causes. They determine how a given end event *could* have happened. One widespread application of deductive systems safety analysis is fault tree analysis.

Fault tree analysis (FTA) postulates the possible failure of a system and then identifies component states that contribute to the failure. It reasons backward from the undesired event to identify all of the ways in which such an event could occur and, in doing so, identifies the contributory causes. The lowest levels of a fault tree involve individual components or processes and their failure modes. This level of the analysis generally corresponds to the starting point in FMEA.

FTA uses Boolean logic and algebra to represent and quantify the interactions between events. The primary Boolean operators are AND and OR gates. With an AND gate, the output of the gate, the event that is at the top of the symbol occurs only if all of the conditions below the gate, and feeding into the gate, coexist. With the OR gate, the output event occurs if any one of the inputs even occurs.

When the probabilities of initial events or conditions are known, the probabilities of succeeding events can be determined through the application of Boolean algebra. For an AND gate, the probability of the output event is the intersection of the Boolean probabilities or the product of the probabilities of the input events or:

Probability (output) = (Prob Input 1) x (Prob Input 2) x (Prob Input 3)

For an OR gate, the probability of the output event is the sum of the "union" of the Boolean probabilities, or the sum of the probabilities of the input events minus all of the products.

Probability (output) = (Prob Input 1) + (Prob Input 2) + (Prob Input 3)
-[(Prob Input 1) x (Prob Input 2) + (Prob Input 2) x (Prob Input 3) + (Prob Input 1) x (Prob Input 1) x (Prob Input 3) + (Prob Input 1) x (Prob Input 2) x (Prob Input 3)]

Where the probabilities of the input events are small (less than 0.1, for example), the probability of the output event for an OR gate can be estimated by the sum of the probabilities of the input events or:

Probability (output) = (Prob Input 1) + (Prob Input 2) + (Prob Input 3)

Figure 15.2 illustrates a simple FTA. The basic events in figure 15.2 are represented as rectangles rather than circles because they could possibly be developed (reduced) further.

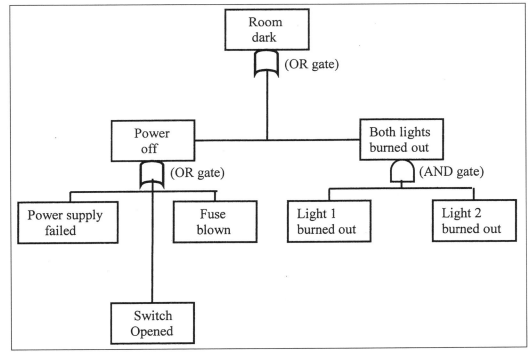

Figure 15.2   Fault tree analysis of light bulbs.

Assume for example, in figure 15.2, that the following probabilities exist:

Probability of power supply failing	= 0.0010
Probability of switch open	= 0.0030
Probability of blown fuse	= 0.0020
Probability of light 1 out	= 0.0300
Probability of light 2 out	= 0.0400

Because the power will be off if the power supply fails or the switch is open or the fuse is blown, the probability of the power being off is the sum of the probabilities of the power supply failing, the switch being open, and the fuse being blown or (0.0010) + (0.0030) + (0.0020) = 0.0060. Both lights will be burned out if light 1 is out and light 2 is out. The probability of both lights being out is (0.0300) x (0.0400) or 0.0012. The room will be dark if the power is off or both lights are burned out, which is (0.0060) + (0.0012) or 0.0072.

The concepts of "cut sets" and "path sets" are useful in the analysis of fault trees. A cut set is any group of contributing elements that, if *all* occur, will *cause* the end event to occur. A path set is any group of contributing elements that, if *none* occur, will *prevent* the occurrence of the end event.

For the example in figure 15.2, the end event will occur if:

- the power supply fails
- the switch is open
- the fuse is blown
- light 1 and light 2 are burned out

Each of the four sets above represents a cut set since, if all of the events in any of the sets occur, the end event (room dark) will occur.

For the example in figure 15.2, the end event will not occur if no events in either of the following sets occur:

- the power supply fails, the switch is open, the fuse is blown, and light 1 is burned out, or
- the power supply fails, the switch is open, the fuse is blown, and light 2 is burned out.

Each of the two sets above represents a path set since, if none of the events in either of the sets occur, then end event (room dark) cannot occur.

Management oversight and risk tree analysis (MORT), originally developed for the U.S. Department of Energy (USDOE) to help conduct nuclear criticality and hardware analysis, is defined as a formalized, disciplined logic or decision tree to systematically relate and integrate a wide variety of safety concepts. As an incident analysis technique, it is a graphical checklist that focuses on three main concerns: specific oversights and omissions, assumed risks, and general management system weaknesses (EG&G 1984).

It is essentially a series of fault trees with three basic subsets or branches:

1. a branch that deals with specific oversights and omissions at the work site
2. a branch that deals with the management system that establishes policies and makes them work
3. an assumed risk branch that acknowledges that no activity is completely free of risk and that risk management functions must exist in any well-managed organization. These assumed risks are those undesirable consequences that have been quantitatively analyzed and formally accepted by appropriate management levels within the organization.

MORT includes about a hundred generic causes and thousands of criteria. The MORT diagram terminates in some 1,500 basic safety program elements needed for a successfully functioning safety program—elements that prevent the undesirable consequences indicated at the top of the tree. MORT has three primary goals:

1. to reduce safety-related oversights, errors, and omissions
2. to allow risk quantification and the referral of residual risks to program organizational management levels for appropriate action

3. to optimize the allocation of resources to the safety program and to organizational hazard control efforts.

Did You Know?

MORT programs and their associated training courses place emphasis on constructing trees for individual program needs and on a set of ready-made MORT trees that can be used for program design, program evaluation, or accident investigation.

- Uses the analytic tree method to systematically dissect an accident.
- Serves as a detailed road map by requiring investigators to examine all possible causal factors (e.g., assumed risk, management controls or lack of controls, and operator error).
- Looks beyond immediate causes of an accident and instead stresses close scrutiny of management systems that allow the accident to occur.
- Permits the simultaneous evaluation of multiple accident causes through the analytic tree (DOE 2009).

REVIEW QUESTIONS

1. What personal qualities indicate the possibility of good management ability?
2. Discuss "the right way" to manage.
3. Discuss "planning," "organizing," "controlling," and "directing."
4. What's the fifth essential element that synergizes planning, controlling, organizing, and directing? Why?
5. How can you ensure compliance in your facility as safety manager through persuasion, not muscle?
6. Discuss participative management.
7. What are some of the drawbacks to collective management?
8. Relate "the camel is a horse that was designed by a committee" to collective management.
9. Discuss behavior-based management verses the three Es of safety.
10. How can using positive reinforcement lead to safe workplace behaviors?
11. What happens when you take away the chocolate chip cookies (i.e., the reward)?
12. What are the ABCs of behavior-based observation? Discuss.
13. Define benchmarking in light of safety management.
14. How can benchmarking be used to improve safety management? Discuss seven different ways.
15. Name and discuss four different potential results of benchmarking.
16. Describe the benchmarking process and the steps.
17. What is total quality management?
18. Discuss how TQM uses the four key concepts.
19. Discuss the primary goals of MORT.
20. What is the difference between inductive and deductive systems safety analysis techniques?
21. When is the appropriate time to apply the PHA technique to a design?

REFERENCES AND RECOMMENDED READING

American Society of Safety Engineers. 1987. *Introduction to system safety*. Des Plaines, IL: American Society of Safety Engineers.

Bureau of National Affairs. 1989. *Occupational safety and health: 7 critical issues for the 1990s* (SSP-136). Arlington, VA.

Cormen, T. H., C. E. Leiserson, R. L. Rivest, and C. Stein. 2002. *Introduction to algorithms*. 2nd ed. New Delhi: Prentice-Hall of India.

Department of Energy. 2009. *Nuclear safety and environment*. http://hs.energny.ov/CSA/CSP/AIP/workbook?Rev2/chpt7/chapt7.htm (accessed January 5, 2009). site has expired.

Department of Labor. 29 CFR part 1910. Federal Register.

EG&G. 1984. *Idaho national engineering laboratory, glossary of terms and acronyms*, Publication 76-45/28. Idaho Falls, ID: EG&G.

EG&G. n.d. *Idaho national engineering laboratory, systems safety course material.* Idaho Falls, ID: EG&G.

Eisma, T. L. 1990. Demand for trained professionals increases educational interests. *Occupational Health and Safety.* Waco, TX: Stevens Publishing.

Ferry, T. 1990. *Safety and health management planning.* New York: Van Nostrand Reinhold.

Firenze, R. J. 1978. *The process of hazard control.* Dubuque, IA: Kendall Hunt.

Firenze, R. J. 1988. *Evaluation and control of the occupational environment.* Cincinnati, OH: National Institute for Occupational Safety and Health.

Firenze, R. J. 1991. *The impact of safety on high-performance/high-involvement production systems.* Melbourne, IA: Creative Work Designs, Inc.

Gloss, D. S., and M. G. Wardle. 1984. *Introduction to safety engineering.* New York: John Wiley and Sons.

Grimaldi, J. V., and R. H. Simonds. 1989. *Safety management.* 5th ed. Homewood, IL: Irwin, 1989.

Gusfield, D. 1997. *Algorithms on strings, trees, and sequences: Computer science and computational biology.* Cambridge, UK: Cambridge University Press.

Hammer, W. 1989. *Occupational safety management and engineering.* 4th ed. Englewood Cliffs, NJ: Prentice Hall.

Hardie, D. T. 1981. The safety professional of the future. *Professional Safety* (May): 17–21.

Hazard, W. G. 1988. Industrial ventilation. In *Fundamentals of industrial hygiene.* 3rd ed., ed. B. Plog. Chicago, IL: National Safety Council.

Health and safety: The importance of wearing helmets. 1989. *USAA Aide Magazine* (October 25, 1989).

Hoover, R. L., R. L. Hancock, K. L. Hylton, O. B. Dickerson, and G. E. Harris. 1989. *Health, safety, and environmental control.* New York: Van Nostrand Reinhold.

Krause, T. R., J. H. Hidley, and S. J. Hodson. 1990. *The behavior-based safety process: Managing involvement for an injury-free culture.* New York: Van Nostrand Reinhold.

Lafore, R. 2002. *Data structures and algorithms in Java.* 2nd ed. Toronto, CA: SAMS.

Lewis, R. 1992. *Sax's dangerous properties of industrial materials.* 8th ed. New York: Van Nostrand Reinhold.

Lyman, W. J., W. R. Reehl, and D. H. Rosenblatt. 1990. *Handbook of chemical property estimation methods.* 2nd printing. Washington, DC: American Chemical Society.

Mathis, T., and T. McSween. 1996. Behavior-based safety: The infamous quick fix. *Industrial Safety and Hygiene News* (February).

McSween, T. 1998. Culture: A behavioral perspective. In *Proceedings of light up safety in the new millennium: A behavioral safety symposium,* 43–49. Des Plaines, IL: America Society of Safety Engineers.

Mitchell, T. M. 1997. *Machine learning.* New York: McGraw-Hill.

NIOSH. 1973. *The industrial environment, its evaluation and control.* Cincinnati, OH: National Institute of Occupational Safety and Health.

NIOSH. 1994. *Pocket guide to chemical hazards.* Washington, DC: U.S. Department of Health and Human Services.

NIOSH. 2009. *An introduction to electrical safety for engineers.* www.cdc.gov/niosh/topics/SHAPE (accessed January 1, 2009).

Nirmalakhandan, N. 2002. *Modeling tools for environmental engineers and scientists.* Boca Raton, FL: CRC Press.

Occupational Health. 1988. *Recognizing and preventing work-related disease.* 2nd ed., eds. B. S. Levy and D. H. Wegman. Boston: Little, Brown and Company, 1988.

Ogle, B. R. 2003. Introduction to industrial hygiene. In *Fundamentals of occupational safety and health.* 3rd ed. Rockville, MD: Government Institutes.

Olishifski, J. B., and B. A. Plog. 1988. Overview of industrial hygiene. In *Fundamentals of industrial hygiene.* 3rd ed., ed. B. A. Plog. Chicago, IL: National Safety Council.

Poynton, C. 2003. *Digital video and HDTV algorithms and interfaces.* New York: Morgan Kaufmann.

Sarkus, D. J. 2001. *The safety coach: Unleash the 7 C's for world class safety performance.* Donora, PA: Championship Publishing.

Simon, H. 1996. *The sciences of the artificial.* 3rd ed. Cambridge, MA: MIT Press.

Skinner, B. F. 1965. *Science and human behavior.* Glencoe, IL: Free Press.

Smith, L. C. 1980. The J programs. *National Safety News* (September).

Sulzer-Azaroff, B., and W. E. Eischeid. 1999. Assessing the quality of behavioral safety initiatives. *Professional Safety* 44 (4): 31–36.

Total Quality Management. 2003. Poway, CA: Total Quality Engineering, Inc.

U.S. Department of Labor. 1995. *Occupational safety and health standards* (part 1910), subpart G, section 1910.95. Washington, DC: OSHA.

USEPA. 2003. *Technology transfer network support center for regulatory air models.* www.epa.gov/scram001/tt22.htm. Site has expired.

Appendix: Chapter Review Answers

Chapter 1
Answers will vary.

Chapter 2
Answers will vary.

Chapter 3
Answers will vary.

Chapter 4

1. Pounds = 1,650 gal x 8.34/gal = 13,761 lb

2. Flow = 2.89 MGD x 1.55 cfs/MGD = 4.48 cfs

3. Flow = $\dfrac{33,444,950 \text{ gal/day}}{1,000,000 \text{ gal/MG}}$ = 33.44 MGD

4. $\dfrac{675 \text{ lb}}{8.34 \text{ lb/gal}}$ = 80.9 gallons

5. Solve for pounds pumped per minute; change to lb/day.
   8.34 lb/gal water x 1.22 SG liquid chemical = 10.2 lb/gal liquid
   40 gal/min x 10.2 lb/gal = 408 lb/min
   408 lb/min x 1440 min/day = 587,520 lb/day

Chapter 5

1. 0.55 = $\dfrac{55}{100}$ = 0.55 = 55%

2. $\dfrac{7}{22}$ = 0.318 = 0.318 x 100 = 31.8%

3. $1.4 = \dfrac{1.4}{100}$, since the weight of 1 L water to 10^6

$\dfrac{1.4}{100} \times 1,000,000 \text{ mg/L}$

14,000 mg/L

4. Grade $= \dfrac{140 \text{ mm}}{22 \text{ m}} \times 100(\%)$

$= \dfrac{140 \text{ mm}}{22 \times 1000 \text{ mm}} \times 100\%$

$= 0.64\%$

5. Three significant figures: 1, 3, and 5.

6. $\dfrac{x \text{ h}}{20 \text{ h}} = \dfrac{4 \text{ gpm}}{10 \text{ gpm}}$

$x = \dfrac{(4)(20)}{10}$

$x = 8 \text{ h}$

7. $(2.3 \text{ ft}^3/\text{s})\,(7.48 \text{ gal/ft}^3)\,(60 \text{ s/min})$

$(\text{ft}^3/\text{s})\,(\text{gal/ft}^3)\,(\text{s/min}) = \left(\dfrac{\text{ft}^3}{\text{s}}\right)\left(\dfrac{\text{gal}}{\text{ft}^3}\right)\left(\dfrac{\text{s}}{\text{min}}\right)$

$= \dfrac{\text{ft}^3}{\text{s}}\,\dfrac{\text{gal}}{\text{ft}^3}\,\dfrac{\text{s}}{\text{min}}$

$= \dfrac{\text{gal}}{\text{min}}$

$(2.3 \text{ ft}^3/\text{s})\,(7.48 \text{ gal/ft}^3)\,(60 \text{ s/min}) = 1032.24 \text{ gal/min}$

8. $1 \text{ MGD} = \dfrac{10^6}{1 \text{ day}}$

$= \dfrac{10^6 \text{ gal} \times 0.1337 \text{ ft}^3/\text{gal}}{1 \text{ day} \times 86,400 \text{ s/day}}$

$$= \frac{133{,}700 \text{ ft}^3}{86{,}400 \text{ s}}$$

$$= 1.547 \text{ cfs}$$

9. Perimeter = 2 x length + 2 x width
       = 2 x 8 in. + 2 x 8 in.
       = 16 in. + 16 in.
       = 32 in.

10. C = π x 6′
     C = 3.14 x 6′
     C = 18.84 ft

11. 24 in. ÷ 2 = 12 in.
     V = H x πr²
     V = 96 in. x π(12 in.)²
     V = 96 in. x π(144 in.²)
     V = 43,407 ft³

12. $45 \dfrac{\text{psi}}{1} \times \dfrac{1 \text{ ft}}{0.433 \text{ psi}} = 104 \text{ ft}$

Chapter 6

1. 0.78125, -1.28125
2. x = -5

Chapter 7

1. 0.2419
2. 0.3746
3. 5.6713
4. 0.5
5. 0.9962

Chapter 8

1. dispersion
2. mean
3. 9.25

Chapter 9

1. 0
2. 1, 0
3. not
4. OR
5. AB + AC

Chapter 10

1. $1,664.00
2. $1,083.00
3. $8,511.11
4. $5,414.71
5. $11,348.15

Chapter 11

Answers will vary.

Chapter 12

1. Potential, kinetic
2. In a steady flow without friction, the sum of the velocity head, pressure head, and elevation head is constant for any incompressible fluid particle throughout its course.
3. (Velocity = 28.8 fps flow = 0.98 ft³/sec or 440 gpm)
   Q = V x A; (20)(0.049) = 0.98 ft³/sec or 440 gpm
4. 660 ft³ x 7.48 gal/ft³ = 4,937 gallons
5. 160 lb - 125 lb = 35 lb; specific gravity = 160/35 = 4.57
6. $$\frac{1,600,000 \text{ gal}}{7.48} = 213,904 \text{ cu ft (rounded)}$$

   Volume = 7.85 x d² x depth

   $$\frac{213,904}{9498.5} = .785 \times 110^2 \times \text{depth}$$
   depth = 22.5 ft
7. pressure = weight x height
   6400 psf = 62.4 lb/cu ft x height
   height = 103 ft (rounded)
8. 35 psi x 144 sq in/sq ft = 5,040 psf
   force = pressure x area
   lb = 5,040 psf x 1.6 sq ft
   lb = 8,064
9. always constant
10. pressure due to the depth of water
11. the line that connects the piezometric surfaces along a pipeline
12. 1500 gpm x 1440 min/day = 2,160,000 gpd = 2.16 MGD
    2.16 MGD x 1.55 cfs/MGD = 3.35 cfs
    Q = A x V
    3.35 = .785 x 1² V
    4.27 ft/sec = V

    $$V_h = \frac{V^2}{2g} = \frac{4.27^2}{64.4}$$
    V_h = 0.28 ft
13. Pressure Head = 110 psi x 2.31 ft/psi = 254.1 ft
14. $$\frac{5.00 \text{ ml/sec}}{3785 \text{ ml/gal}} = 0.0013 \text{ gal/sec}$$

    $$\frac{0.0013 \text{ gal/sec}}{7.48 \text{ gal/cu ft}} = 0.00018 \text{ cfs}$$

    Q = A x V

$0.00018 = .785 \times .33^2$

$V = 0.002$ ft/sec

$$V_h = \frac{V^2}{2g} = \frac{0.002^2}{64.4} = 6.2 \times 10^{-8}$$

15. $Q = A \times V$
    $1.46 = .785 \times .5^2 V$
    7.44 ft/sec $= V$

$$V_h = \frac{V^2}{2g} = \frac{7.44^2}{64.4}$$

$V_h = 0.86$ ft

16. pressure energy due to the velocity of the water
17. a pumping condition in which the eye of the impeller of the pump is above the surface of the water from which the pump is pumping
18. the slope of the specific energy line
19. h = (2.31 ft/psi)(P)
    = (2.31 ft/psi)(80 psi)
    = 185 ft
20. P = (0.433 psi/ft)(h)
    = (0.433 psi/ft)(140 ft)
    = 61 psi
21. P = [(0.433 psi/ft)(S.G. of oil)] (h)
    = [(0.433 psi/ft) (0.9)](90 ft)
    = (0.390 psi/ft)(90 ft) = 35.1 psi
22. pressure expressed in unit of feet of water instead of psi

Chapter 13

1. to convert a source of applied voltage to a load resistance with a minimum *IR* voltage drop in the conductor so that most of the applied voltage can produce current in the load resistance
2. resistivity
3. circular mil
4. conductivity
5. allowable power loss in the line; permissible voltage drop in the line; the current-carrying capacity of the line; conductor installed where ambient temperature is relativity high.
6. negative
7. the two are directly proportional. As flux density increases, field strength also increases.
8. the type of material and the flux density
9. the lag between magnetization and the forces that produce it
10. North Pole
11. 141.4 volts (200 x 0.707)
12. 846 bolts (600 volts x 1.41)
13. number of turns in the coil; length of the coil; cross-sectional area of the coil; permeability of the core
14. decrease
15. 4 milliseconds
16. area of plates, distance between plates, dielectric constant

17. 120 µf
18. apparent power
19. 1
20. effective values and phase angles

Chapter 14

Answers will vary.

Chapter 15

Answers will vary.

Index

orifice flow, 213; flowmeter, 209
orthogonal matrix, 95
OSHA Form 300, 34
ototoxic, 155
ototraumatic, 155
overhanging beam, 147
overload, 31
oxidation, 34
oxidizer, 34
oxygen deficient atmosphere, 34

parallel circuits, 242–52
parallelogram, 131
particulate matter, 62
parts per million (ppm), 58–59
passive infrared sensors, 312
password, 319; authentication protocol (PAP), 319;
    cracking, 319; sniffing, 319
patch, 319
peak amplitude, 264
peak-to-peak amplitude, 264
penetration, 319; testing, 319
percent, 67–70
performance standards, 34
perimeter, 82
perimeter-based security, 319
period, 262
permeability, 258
permissible exposure limit (PEL), 34
permitted worker, 162
personal fire walls, 319
personal protective equipment (PPE), 34
phase, 267
phase angle, 218, 267; difference, 267; relationships, 267
phasor, 268
photon, 162
physical security, 293
piezometric surface, 197–99
piloted ignition, 181
PIN, 319
pipe diameter, 205; head pressure, 211; length, 205;
    velocity pressure, 212
piping hydraulics, 208
plain text, 319
planning, 336
plastic limit, 171
pneumatics, 144
Poisson's ratio, 140
polarity, 254
policy, 319
polymorphism, 319
population, 106
portable barrier, 306; computing, 323
positive reinforcement, 338
potential energy, 193, 200
pound force, 129
power, 233, 240–42, 241–52
powers and exponents, 71–73
preindustrial safety, 20

preliminary assessment, 34; hazards analysis, 343–43
presbycusis, 155
present worth, 121
pressure, 34, 50, 89–90, 188, 192; calculation, 210
prime number, 65
product, 65
product-over the sum method, 250
product safety, 12
program policy, 319
promiscuous mode, 319
properties of materials, 137–52
proprietary information, 319
protective relays, 219
protocol, 319

quadratic equation, 97–100

rad, 162
radiant heat, 34
radiation, 34, 162; dose, 51, 161; workers, 162
radioactive decay, 162; equations, 166
radioisotope, 163
radius of curvative (beam), 149
ramp, 136
range, 106
rank, 96
ratio, 74
RC series circuits, 284; time constant, 285
reactions, 150
reactive, 34
reactivity hazard, 34
real ear attenuation, 153
reciprocal method, 249
reconnaissance, 320
registry, 320
regulation, 6
rem, 163
remote access, 320
reportable quantity (RQ), 35
resistance, 230, 236–37, 246–50, 266–67
resistivity, 222
resolution of forces, 128–31
Resources Conservation and Recovery Act (RCRA), 35
resource exhaustion, 328
restrained beam, 147
restricted area, 163
reverse engineering, 320; hookup, 320
rim speed, 143
risk, 35, 320; analysis, 320; assessment, 35;
    characterization, 35; management, 35
RL series circuit, 273–74
roentgen, 163
rogue program, 320
roughness, 205
router, 320
rupture, 138

safety, 35
safety engineering, 3–16

About the Authors

Frank R. Spellman is assistant professor of Environmental Health at Old Dominion University. He has extensive experience in environmental science and engineering, in both the military and the civilian communities. He is the author or coauthor of more than sixty books, including *Physics for Nonphysicists* (Government Institutes, 2009) and *Occupational Safety and Health Simplified for the Chemical Industry* (Government Institutes, 2009).

Nancy E. Whiting is a freelance technical writer in water/wastewater issues, environmental science, and education. She is the coauthor, with Frank Spellman, of *Safety Engineering*, 2nd ed. (Government Institutes, 2005) and *Environmental Science and Technology*, 2nd ed. (Government Institutes, 2006).